BATTLEGROUNDS

Also by H. R. McMaster

Dereliction of Duty: Johnson, McNamara, the Joint Chiefs of Staff, and the Lies That Led to Vietnam

H. R. McMASTER

BATTLEGROUNDS

THE FIGHT TO DEFEND
THE FREE WORLD

WILLIAM
COLLINS

William Collins
An imprint of HarperCollins*Publishers*
1 London Bridge Street
London SE1 9GF

WilliamCollinsBooks.com

First published in Great Britain in 2020 by William Collins
First published in the United States by Harper in 2020

1

A catalogue record for this book is
available from the British Library

ISBN 978-0-00-841040-7

Typeset in Dante MT Std
Printed and bound in Great Britain by
CPI Group (UK) Ltd, Croydon

For Katharine, Colleen, and Caragh—the real KCC detectives, who inspired much more than bedtime stories, including this book

CONTENTS

PREFACE

THIS IS not the book that most people wanted me to write. Friends, agents, editors, and even family, asked me to write a tell-all about my experience in the White House to confirm their opinions of President Donald Trump. Those who supported the president would have liked me to depict him as an unconventional leader who, despite his brash style, made decisions and implemented policies that advanced American interests. Those who opposed the president wanted an account to confirm their judgment that he was a bigoted narcissist unfit for office. And they wanted me to write it immediately, so that the book might influence the outcome of the 2020 presidential election. Although writing such a book might be lucrative, I did not believe that it would be useful or satisfactory for most readers. The polarization of America's polity and that of other free and open societies is destructive, and I wanted to write a book that might help transcend the vitriol of partisan political discourse and help readers understand better the most significant challenges to security, freedom, and prosperity. I hoped that improved understanding might inspire the meaningful discussion and resolute action necessary to overcome those challenges.

SPECIAL CHARACTERS

Chapter 3: An Obsession with Control
中国: "Middle Kingdom"

分久必合, 合久必分: "After a long split, a union will occur; after a long union, a split will occur"

天下: "All under heaven"

Chapter 4: Turning Weakness into Strength
名不正, 则言不顺; 言不顺, 则事不成: "If names cannot be correct, then language is not in accordance with the truth of things; and if language is not in accordance with the truth of things, affairs cannot be carried on to success"

擦枪走火: "to shoot accidently while polishing a gun"

Chapter 12: Making Him Safer Without Them
적화통일: "red-colored unification"

돈주: "masters of money"

Introduction

A great deal of intelligence can be invested in ignorance when the need for illusion is deep.

—**SAUL BELLOW**

ON FRIDAY, February 17, 2017, I was in my hometown of Philadelphia on the way to the Foreign Policy Research Institute, a think tank. The purpose of the meeting was to discuss the findings of a study I had commissioned on Russia's annexation of Crimea and invasion of Ukraine in 2014. As the lieutenant general director of the clunkily named Army Capabilities Integration Center, my job was to design the future army. To fulfill that responsibility, I sought to understand how Russia was combining conventional and unconventional military capabilities along with cyber attacks and information warfare—what we were calling Russia new-generation warfare (RNGW). The study recommended how to improve the future army's ability to deter and, if necessary, defeat any forces that employed similar capabilities against the United States or our allies. We modeled the effort on General Donn Starry's study of the 1973 Arab-Israeli War. Starry's findings helped drive a renaissance in the post–Vietnam War army based on changes in fighting doctrine, training, and leader development. It was clear to me that Russia, China, and other nations had studied the U.S. Army after the lopsided U.S. victory over Saddam Hussein's armed forces in the 1991 Gulf War and the initially successful U.S. military campaigns during the 2001 invasion of Afghanistan and the 2003 invasion of Iraq. As a trained historian as well as a soldier, I believed that the old saying "The military is always prepared to fight the last war" was wrong. Militaries that encountered the greatest difficulties at the onset of war studied their recent past only superficially.[1] Learning from history, I believed, was essential if the U.S. military were to maintain its competitive advantages over potential enemies.

I intended to begin the discussion at the institute with a description of how RNGW combined disinformation, denial, and disruptive technologies for psychological as well as physical effect. Russian president Vladimir Putin and his generals wanted to accomplish their objectives below the threshold of what might elicit a military response from the United States and countries of the North Atlantic Treaty Organization (NATO). RNGW seemed to be working, and we were likely, I thought, to see more of it. The stakes were high. Russian aggression in the last decade had taken many forms, from cyber attacks to political subversion to assassination and the use of military power such as the invasion of Georgia in 2008. Russia had changed the borders of Europe by force for the first time since the end of World War II. It seemed likely that Putin, emboldened by perceived success, would become even more aggressive in the future.

It was warm for February. I was enjoying the walk down Walnut Street when my phone rang, displaying a partially blocked Washington, DC–based number. It was Katie Walsh, the White House deputy chief of staff. She asked if I could travel to Florida that weekend to interview with President Trump for the position of assistant to the president for national security affairs. I said yes and called my wife, also named Katie, as I walked the last block. Katie was used to phone calls that suddenly changed our lives. This was one of them.

I had scheduled a premeeting with the *Philadelphia Inquirer* reporter Trudy Rubin. Trudy was always ahead of other analysts in her understanding of the complex problem set in the Middle East. I benefited from our conversations about the region. She predicted many of the difficulties that the United States encountered in the second Iraq War and characterized our unpreparedness for those challenges as "willful blindness."[2] We both agreed that while many often debated whether the United States should have invaded Iraq, the better question was who thought it would be easy and why. Trudy was about to return to Syria to report on the humanitarian catastrophe associated with the Syrian Civil War and the rise of the terrorist group Islamic State in Iraq and Syria, or ISIS. The most difficult part of that campaign, we agreed, would be how to get to a sustainable political outcome in Syria and Iraq that led to the enduring defeat of ISIS and an end to the humanitarian catastrophe across the Middle East. I told her in confidence about the unexpected phone call I had just received. She replied that she hoped I would be selected. Trudy was not a supporter of President Donald Trump, but he *was* the elected president, and there *was* work to be done. She and I both felt that, in recent years, the balance of power and persuasion had shifted against the United States and other free and open societies. Much of that shift, we believed, had been self-inflicted due to failures to understand fully the emerging challenges to American security, prosperity, and influence.

Service in our army gave me the opportunity to work alongside dedicated and courageous men and women in our armed forces, intelligence agencies, and diplomatic corps to implement the policies and strategies that came from Washington, DC. I would soon enter my thirty-fourth year of service as an officer, and I was considering retirement. I felt privileged to have served, especially in command positions. I had spent nearly half my career overseas and over five years in combat. I would look back fondly on the tremendous, intangible rewards of service, especially being a part of endeavors much larger than myself and being a member of teams that took on the quality of a family, in which the man or woman next to you was willing to give everything, even their own life, for you. I was reluctant to retire because I felt a sense of duty to my fellow servicemen and women, many of whom were still serving in battlegrounds overseas.

Service in combat was rewarding, but the experience was also difficult and sometimes frustrating. It was difficult because one bears witness to the horror of war and the sacrifices of young men and women who fight courageously and selflessly for our nation and for one another. It was frustrating because of the wide gap between the assumptions on which some policies and strategies were based and the reality of situations on the ground in places like Iraq and Afghanistan. Serving as national security advisor might give me an opportunity to help a new, clearly unconventional president challenge assumptions and close gaps between reality overseas and fantasy in Washington. Trudy knew that I was apolitical; in the tradition of Gen. George C. Marshall (the architect of victory in World War II), I had never even voted. If selected, I would do my duty under President Trump as I had under five other presidents.

⋆ ⋆ ⋆

I BELIEVED we were at the end of the beginning of a *new* era. At the end of the *last* era, the United States and other free and open societies had reason to be confident. The Cold War ended in victory over Communist totalitarianism. The Soviet Union collapsed. Then, during the 1991 Gulf War, America put together a broad international coalition and demonstrated tremendous military prowess to defeat Saddam Hussein's army and free Kuwait. But after the end of the Cold War, America and other free and open societies forgot *that they had to compete* to keep their freedom, security, and prosperity. The United States and other free nations were confident—overconfident. Overconfidence led to complacency. I bore witness to that growing confidence.

In November 1989, our cavalry regiment was on patrol near Coburg, West Germany, the town where Martin Luther translated the Bible into German in the sixteenth century. As a captain in the Second Armored Cavalry Regiment, I saw the need to compete as obvious. Our regiment patrolled a stretch of the Iron Curtain that divided democracies and dictatorships in Europe.[3] It was really an iron *complex,* one designed to keep the subjugated peoples in and freedom out. Fortifications began well east of the actual border between the two German states—with a ten-foot-tall, steel-reinforced fence covered in electric trip wires. Then there was a road. East German border guards drove their jeeps along that road, monitoring the soil next to it for footprints. A steep ditch prevented would-be escape vehicles from plowing through. Beyond the ditch stood two more fencerows, separated by a one-hundred-foot-wide minefield. Those who made it past the mines then had to cross a three-hundred-foot-wide no-man's-land. Some of our sergeants had seen East German guards shoot unarmed civilians there. It was a formidable system. But it was artificial. And then, on November 9, 1989, it collapsed. A confused East German Politburo member announced that East Germans could use all border crossings to "permanently exit" the nation. With people gathering at the gates near Coburg, guards stepped aside and threw the gates open. Hundreds, then thousands, then tens of thousands of East Germans flooded across. Scouts from Eagle Troop, on patrol that day, received countless hugs, bouquets of flowers, and bottles of wine. There were tears of joy. Meanwhile, Berliners were celebrating as they chiseled away at the wall that had divided them since 1961. The wall fell. The East German government withered away. The Soviet Union broke apart. We had won the Cold War.

But then came a hot war far away from the Iron Curtain. In 1989, Saddam Hussein's first decade as Iraq's dictator was coming to a close. He should have been fatigued. In 1980 he had started a disastrous eight-year war with Iran that killed more than 600,000 people. Since seizing power in 1979, he had employed a Stalinist model of repression, murdering more than another million of his own people in a country of 22 million, including an estimated 180,000 Kurds in a genocidal campaign in which he used poison gas to massacre entire villages of innocent men, women, and children. But in 1990, Saddam felt more underappreciated than fatigued. Had he not defended the Sunni Muslim and Arab world against the scourge of Iran's Shia Islamist revolution? Did not Kuwait, Saudi Arabia, and other Arab states owe him a debt of gratitude—and cash to cover the cost of that war?

Saddam's tanks rumbled toward Iraq's southern border in July 1990, and on August 2, the U.S. ambassador to Iraq was in London when the first of more than three hundred thousand Iraqi troops poured into Kuwait to make that small but wealthy nation Saddam's nineteenth province. President George H. W. Bush and his team got a coalition of thirty-five nations to agree that the annexation "would not stand."

Those same troopers who were patrolling the East-West German border in November 1989 arrived in Saudi Arabia almost exactly one year after they watched the Iron Curtain part. Three months later, Eagle Troop was leading the so-called left hook, a massive envelopment attack, to crush Saddam's Republican Guard and kick the door open to Kuwait with a blow from the western desert.

As our troop moved out on February 26, heavy morning fog dissipated. It was replaced by high winds and blowing sand. Visibility was limited. Our scout helicopters were grounded. It was just after 4 p.m. We moved in formation. One scout platoon, Lt. Mike Petschek's First Platoon, led with six Bradley armored fighting vehicles, which carry a scout squad and were armed with a 25 mm chain gun and a TOW antitank missile launcher. The other scout platoon, Lt. Tim Gauthier's Third Platoon, moved along our southern flank. Our tanks moved behind the lead scouts in a nine-tank wedge, with my tank in the center. Lt. Mike Hamilton's Second Platoon was to my tank's left and Lt. Jeff DeStefano's Fourth Platoon was to my tank's right. Our 132 troopers were well trained and confident, in their equipment and in one another—men bound together by mutual trust, respect, affection. As a twenty-eight-year-old captain, I was proud to command that extraordinary team.

The troop was not very high-tech by twenty-first-century standards. We had three of these new devices called Global Positioning Systems, or GPS. But given that they worked only sporadically, we navigated mainly by dead reckoning in the flat, featureless desert. Because the troop had no maps, leaders did not know that they were paralleling a road that ran through a small abandoned village and then into Kuwait. We also did not know that we were entering an old Iraqi training ground recently reoccupied by a Republican Guard brigade and an armored division. Their mission was to halt our advance.

The Iraqi brigade commander, Major Mohammed, knew the ground well. Mohammed had attended the Infantry Officer Advanced Course at Fort Benning, Georgia, in the 1980s, when the United States was cultivating an ill-conceived relationship with Iraq to balance against Iran. Mohammed's defense was sound. He fortified the village with anti-aircraft guns and put his infantry in protected positions. He took advantage of an imperceptible rise in the terrain that ran perpendicular to the road running east to west through the village to organize a "reverse slope" defense. He built two engagement areas, or "kill sacks," on the eastern side of the ridge, emplaced minefields, and dug in approximately forty tanks and sixteen BMPs, Russian-made infantry fighting vehicles, on the back side of the ridge. His plan was to destroy us piecemeal as we moved across the crest. Hundreds of Iraqi infantry occupied bunkers and trenches between the armored vehicles. He positioned his reserve of eighteen more T-72 tanks and his command post along another subtle ridgeline farther east.

At 4:07, Staff Sergeant John McReynolds's Bradley drove on top of an Iraqi bunker positioned to provide early warning. Two enemy soldiers emerged and surrendered. McReynolds's wingman, Sgt. Maurice Harris, was scanning into the village through the blowing sand when his Bradley came under fire. As Harris returned fire with his 25 mm cannon, Lieutenant Gauthier moved forward and fired a TOW antitank missile. Thus began twenty-three minutes of furious combat.

As our tanks fired nine high-explosive rounds simultaneously into the village, we received permission to advance to the 70 Easting, a north–south running grid line on a map. We switched to a tanks lead formation. I instructed Second and Fourth Platoons to "follow my move" and we passed through the scouts' Bradleys. As our tank came over the crest of that imperceptible rise, our gunner, Sgt. Craig Koch, and I identified the enemy simultaneously: eight T-72 tanks in prepared positions faced us at close range. Koch announced, "Tanks direct front." The crew acted as one. The gun recoiled, and the breech dropped. The enemy tank exploded in a huge fireball. Pfc. Jeffrey Taylor loaded a tank-defeating "sabot" round, which thrusts a fourteen-pound depleted uranium dart out of the gun tube at two kilometers a second. He armed the gun and yelled, "Up!" as he threw his body against the turret wall to get out of the gun's recoil path. Our tank crew destroyed the first three tanks in about ten seconds. When our other eight tanks crested the rise, they joined in the assault. In about a minute, everything in the range of our guns was in flames. Our tank driver, Spec. Chris Hedenskog, informed me, "Sir, we just went through a minefield." He knew that it would be dangerous to stop right in the middle of the enemy's kill sack, the area in which all his tanks could concentrate their fire. We had a window of opportunity to shock the enemy and take advantage of the first blows we had delivered, to turn physical advantage into psychological advantage. So, our tanks drove around the antitank mines, with the Bradleys and other vehicles following in our tracks. We ran over antipersonnel mines, but they popped harmlessly. Our training was paying off. As McReynolds recalled, "We did not have to be told what to do; it just kinda came natural."

Just as we cleared the western defensive positions, our executive officer, John Gifford, radioed, "I know you don't want to hear this, but you're at the limit of advance; you're at the 70 Easting." I responded, "Tell them we can't stop. Tell them we have to continue this attack. Tell them I'm sorry." Stopping would have allowed the enemy to recover. I felt that we had the advantage and had to finish the battle rapidly. The army's cavalry culture encourages initiative, and the stakes were too high not to take advantage of the hard blow we had just delivered.

We crested a second rise and entered the reserve's circular perimeter. Iraqi tank commanders were trying to deploy against us. They were too late. We destroyed all eighteen tanks at close range. Then we stopped. There was nothing left to shoot. Our fire support officer, Lt. Dan Davis, called in a massive artillery strike on fuel and ammo stocks farther east. There was some more fighting to do, but the main attack had lasted twenty-three minutes.

Eagle Troop destroyed a much larger enemy force that had all the advantages of the defense and took no casualties. Our fight was a lopsided victory in a larger battle and war that were lopsided victories. As confidence grew based on the military victory in the Gulf War, analysts undervalued the qualitative advantage of U.S. forces and the narrow political objective of simply returning Kuwait to the Kuwaitis. They assumed that future enemies would repeat Saddam's mistake of trying to fight the U.S. and coalition military forces on our own terms rather than asymmetrically. And, as many reflected on the victory over the Soviet Union in the Cold War, they forgot how the United States and its allies and partners had competed based on a clear understanding of their adversary, what was at stake, and the long-term strategy designed to ensure their security, promote prosperity, and extend their influence.

★ ★ ★

IN RETROSPECT, what those cavalry troopers experienced in Coburg, Germany, and what would become known as the Battle of 73 Easting in the Iraqi desert, marked the end of an era.[4] It was then, in the 1990s, that American leaders, flush with victories in the Cold War and the Gulf War, forgot that the United States had to compete in foreign affairs. Coburg was also the birthplace of Hans Morgenthau, who fled the Nazis in 1937 and became one of the fathers of the discipline of international relations. In 1978, in his last coauthored essay with Ethel Person, titled "The Roots of Narcissism," Morgenthau lamented preoccupation with self in foreign policy because it led to alienation from other nations and aspirations that exceeded the limits of ability. It was there in Coburg, near Morgenthau's birthplace, that American confidence grew as the Cold War ended and the world entered what the political analyst Charles Krauthammer called "the unipolar moment." America's stature as the only superpower encouraged narcissism, a preoccupation with self, and an associated neglect of the influence that others have over the future course of events. Americans began to define the world only in relation to their own aspirations and desires.[5]

Over-optimism and a preoccupation with self inspired three flawed assumptions about the new, post–Cold War era. First, many accepted the thesis that the West's victory in the Cold War meant "the end of history," what political philosopher Francis Fukuyama described as "the universalization of Western liberal democracy as the final form of human government."[6] Although Fukuyama warned that ideological consensus in favor of democracy was not a foregone conclusion, many assumed that an arc of history guaranteed the primacy of free and open societies over authoritarian and closed societies, and of free-market capitalism over authoritarian, closed economic systems. Ideological competition was finished.

Second, many assumed that old rules of international relations and competition were no longer relevant in what President George H. W. Bush hoped would be "a new world order—a world where the rule of law, not the rule of the jungle governs the conduct of nations." The post–Cold War world was unipolar. Russia was in disarray after the collapse of the Soviet Union. China's economic miracle was just beginning, and Chinese Communist Party leaders adhered to paramount leader Deng Xiaoping's directive to hide their capabilities and bide their time. An emerging condominium of nations would vitiate the need to compete; all would work together and through international organizations to solve the world's most pressing problems.[7] Great power competition was passé.

Third, many asserted that American military prowess demonstrated during the 1991 Persian Gulf war manifested a revolution in military affairs (dubbed RMA) that would allow the U.S. military to achieve "full-spectrum dominance" over any potential enemy. If any adversary had the temerity to challenge a technologically dominant U.S. military, the war would result in a rapid, decisive U.S. victory.[8] Military competition was over.

Those three assumptions underpinning U.S. policies not only were over-optimistic, they also led to complacency and hubris. *Hubris,* an ancient Greek term defined as extreme pride leading to overconfidence, often results in misfortune. In Greek tragedies, the hero vainly attempts to transcend human limits and often ignores warnings that predict a disastrous fate. In the case of the new, post–Cold War era, warnings that might have drawn into question the three assumptions I've just outlined went unheeded by too many in the U.S. policy, political, and military establishments.

First, autocracy was making a comeback. By the end of the 1990s, market-oriented reforms failed in Russia, resulting in the election of Vladimir Putin, a little-known director of the Russian Federal Security Service, or FSB (the successor organization to the KGB). Writing in the *Hoover Digest* in April 2000, David Winston, a strategic advisor to congressional leadership, warned that the newly elected Russian president "will be strongly tempted to revert to the traditional paths of autocracy and statism" and "may see both the fate of Russia and his rule through the traditional prism of military prowess and conquest." But then, autocracy had never really gone away. Despite many predictions of its imminent collapse or implosion, the despotic regime in North Korea adapted to the loss of aid from the defunct Soviet Union, endured a devastating famine, extorted money and goods from the West and South Korea in exchange for a weak nuclear agreement, and transitioned the dictatorship from Kim Il-sung, known as the "Great Leader" since 1948, to his son Kim Jong-il, the "Dear Leader." Meanwhile, a nascent reform movement in Iran was stifled as the Islamist revolutionaries tightened their grip on their theocratic dictatorship.

Second, a new great power competition was emerging. China had paid close attention to the 1991 Gulf War and was deeply embarrassed by the 1996 Taiwan Strait Crisis, during which the United States responded to Chinese missile threats meant to intimidate Taiwan with a massive show of force. The two U.S. aircraft carrier groups that converged on the strait exposed the inferiority of the Chinese People's Liberation Army (PLA) Navy compared to the U.S. fleet. As China's economy grew, so did the PLA. And as their military grew, China began to flex its muscles. On April 1, 2001, fighter pilot Wang Wei maneuvered his PLA Navy J-8 fighter aggressively over the South China Sea in an effort to intimidate the crew of a U.S. Navy EP-3 signals intelligence aircraft. After two passes at the U.S. aircraft, he misjudged his approach, colliding with its nose and propeller. The J-8 broke into pieces, and the U.S. aircraft made an emergency landing on Hainan Island. Wang Wei's body was never recovered. The Chinese detained the twenty-four U.S. crew members for eleven days.[9] The demonstration of U.S. military prowess in the Gulf War and the Taiwan Strait Crisis, as well as increasing tension in the South China Sea, spurred China to undertake the largest peacetime military buildup in history.

Third, as China began to challenge so-called American military dominance, increasingly potent jihadist and Iranian state-sponsored terrorist organizations attacked asymmetrically, avoiding military strength and exploiting weakness. The jihadist terrorist movement grew after the Afghan War of the 1980s and the Gulf War. Its leaders used a perverted interpretation of Sunni Islam to inspire recruits and rationalize violence against the "far enemies," the United States and Europe, and the "near enemies," Israel and Arab monarchies. Mass murder of the defenseless was the preferred tactic. On February 26, 1993, Ramzi Yousef, a Kuwaiti-born Pakistani terrorist who attended an Al-Qaeda training camp in Afghanistan, drove with his Jordanian co-conspirator into the parking garage underneath the North Tower of the World Trade Center in New York City. After building their weapon in a Jersey City apartment, they had packed the 1,200-pound bomb into a yellow Ryder van. Six people were killed and more than a thousand injured. Yousef had hoped that his explosion would topple Tower 1, which would then fall into Tower 2 and kill the occupants of both buildings, which he estimated to be about 250,000 people. Three years later, in 1996, Hezbollah terrorists (with Iranian backing and support) attacked U.S.

military forces housed in the Khobar Towers in Dhahran, Saudi Arabia, killing 19 Americans and wounding 372. In April 1998, Al-Qaeda issued a fatwa (a ruling on a point of Islamic law) from its safe haven in Afghanistan calling for the indiscriminate killing of Americans and Jews everywhere. Then, in August of that year, the terrorist organization turned words into action with simultaneous bombings of U.S. embassies in Nairobi, Kenya, and Dar es Salaam, Tanzania, killing 224 people and wounding more than 5,000. Twelve of those killed in Kenya were U.S. citizens. But Al-Qaeda was not finished. On October 12, 2000, the U.S. Navy destroyer *Cole* was docked in Aden, Yemen, for refueling. At around 11:18 a.m., a fiberglass boat laden with C4 explosives sped toward the port side and exploded on impact, blowing a forty-by-sixty-foot hole in the ship's port side and killing seventeen sailors.[10] By the turn of the century, the director of central intelligence, Adm. James Woolsey's observation in 1993 that "Yes, we have slain a large dragon, but we live now in a jungle filled with a bewildering variety of poisonous snakes,"[11] seemed particularly prescient. But in the new century, the free and open societies of the world would confront both.[12]

Those and other harbingers of an emerging geopolitical landscape much different from the idealized new world order might have inspired a fundamental reassessment of U.S. foreign policy and national security strategy and sparked a questioning of the assumptions underpinning the optimistic view of the post–Cold War world. They did not. Indeed, President Bill Clinton wrote the following in the preface to the December 2000 National Security Strategy report:

> **As we enter the new millennium, we are blessed to be citizens of a country enjoying record prosperity, with no deep divisions at home, no overriding external threats abroad, and history's most powerful military. Americans of earlier eras may have hoped one day to live in a nation that could claim just one of these blessings. Probably few expected to experience them all; fewer still all at once.[13]**

At the turn of the century, the United States was therefore set up for a rude awakening of tragic proportions. Like Icarus of the ancient Greek legend, U.S. leaders disregarded admonitions against over-optimism and complacency. Icarus's father instructed him to fly neither too low, lest the sea's dampness clog his wings, nor too high, lest the sun's heat melt them. But Icarus flew too close to the sun, his wings melted, and he tumbled into the sea and drowned. Before the mass murder attacks of September 11, 2001, America was flying too high.

In the new century, three shocks and disappointments undermined American confidence. First, the September 11, 2001, Al-Qaeda mass murder attacks in New York, Washington, and over a field in Pennsylvania hit like a sudden earthquake. The lives of nearly three thousand innocents were lost; many more suffered physical and psychological wounds. The attacks inflicted an estimated $36 billion in physical damages alone, with even higher costs accumulated when one considers the broader effect the attacks had on the American and global economies.[14] Second, the unanticipated length and difficulty of wars in Iraq and Afghanistan, and the cost of those wars in blood and treasure, came like slow, rolling aftershocks from 9/11. Third, the 2008 financial crisis had the effect of a tsunami earthquake. It began with subterranean rumblings caused by subprime mortgages and unregulated use of derivatives (contracts based on overvalued homes and bad loans). When the tidal wave hit, it created the worst economic disaster since the Great Depression of 1929. Housing prices fell 31 percent, more than during the Depression. The U.S. Treasury disbursed nearly $450 billion to banks to stimulate the economy, and approximately $360 billion to Freddie Mac, Fannie Mae, and AIG.[15] The crisis passed, but two years later, unemployment was still above 9 percent and an unknown number of discouraged workers gave up looking for work.

* * *

OVER THE seven years following the 9/11 attacks, optimism and confidence eroded and, after 2008, began to give way to pessimism and resignation. In 2009, a new president implemented a foreign policy based mainly on his opposition to the Iraq War and animated by a worldview skeptical of American interventions and activist foreign policy abroad. In a June 2013 speech during which he announced the planned withdrawal of 33,000 troops from Afghanistan, President Barack Obama cited the cost of the war and the "rising debt and hard economic times" that followed the financial crisis. He stated that "the tide of war is receding" and that it was "time to focus on nation building here at home."[16] He saw the war in Iraq as part of a broader historical pattern of U.S. interventions. After a retrospective interview with President Obama in the waning days of his second term, *Atlantic* reporter Jeffrey Goldberg observed that President Obama "consistently invokes what he understands to be America's past failures overseas as a means of checking American self-righteousness." The president and many of those who served him were sympathetic to the New Left interpretation of foreign affairs, one that considers so-called Western capitalist imperialism as the primary cause of the world's problems. "We have history," President Obama said. "We have history in Iran, we have history in Indonesia and Central America. So we have to be mindful of our history when we start talking about intervening, and understand the source of other people's suspicions."[17] An underlying premise of the New Left interpretation of history is that an overly powerful America is more often a source of, rather than part of the solution to, the world's problems. To return to the Icarus analogy, under the Obama administration, we began to fly too low.

Across multiple administrations, U.S. foreign policy and national security strategy has suffered from what we might derive from Morgenthau's essay "Strategic Narcissism": the tendency to view the world only in relation to the United States and to assume that the future course of events depends primarily on U.S. decisions or plans. The two mind-sets that result from strategic narcissism, overconfidence and resignation, share the conceit of attributing outcomes almost exclusively to U.S. decisions and undervaluing the degree to which others influence the future. The over-optimism that energized U.S. foreign policy under the George W. Bush administration contributed to an underappreciation of the risks of action, such as the invasion of Iraq in 2003. The pessimism about the efficacy of U.S. engagement abroad that influenced U.S. foreign policy under the Barack Obama administration led to an underappreciation of the risks of inaction, such as the complete withdrawal of U.S. forces from Iraq in 2011 or the decision to forgo military reprisals for the Assad regime's mass murder of Syrian civilians with chemical weapons in 2013. Both forms of strategic narcissism were based mainly on wishful thinking and the definition of problems as one might like them to be as a way to avoid harsher realities. I experienced the effect of strategic narcissism up close. I was often on the receiving end of ill-conceived plans disconnected from the problems they were ostensibly meant to address. That is because strategic narcissism leads to policies and strategies based on what the purveyor prefers, rather than on what the situation demands. The assumptions that underpin these policies and strategies often go unchallenged as they provide a deceptive rationale for folly.

As I got on the flight to Palm Beach, Florida, to be interviewed by a man I had never met, I thought that, if given the opportunity, I would try to help restore America's strategic competence. And I thought that the first step might be to begin with historian Zachary Shore's concept of strategic empathy, what Shore describes as "the skill of understanding what drives and constrains one's adversary,"[18] as a corrective to strategic narcissism. During the interview at Mar-a-Lago, President Trump seemed sympathetic to my observation that the United States had not competed effectively in recent years and that, as a result, determined adversaries had gained strength and our power and influence had diminished. As defense expert Nadia Schadlow observed in her 2013 essay "Competitive Engagement," "being successful in a competition requires knowing and understanding both one's competitors and oneself."[19] I began with those tasks when I assumed my duties as national security advisor just three days after the call I received in Philadelphia. I asked Schadlow to join me as senior director for national security strategy to develop options that would enhance America's ability to compete more effectively and shift the balance back in favor of the United States and the free and open societies of the world.

There was a lot of work to do. Two days after I arrived in Washington, I held an "all hands" meeting in which I shared with the national security staff my view that our strategic competence had eroded based, in part, on our narcissistic approach to foreign policy and national security strategy. Our job was to provide the president with options and integrated strategies that combined elements of national power with efforts of like-minded partners to make progress toward clearly defined goals. The work, however, should begin with identifying challenges and understanding them on their own terms and from the perspective of "the other." I asked our team not only to map the interests of rivals, adversaries, and enemies, but also to consider the emotions, aspirations, and ideologies that drive and constrain them. The options we developed, if approved, would become integrated strategies. I insisted that these strategies must identify not only goals, but also our assumptions—especially assumptions concerning the degree of agency and control that we and our partners could expect in order to make progress toward those goals. The strategies needed to be logical with regard to the means employed and the desired ends. We would also work hard to describe what was at stake and to explain why accomplishing those ends was worth the risks and potential cost in treasure and, especially, blood. I then laid out what I saw, from my more than three decades in the military and from studying national security as a historian, as the four categories of challenges to national and international security. These would be our priorities as we developed integrated strategies for the president.

First, great power competition was back with a vengeance, highlighted by Russia's annexation of Crimea, invasion of Ukraine, intervention in Syria, and the sustained campaign of political subversion against the United States and the West. And it was clear that China under Chairman Xi Jinping was no longer hiding its capabilities and biding its time as the People's Liberation Army accelerated island building in the South China Sea, tightened control of its population internally, and extended its diplomatic, economic, and military influence internationally.

Second, the threat from transnational terrorist organizations was greater than it was on September 10, 2001. Terrorist groups were increasing their technological sophistication and lethality. They were also growing in magnitude due to slick recruiting and the perpetuation of conflict in and around the two epicenters of the war in Afghanistan and the war in Syria.

Third, hostile states in Iran and North Korea were becoming more dangerous. The new dictator in Pyongyang aggressively pursued nuclear weapons and long-range missiles. An old dictator in Tehran expanded support for terrorists and militias across the Middle East and beyond in a way that prolonged destructive wars and increased the threat to Israel, Arab states, and U.S. interests in the Middle East.

Fourth, new challenges to security were emerging in complex arenas of competition from space to cyberspace to cyber-enabled information warfare to emerging disruptive technologies. Moreover, a range of interconnected long-range problems demanded an integrated effort now including the environment, climate change, energy, and food and water security.

As we began to frame those challenges as the first step toward developing integrated strategies, we paid particular attention to improving our competence. We emphasized the importance of history. Ignorance or misuse of history often led to the neglect of hard-won lessons or the use of simplistic analogies that masked flaws in policy or strategy. Understanding the history of how challenges developed would help us ask the right questions, avoid mistakes of the past, and anticipate how "the other" might respond.

Supposition about the future should begin with an understanding of how the past produces the present. Policies and strategies must be based on the recognition that rivals and enemies will influence the future course of events. How "the other" responds will depend, in part, on their own interpretation of history. As former secretary of state and national security advisor Henry Kissinger observed, all states "consider themselves as expressions of historical forces . . . what really happened is often less important than what is thought to have happened."[20] And more than 2,500 years ago, the Chinese military theorist Sun Tzu wrote, "If you know the enemy and know yourself you need not fear the results of a hundred battles."[21] So, in order to overcome strategic narcissism, we must strive to understand our competitors' view of history as well as our own.

Still, it does no good to improve our strategic competence if the United States and our partners do not possess the confidence to overcome new and pernicious threats to our free and open societies. To rebuild and sustain that confidence requires communicating clearly what is at stake and describing how the proposed strategies are designed to achieve sustainable outcomes at acceptable costs. This is what British prime minister Winston Churchill described as "an all-embracing view which presents the beginning and the end, the whole and each part, as one instantaneous impression retentively and untiringly held in the mind."[22] None of the competitions discussed in this book will be resolved quickly; strategies, while remaining flexible and adaptable to changing conditions, must be sustained over time. Consistency and will are, therefore, important dimensions of strategic competence.

But our will is diminished. As our foreign policies swung from over-optimism to resignation, identity politics interacted with new forms of populism. That interaction divided us and diminished confidence in our democratic principles, institutions, and processes. We might apply empathy to ourselves as well as to the other and, as we discuss the challenges we face, seek common understanding, and work together to secure freedom and prosperity for future generations. It is my hope that this book might contribute to those discussions.

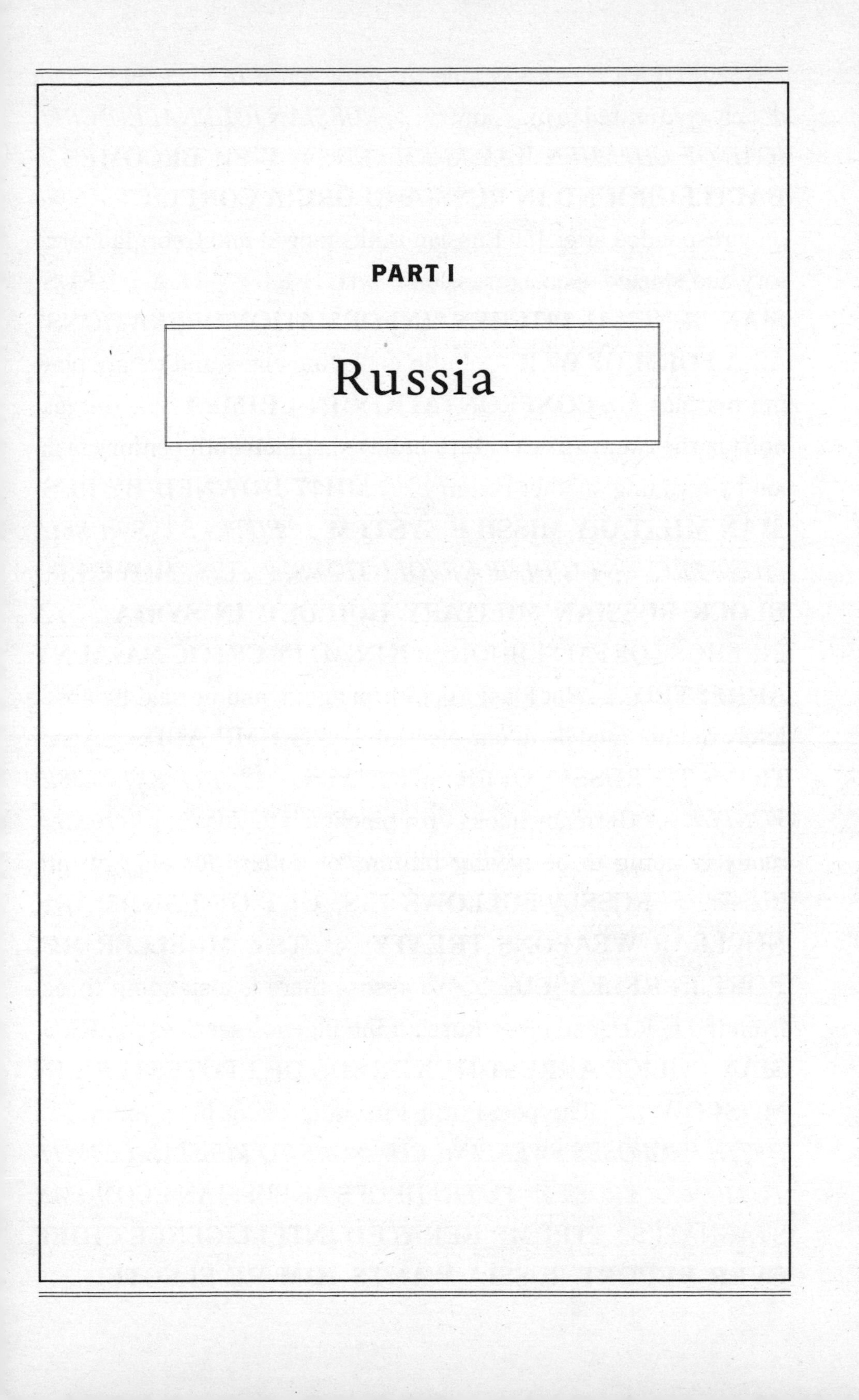

PART I

Russia

MASKIROVKA . . . I was able to get a sense of his soul; a man deeply committed to his country . . . *RUSSIAN JOURNALIST CRITICAL OF CHECHEN WAR IS KILLED* . . . **WEB BECOMES A BATTLEGROUND IN RUSSIA-GEORGIA CONFLICT** . . . We only responded after 150 Russian tanks moved into Georgian territory and started open aggression . . . ПЕРЕГРУЗКА . . . **RUSSIAN GENERAL PITCHES "INFORMATION OPERATIONS" AS A FORM OF WAR** . . . Putin is playing chess and we are playing marbles . . . **CONFRONTATION IN CRIMEA** . . . You just don't in the twenty-first century behave in nineteenth-century fashion by invading another country . . . **MH17 DOWNED BY RUSSIAN MILITARY MISSILE SYSTEM** . . .*PUTIN SAYS RUSSIA MUST PREVENT "COLOR REVOLUTIONS"* . . . **U.S. MOVES TO BLOCK RUSSIAN MILITARY BUILDUP IN SYRIA** . . . A FIREHOSE OF FALSEHOOD . . . **KREMLIN CRITIC NAVALNY ARRESTED** . . . But I just asked him again, and he said he absolutely did not meddle in our election . . . **TRUMP ADDS SANCTIONS TO RUSSIA OVER SKRIPALS** . . . *JOINT EXERCISE: VOSTOK* . . . Germany hooks up a pipeline into Russia, where Germany is going to be paying billions of dollars for energy into Russia . . . **RUSSIA FOLLOWS U.S. OUT OF LANDMARK NUCLEAR WEAPONS TREATY** . . . **THE MUELLER REPORT IS RELEASED** . . . We assess there is a standing threat from the G.R.U. and other Russian intelligence services . . . **RUSSIAN POLICE ARREST HUNDREDS OF PROTESTERS IN MOSCOW** . . . The post-Putin Russia is being born today . . . *PUTIN PROPOSES SWEEPING CHANGES TO RUSSIAN CONSTITUTION* . . . TRUST IN PUTIN DROPS AS RUSSIAN ECONOMY STAGNATES . . . **TRUMP BERATED INTELLIGENCE CHIEF OVER REPORT RUSSIA WANTS HIM RE-ELECTED** . . .

Map copyright © 2020 Springer Cartographics LLC. Based on United Nations map.

CHAPTER 1

Fear, Honor, and Ambition: Mr. Putin's Campaign to Kill the West's Cow

The more powerful enemy can be vanquished only by exerting the utmost effort, and most thoroughly, carefully, attentively, and skillfully making use without fail of every, even the smallest, "rift" among the enemies.

—V. I. LENIN

GENEVA IS the ideal city for a confidential diplomatic meeting. It is easy to blend in there. The city hosts more than three thousand official meetings annually, attended by more than two hundred thousand delegates. Government airplanes flow in and out of the airport. Convoys of black limousines and SUVs crisscross the city. Officials of friendly and not-so-friendly nations arrive at each other's consulates, shake hands, and sit across from one another at long conference tables. At the U.S. Mission to the United Nations and other international organizations, my meeting in February 2018 with Nikolai Patrushev, the secretary of the Security Council of Russia, fell into the not-so-friendly category.

Patrushev asked to meet me soon after I became national security advisor in early 2017. I agreed. I thought it important to open a routine channel of communication between the White House and the Kremlin below the level of Presidents Vladimir Putin and Donald Trump. Russia is, of course, a nuclear power, and a strained relationship is better than no relationship, if for no other purpose than to prevent misunderstandings that might increase the chance of war. There was much to discuss.

By 2017, it was clear that Russia was pursuing an aggressive strategy to subvert the United States and other Western democracies. Russian cyber attacks and information warfare campaigns directed against European elections and the 2016 U.S. presidential election were just one part of a multifaceted effort to exploit rifts in European and American society through propaganda, disinformation, and political subversion. As social media began to polarize the United States and other Western societies and pit communities against each other, Russian agents conducted cyber attacks and released sensitive information. Although Russian leaders routinely denied responsibility, the Kremlin was reportedly directing a sophisticated campaign.[1] Russia also used cyber attacks and malicious cyber intrusions to create vulnerabilities in critical infrastructure, such as in the energy sector. For example, by early 2018, the United States knew that Russia had conducted the NotPetya cyber attack that first infected Ukraine's government agencies, energy companies, metro systems, and banks.[2] It spread later to Europe, Asia, and the Americas, costing ten billion dollars in losses and damages around the world.[3]

Having studied the evolution of Russia new-generation warfare (RNGW) for years, I looked forward to talking with Patrushev to understand better the motivations behind this pernicious form of aggression that combined military, political, economic, cyber, and informational means. The day after our meeting with Patrushev, I gave a speech at the Munich Security Conference pledging that "the United States will expose and act against those who use cyberspace, social media, and other means to advance campaigns of disinformation, subversion and espionage." During my year as national security advisor, we had worked hard to impose costs on Russia. I hoped to convince Patrushev of the dangers associated with Russia's continued implementation of a strategy that pushed our two nations along a path toward worsening relations and potential conflict.

The potential for conflict with Russia was growing. The civil war in Syria was a particular concern. In March 2019, Russian general Valery Gerasimov cited the Syrian Civil War as a successful example of Russian intervention to "defend and advance national interests beyond the borders of Russia."[4] The war was a humanitarian catastrophe. Russia had supported the Syrian regime of Bashar al-Assad since the beginning of the conflict in 2011. In August 2013, the Syrian regime used poison gas to kill more than fourteen hundred innocent civilians, including hundreds of children, but it was not its first use of chemical weapons, nor would it be the last. From December 2012 to August 2014, the Syrian regime used them against civilians at least fourteen times. Despite President Barack Obama's declaration in 2012 that the use of these heinous weapons to murder civilians was a red line, the United States did not respond. President Putin likely concluded that America would not react to aggression. By the end of spring 2014, an emboldened Putin had annexed Crimea and invaded Eastern Ukraine. And then, in September 2015, Russia intervened directly in the Syrian Civil War to save Assad's murderous regime. After another massacre with nerve agents at Khan Shaykhun in April 2017, President Trump ordered the U.S. military to strike Syrian facilities and aircraft with fifty-nine cruise missiles.[5] By 2018, Russian-supported forces fighting for Assad's regime were converging with American-supported forces fighting the terrorist group ISIS. When I met Patrushev, the danger of a direct clash between Russians and Americans on the ground in Syria was not only more likely—it had already happened.[6]

On February 7, 2018, the week prior to the Geneva meeting, Russian mercenaries and other pro-Assad forces reinforced with tanks and artillery attacked U.S. forces and the Kurdish and Arab militiamen they were advising, in northeastern Syria. The mercenaries were from the company owned by Yevgeny Prigozhin, a Russian oligarch known as "Putin's cook," a man indicted by U.S. Special Counsel Robert Mueller and sanctioned by the Trump administration for his role in sowing disinformation during the 2016 U.S. presidential election.[7] It was an ill-conceived and poorly executed attack. U.S. forces and their Syrian Democratic Forces partners killed more than two hundred Russian mercenaries while suffering no casualties.[8] Eager to suppress negative news prior to the forthcoming presidential election, the Kremlin lied about the number of casualties suffered. Putin wanted to win the election by the widest possible margin. News of a costly defeat brought on by Russia's need to finance reconstruction of a country it had helped destroy would not help achieve this. The ultimate purpose of the Russian-led attack was to seize control of an old Conoco oil plant that promised to generate revenue and defray the costs of the war and reconstruction. No battle like that between Russians and Americans had ever occurred, even during the height of the Cold War.

A year had passed since Patrushev suggested we meet. I had delayed in deference to Secretary of State Rex Tillerson, who wanted to make a personal assessment of Russia's intentions first. Tillerson had hoped that his preexisting relationship with President Putin and Foreign Minister Sergey Lavrov, which he developed as chief executive officer of ExxonMobil, might deliver some improvement in U.S.-Russian relations. He wanted to offer Putin an "off ramp" in Ukraine and Syria based on the assumption that those interventions, including U.S. and European economic sanctions imposed on Russia, might entice Lavrov to negotiate an eventual Russian withdrawal. In Lavrov's case, it was not clear that he could deliver even if the possibility for improved relations existed.

Lavrov's approach to foreign policy was old-Soviet style, reflexively anti-Western and suspicious of new initiatives. Lavrov invariably accused the United States and the West of instigating the 2003 Rose Revolution in Georgia, the 2004 Orange Revolution in Ukraine, and the 2005 Tulip Revolution in Kyrgyzstan, as well as large-scale protests in Russia in 2011. It seemed that Lavrov had neither the independence of mind to come up with solutions nor the latitude to make basic decisions. By early 2018, it was clear that Tillerson's valiant efforts to find areas of cooperation with Russia had foundered. It was past time to establish a direct channel of communication between the White House and the Kremlin, other than the occasional phone calls and meetings between Trump and Putin. Since Putin had centralized power in an unprecedented way, even for a country with a long history of authoritarianism, it was important to have a relationship with someone close to Putin himself. Patrushev, Putin's right-hand man, who occupied a position that is the Russian equivalent of national security advisor, was the ideal candidate.[9]

No one on our team believed that the Geneva meeting would solve our problems with Russia. Events of the following month confirmed that belief. Soon after our meeting, Russia used a banned nerve agent in an attempted murder of a former intelligence official in Salisbury, United Kingdom, and Putin made a chest-thumping speech in which he announced new nuclear weapons. We hoped, however, that this new channel of communication between the White House and the Kremlin might lay a foundation for some bilateral diplomatic, military, and intelligence engagement with Russia across both governments. Discussions between the U.S. National Security Council staff and the Secretariat of the Security Council of Russia had existed under prior administrations. We could foster a common understanding of each nation's interests and an awareness of where those interests diverged or converged. The two countries might then manage their differences and find some areas for cooperation. Mapping our interests might be a first step toward avoiding costly competitions or dangerous confrontations like the recent clash in Syria. At the very least, we might prepare more fully for the president's meetings with Putin to secure favorable outcomes.

I traveled with Dr. Fiona Hill, the National Security Council's senior director for Europe and Russia, and Mr. Joe Wang, director for Russia. During our long flight on the "big blue plane," as we referred to the air force Boeing 757, we discussed Vladimir Putin, Russian policy, and the man whom I would soon meet, Nikolai Patrushev. Fiona is one of the foremost experts on Russia under Putin. In her book *Mr. Putin*, coauthored with Clifford Gaddy, she observed that "Putin thinks, plans, and acts strategically." She also observed, however, that "for Putin, strategic planning is contingency planning. There is no step-by-step blueprint." Our other travel companion, Joe, a bright young State Department civil service officer of ten years, judged that prospects for a near-term improvement in U.S.-Russian relations were dim mainly due to Mr. Putin's need for an external foe to prevent internal opposition. This need to direct the Russian people's attention away from internal problems drove an increasingly aggressive foreign policy, while the need to generate support for that foreign policy amplified rhetoric designed to conjure the external enemy as a menace. In his March 2018 speech announcing new nuclear weapons, Putin even showed "automated videos depicting" nuclear warheads descending toward the state of Florida.

On the plane, I recalled what I knew about Patrushev. He and Putin had a lot in common. Both entered the KGB in the 1970s. Patrushev succeeded Putin as director of the FSB from 1999 to 2008. Putin, Patrushev, and other prominent former KGB officers who moved into influential Kremlin positions after the 2000 Russian presidential elections believed that they were the ultimate patriots. Putin trusted and relied on Patrushev. Both men understood that, especially in Russia, knowledge is power. Their base of knowledge allowed them to form a protection racket that propelled Putin to the pinnacle of power and kept him there for more than two decades. The future Russian president's climb began in the late 1990s, when he was head of the GKU, the government's inspectorate charged with uncovering fraud and corruption in government and federal agencies. He used that position to build dossiers on Russian oligarchs, powerful businessmen who had accumulated great wealth during the era of Russian privatization in the 1990s. He detailed their finances and business transactions. Putin had dirt on everyone. Because the rule of law had broken down in Russia, the oligarchs regarded him as an arbiter whose persuasive power derived from holding them hostage. Putin prevented infighting that might have collapsed the corrupt system and crushed all of them. When he became the head of the FSB in July 1998, he named Patrushev as head of a new Directorate of Economic Security. He and Patrushev then used their skills as KGB case officers to collect and monopolize information. In exchange for respecting the oligarchs' property and allowing them to amass wealth, Putin expected them to act as his agents, use their business activity to promote Russian interests abroad, and comply with direction from him, their case officer, and protector.[10]

Fiona, Joe, and I landed in Geneva in the early morning of February 16, 2018. Ted Allegra, an experienced diplomat and gracious host who was the chargé d'affaires ad interim of the U.S. Mission in Geneva, greeted us. We held an informative video telephone conference with the U.S. ambassador to Russia, Jon Huntsman. Huntsman, a wise statesman, politician, and businessman who had served previously as governor of Utah, U.S. ambassador to Singapore, and, most recently, U.S. ambassador to China, worked daily in a difficult and hostile environment. The ambassador was supportive of the Patrushev meeting and the opening of a channel between Patrushev's Secretariat and the NSC staff. He described how Russian harassment of embassy officials had intensified in recent months. But he took a long view of U.S.-Russian relations and felt that we should lay the groundwork for improved relations. I met Mr. Patrushev outside the U.S. consulate. As he exited the limousine, he evinced the self-assurance one might expect from an old KGB official. Two of his senior staff, a deputy secretary of his Security Council and a senior aide responsible for the U.S.-Russia relationship, along with a staff officer serving as a note taker, accompanied him.

After introductions, I offered coffee to the Patrushev delegation—none of them touched the light refreshments we had on hand—and we sat down across from one another. I welcomed the delegation and, after mentioning my interest in Russian history and literature, reviewed the purpose of the meeting and the sustained dialogue that was meant to follow it: to develop mutual understanding of our interests. I asked Mr. Patrushev to begin. He spoke for the better part of an hour. His version of the Kremlin's view of the world revolved around three main points. First, he portrayed Russia's annexation of Crimea and invasion of Ukraine as defensive efforts to protect ethnic Russian populations from what he described as Ukrainian far-right extremists and U.S. and European attempts to engineer a pro-European Union and, therefore, an anti-Russian government in Kiev. Second, he described the expansion of NATO countries and the rotation of NATO forces into areas that Russia considered traditional spheres of influence as threatening. Third, he argued that the United States, its allies, and its partners had increased the terrorist threat across the greater Middle East through ill-conceived interventions in Afghanistan, Iraq, and Libya.[11] Finally, perhaps in anticipation of my comments, he flatly denied attacking the 2016 U.S. presidential election or attempting to subvert Western democracies. None of what Patrushev said was surprising, and I did not want to waste time rebutting his assertions or denials. Instead, I endeavored to elevate the discussion to generate mutual understanding of our vital interests in four areas.

First, I noted that both our countries were interested in the prevention of a direct military conflict. Russia's annexation of Crimea and invasion of eastern Ukraine was particularly dangerous to peace, not only because it was the first time since World War II that borders within Europe were changed by force, but also because Russia's continued use of unconventional forces in Ukraine or elsewhere in Europe could escalate.[12] One of the historical parallels to Russia's invasion of Ukraine was Austria-Hungary's invasion of Serbia in 1914, which triggered World War I. World War I was a powerful analogy because it was a war in which none of the participants would have engaged had they known the price they would pay in treasure and, especially, blood. Many people wanted war, but no one got the war he or she wanted. Moscow and Washington both needed to acknowledge the risk that the next Russian attempt might trigger a military confrontation, even if Russia intended to act below the threshold of what might elicit a military response from NATO. I wanted Patrushev to see the U.S. and European Union sanctions on Russia for invading Ukraine and annexing Crimea as more than punitive; they were meant to deter Russia from future actions that could lead to a destructive war. I thought that Patrushev might agree that we were in a dangerous, transitional period. Communicating the United States' vital interests and our determination to counter Russian aggression would disabuse Kremlin leaders of any belief that they could exploit perceived American complacency and wage new-generation warfare without risk.

I also wanted Patrushev to understand that the United States was awake to the danger of Putin's playbook and, in particular, RNGW. Russia's actions in Crimea and Ukraine were analogous to a long-standing Russian military strategy known as *maskirovka*, or the use of tactical deception and disguise. Like *maskirovka*, Putin's playbook combined disinformation with deniability. The new playbook added disruptive technologies and the use of cyberspace to enable conventional and unconventional military forces. And, where possible, the Kremlin fostered economic dependencies to coerce weak states and deter a response to aggression. The parallels to previous dangerous periods, not only in the nineteenth but particularly in the twentieth century, were striking. In recent years, Russia had acted aggressively, counting on American complacency based on the self-delusion that great power competition was a relic of the past. Consider Secretary of State John Kerry's comments: "You just don't in the twenty-first century behave in nineteenth-century fashion by invading another country on completely trumped-up pretext."[13] I thought it important to let Patrushev know that we were prepared to compete and would no longer be absent from the arena.

Second, both our nations sought to preserve our sovereignty or the ability to shape our relationships abroad and govern ourselves at home. Russia's sustained campaign of disinformation, propaganda, and political subversion was a direct threat to our sovereignty and that of our allies. I suggested that it was in the Kremlin's interest to stop this activity because Russian actions would unite Americans and other Western societies against Russia. Their recent efforts to influence election outcomes had failed or backfired. For example, Russian disinformation aimed against Emmanuel Macron in France during the 2017 presidential election increased support for the candidate and probably helped Macron win the presidency. Another example of Russia's heavy-handed tactics backfiring was the failed October 2016 coup in Montenegro that intended to prevent that country's accession to NATO. Russia's meddling actually accelerated Montenegro's admission to NATO and its application for membership in the European Union. Finally, Russian efforts to convince the Trump administration to lift economic sanctions in 2017 failed as the administration instead sanctioned more than one hundred individuals and companies in response to Russia's continued occupation of Crimea and aggression in Eastern Ukraine. More sanctions would follow under the Global Magnitsky Human Rights Accountability Act.[14] During my conversation with Patrushev, I joked that Russia's efforts to divide Americans and meddle in our election made the imposition of severe sanctions on Russia the only subject that united Congress. In fact, the first major foreign policy legislation to emerge from the U.S. Congress after President Trump took office was a sanctions bill on Russia, the Countering America's Adversaries Through Sanctions Act, which passed in the Senate in a 98–2 vote after flying through the House by a 419–3 margin.[15] At this, Patrushev cracked a smile, perhaps to acknowledge that we both were very much aware of Russia's subversive activities.

Third, both nations must protect our people from jihadist terrorist organizations. That is why it did not seem to be in Russia's long-term interests to provide weapons to the Taliban in Afghanistan or to spread disinformation that the United States supported terrorist groups. Such actions strengthened organizations that posed a common threat to both our countries. Moreover, Russia's support for Iran, Iran's proxy militias, and Bashar al-Assad's forces in their brutal campaign in Syria not only perpetuated the humanitarian and refugee crisis, but also fueled a broader sectarian conflict that strengthened jihadist terrorists like ISIS and Al-Qaeda. These terrorist organizations draw strength from the fear that Iranian-backed Shia militias generate among Sunni communities, allowing them to portray themselves as patrons and protectors of those communities. I hoped that Patrushev might see that Russia's support for Iran only reinforced jihadist terrorist recruitment and support among Sunni Muslim populations.

Finally, I raised the subject of how Russia seemed to act reflexively against the United States even when cooperation was in its interests. I used the case of Russia's circumvention of UN sanctions against North Korea as an example. In addition to the direct threat of North Korean nuclear missiles to Russia itself, a nuclear-armed North Korea might lead other neighboring countries like Japan to conclude that they needed their own nuclear weapons. Moreover, North Korea had never developed a weapon that it did not try to sell. It had already tried to help Syria develop an Iranian-financed nuclear program in an effort that was only thwarted by a September 2007 Israeli strike on the nuclear reactor under construction near Dayr al-Zawr, Syria. Ten North Korean scientists were reportedly killed in the strike.[16] What if North Korea sold nuclear weapons to terrorist organizations? What nation would be safe?

Patrushev listened but showed no discernible reaction. After a break, we agreed to charge our teams with mapping our interests and preparing materials for presentation to Presidents Trump and Putin in advance of their next meeting. I departed Geneva convinced of the importance of our work. I realized, however, that relations were unlikely to improve due to Putin's motivations, his objectives, and the strategy he was pursuing.

When he assumed the presidency at the turn of the century, Putin worked to strengthen the system that had put him there. His overarching goal was to restore Russia's status as a great power. He would be patient, estimating that Russia would need fifteen years to build strength before it was ready to challenge the West.[17] Indeed, approximately fifteen years later, he annexed Crimea, invaded Ukraine, and intervened in the Syrian Civil War.

* * *

OUR EFFORT to map Russian and U.S. interests as a way of managing our relationship addressed only one dimension of the challenge before us. That is because Putin, Patrushev, and their colleagues in the Kremlin are motivated as much by emotion as by calculations of interest. As the Athenian historian and general Thucydides concluded twenty-five hundred years ago, conflict is driven by fear and honor as well as interest.[18] President Putin and those, like Patrushev, whom he brought with him into the Kremlin were shaken by the collapse of the Union of Soviet Socialist Republics (aka USSR, aka Soviet Union) and feared the possibility of a "color revolution" in Russia. They were proud men whose sense of honor had been insulted by the West's victory in the Cold War and whose livelihoods depended on the Soviet system. Putin described the breakup of the empire and the end of Soviet rule in Russia as "a major geopolitical disaster of the century," one that not only was a "genuine drama" for those who suddenly found themselves outside Russia, but also caused problems that "infected Russia itself."[19] The breakup resulted in the loss of half of the former Soviet population and almost a quarter of its territory. At the height of Soviet dominance, Russian influence extended as far west in Europe as East Berlin. Since the USSR's collapse, Russia had lost control of nearly all of Eastern Europe. Ethnic Russians were scattered across the newly independent successor states of the USSR, such as Ukraine, Georgia, and Estonia. Russians who remember Soviet greatness, including Putin, Patrushev, and their KGB colleagues, watched as their former vassal states, unshackled from Communist authoritarian control, liberalized and eagerly joined other free and open societies under the European Union and North Atlantic Treaty Organization.

The Soviet Union had a truly global reach, penetrating into Asia, Africa, and Latin America. Throughout Communist states, Russians were often looked up to as big brothers in an ideological war—or at least that was how many Russians imagined their Communist brethren viewed them. But then men like Putin and Patrushev saw their mighty empire, one of two global superpowers, fall to the status of a struggling regional power—and it stung. Once Putin achieved the presidency, he set about restoring Russia's lost grandeur, a process that is still under way. Above all, Putin fears an internal threat to the kleptocratic political order he has built with the oligarchs and his cronies in the KGB. To allay fear and restore honor, he has consolidated his base of power internally and gone on the offensive against Europe and the United States.

By the time Putin became president, the cheerfulness associated with the prospect of transforming post-Soviet Russia into a successful state with a booming economy had given way to gloom. In the 1990s, Russian efforts to transition to a market economy proved unable to overcome the complete collapse of the Communist system. Greedy apparatchiks, members of the Soviet bureaucratic political apparatus, were empowered in the wake of that collapse. Because market reforms threatened their grip on power, Russian politicians led a backlash against free-market reformers. The failure to either establish an adequate legal framework or to eliminate Soviet-era bureaucracy made the transition to a market economy even more difficult. The final straw was the financial crisis of 1998, when the Russian ruble lost two thirds of its value. The failure of market reforms and the rise of the oligarchs created a system that was not only fragile, but also ideal for Putin and those who retained political control through the post-Soviet transition to consolidate their power. As journalist (and later Canadian foreign minister) Chrystia Freeland observed, Russia was "an ex-KGB officer's paradise." Under Boris Yeltsin's government, the Siloviki (hard-line functionaries of the Soviet-era Ministry of the Interior, the Soviet Army, and the KGB) comprised only 4 percent of the government. Under Putin, it grew to 58.3 percent. Fear of losing control as the post-Soviet economy and social structure were collapsing propelled the Siloviki into power. And Putin, Patrushev, and their Siloviki colleagues wanted Russia to be feared again.[20]

Putin, Patrushev, and the Siloviki did not view U.S. assistance in the post-Soviet period as it was intended. The United States wanted to assist Russia with the traumatic transition and reduce dangers and complications. Under the Freedom Support Act, the United States aimed to increase security by dismantling nuclear weapons through the Cooperative Threat Reduction program, provide food aid, and support Russia's transition with $2.7 billion in appropriated funds and technical assistance. However, Putin and the Siloviki viewed U.S. assistance as an affront to Russian sovereignty and an effort to exploit Russian weakness. In their telling, their former enemy lorded its Cold War victory over Russian heads, insisting on reforms that left their nation in economic meltdown. In a 2015 speech to leaders of the FSB, Putin stated that "Western special services continue their attempts at using public, nongovernmental and politicized organizations to pursue their own objectives, primarily to discredit the authorities and destabilize the internal situation in Russia."[21]

Putin made a strong debut as president. Recognizing the connection between foreign policy and popularity at home, he ferociously prosecuted a war in Chechnya against separatists and terrorists who had conducted a series of attacks against Russian civilians. That war, which caused an estimated twenty-five thousand civilian deaths between 1999 and 2002, inspired only praise in Russia and weak statements of disapproval from the West. But Putin's fears and suspicions of unrest and opposition grew as he witnessed the so-called color revolutions in Georgia in 2003, in Ukraine in 2003 and 2004, and in Kyrgyzstan in 2005, which toppled undemocratic regimes and led to the election of new presidents. He vowed that protests like those would never hit Russia: "For us, this is a lesson and a warning, and we'll do everything so it never happens in Russia." But they did. His anxiety may have peaked in 2012–2013, during widespread protests sparked by a rigged election in which he "won" 63.6 percent of the vote, according to Russian media.[22] Protests returned in 2017 and 2018, concerning corruption and an increase in the retirement age. Then, in the summer of 2019, massive protests erupted in Moscow following the removal of opposition candidates from the ballot in the Moscow city Duma elections.

Although the color revolutions and the protests in Moscow

were based on the populations' desire for freedom and improved governance, Putin saw U.S. and European hands behind them.[23] Fear and the sense of lost honor were mutually reinforcing and would continue to drive his foreign ambition. To protect himself from internal opposition and restore Russia to greatness, Putin revived Russia's nationalist mission. He portrays Russia as unafraid and designs his foreign policy to intimidate neighbors and subvert Western democracies.

★ ★ ★

RNGW BECAME Putin's playbook for surviving while weakening competitors. Russia does not have the power to compete directly with the United States and its allies in Europe and Asia. By all measures, the combined economies of the United States and European nations dwarf Russia's economy. The European Union and the United States had a combined gross domestic product (GDP) of $36.5 trillion in 2017; Russia's GDP was a meager $1.5 trillion. Russia's GDP per capita in 2017 was approximately $10,750, roughly one sixth of the U.S. GDP per capita and ranking below far less powerful countries such as Chile, Hungary, and Uruguay. Russia's economy is also woefully undiversified, with oil and gas products comprising 59 percent of all exports, leaving it vulnerable to shifting oil prices in a year that saw a 48.1 percent decline in crude oil prices. That same year, in 2014, the Russian ruble declined 45.2 percent relative to the U.S. dollar. Though the Russian economy improved under Putin, with real income doubling between 1999 and 2006 due in large measure to rising oil prices, Putin's creation of institutions and a system to maintain his exclusive control impeded economic growth and modernization. Sanctions in response to Russian aggression did not help. But corruption is the greatest impediment to investment and economic growth; Russia ranks 135th globally on the Corruption Perceptions Index.[24]

Nor are demographic trends in Russia's favor. In the past three decades, due to declining fertility rates, Russia's population dropped from 148 million in 1991 to 144 million in 2018. In addition to a fall in the birth rate, despite government incentives like payments to mothers and childcare services, declining migration also contributed to the population drop. Russia's population is expected to fall to 132.7 million by 2050. Health is also poor due to risky behaviors such as excessive drinking and smoking. In 2019, life expectancy in Russia was seventy-two years, with a one-in-four chance of a Russian man dying before the age of fifty-five. This is equivalent to life expectancy in developing nations such as Nepal and Bhutan.[25]

But, consistent with what they did as KGB case officers, Putin and Patrushev seemed less interested in building Russia up than in tearing other nations apart. In an old Russian joke, a farmer has only one cow and hates his neighbor because he has two. A sorcerer offers to grant the envious farmer a single wish, anything he wants. "Kill one of my neighbor's cows!" he demands.[26] We might think of President Putin as the farmer with only one cow. To kill his neighbors' cows, Putin employs sophisticated strategies designed to achieve objectives below the threshold that might elicit a concerted response from either the targeted state or others, such as the NATO alliance.

Rather than build Russia up to a position of predominance, Putin wants to drag others down, weaken rival states, and unravel alliance networks that give those states strategic advantages. In a 2013 article, Valery Gerasimov, the chief of Russia's General Staff (the Russian equivalent of the U.S. chairman of the Joint Chiefs of Staff), argued that "the very 'rules of war' have changed." He added that "non-military means of achieving political and strategic goals have grown and, in many cases, exceeded the power of force of weapons in their effectiveness [*sic*]."[27] These comments became known as the "Gerasimov Doctrine." Whether under the moniker of Russia new-generation warfare, the Gerasimov Doctrine, or hybrid warfare, Putin's playbook combines disinformation and deniability with the use of disruptive technologies to target states' strengths and exploit their weaknesses. It also entails cultivating economic dependencies, especially on Russian-supplied energy, while brandishing and using improved, unconventional, conventional, and nuclear military capabilities. With this playbook, Putin aims to kill his neighbor's cow and restore Russia's relative power.

* * *

TO UNDERSTAND the sophistication of the Kremlin's strategy, it is worth examining more closely one of the issues I raised with Patrushev: Russian interference in other nation's domestic politics. Efforts in 2016 to subvert democracy in Ukraine, Montenegro, and the United States demonstrated the range of actions available in Putin's playbook.

Russian propaganda has been described as a "firehose of falsehood" that spews rapid, continuous, and repetitive disinformation.[28] Typically, successful disinformation and propaganda campaigns prioritize consistency in messaging. Russia under Putin, however, abandoned consistency because the aim was not to make audiences believe something new but to question just about everything they heard; the purpose was to disrupt, divide, and weaken societies that Putin saw as competitors. The Kremlin uses many tools to fulfill that purpose, such as direct financial support of fringe political parties at the opposite ends of the political spectrum. Russian disinformation is designed to shake citizens' belief in their common identity and in their democratic principles, institutions, and processes by manipulating social media, planting false stories, and creating false personas. Russia also uses media arms such as the television network RT (formerly Russia Today) and the news agency Sputnik to broadcast a steady stream of disinformation. RT has a $300 million annual budget for broadcasting propaganda that looks like legitimate news in multiple languages. The network has more subscribers on YouTube than Fox News, CBS News, or NBC News.[29] RT and other Kremlin-sponsored media emphasize conspiracy theories designed to raise doubts over the reliability of real reporting as well as the virtue and effectiveness of democratic governance. Furthermore, as part of a number of influence mechanisms known as "active measures," even Russian leaders often make patently false statements to reinforce these disinformation efforts. The repetitive nature of these narratives is intended to portray a certain point of view as a popular perception. Additionally, the targets of this information are no longer limited to audiences at home and in the West, but are expanding into regions such as Africa, in an effort to establish Russia as a global superpower by fostering wide influence.[30]

Corrupting elections is only one part of the broader effort to kill the cow of confidence in democratic processes and institutions. After multiple attempts at engineering election outcomes across a decade, the Kremlin and its intelligence arms, the Main Directorate of the General Staff of the Armed Forces (GRU) and the Foreign Intelligence Service (SVR), improved and modified their objectives and approach. In 2004, Russian support for Viktor Yanukovych's presidential candidacy in Ukraine not only consisted of over three hundred million dollars in funding but also probably included the poisoning of his main opponent, Viktor Yushchenko. In a runoff election between the two candidates, efforts to assist Yanukovych included ballot stuffing and the busing of voters from one polling station to another to cast multiple ballots. The GRU and SVR succeeded in throwing the election to Yanukovych, but their brazen actions jump-started what became known as the Orange Revolution. Ukrainians protested against the rigged elections. The Ukrainian Supreme Court ruled the runoff results invalid, and Yushchenko, still in pain and permanently disfigured from the poisoning, was elected in a new runoff election held under the watchful eyes of more than thirteen thousand foreign observers.[31]

Five years later, in 2009, the Kremlin succeeded in engineering another Pyrrhic victory in Moldova for an anti-European Union party.[32] That party could not form a government, however, and a European-friendly party won in fresh elections later that year. Still, the GRU kept refining its methods. Returning to Ukraine in 2010, Russia finally succeeded in helping Viktor Yanukovych secure a victory in the Ukrainian presidential election. But in

2014, it failed to get him reelected, despite a massive effort that combined election influence operations with cyber warfare (including coordinated cyber attacks to fake vote totals and infect election servers).[33] Amid protests and with the country on the brink of a civil war, Yanukovych was ousted, a move the Kremlin denounced as a "coup." Following that, the pro-European Union candidate Petro Poroshenko won in a special election. After 2015, Russia expanded attacks on democratic elections in NATO and European Union countries and, in 2016, the U.S. presidential election. Although the Kremlin often fell short of the outcomes it wanted, efforts to subvert democratic processes in the United Kingdom, Germany, the Netherlands, Montenegro, Italy, Bulgaria, Austria, Spain, Malta, France, and the Czech Republic contributed to its principal goals of weakening citizens' confidence and polarizing their societies.[34]

The Kremlin has learned to tailor disinformation campaigns to the target country. In smaller countries less able to resist Russian efforts, such as Montenegro, a country of 640,000 people on the Adriatic Sea, the Kremlin made audacious attempts to determine an election outcome. In large, more distant countries like the United States, Russia prioritized undermining confidence in democratic institutions and processes. Disinformation and sabotage in Montenegro and in the United States, both in 2016, reveal how the Kremlin customizes its campaigns based on opportunities, risks, and the ability of the target country to resist and retaliate.

"The smaller we are, the more vulnerable we are," Montenegrin president Milo Dukanovic observed. In 2016, the Kremlin applied a broad range of new-generation warfare capabilities to prevent Montenegro from joining NATO and the European Union. Montenegro would be the last piece completing the jigsaw puzzle representing NATO's control of the Adriatic coast. In the run-up to the October parliamentary election, Russian state entities directed funds to the parties challenging Mr. Dukanovic's candidacy for president. Russia spread disinformation on social media and conducted cyber attacks against government and news websites. Russian agents felt especially emboldened in

Montenegro, a small country with a large Slavic ethnic minority that tends to feel a kinship with ethnic Russians. If the election results did not go the Kremlin's way, Russian agents were ready to initiate a violent coup d'état, assassinate Dukanovic, and install a pro-Russia government. On Election Day, Dukanovic was the projected winner.[35]

The coup failed. The night before, Montenegrin authorities arrested twenty Serbian and Montenegrin citizens. After the arrests, Serbia's prime minister, Aleksandar Vucic, revealed that Serbian law enforcement had uncovered the plot and passed the information on to the Montenegrin authorities.[36] Vucic soon received a visit from an angry Nikolai Patrushev. Serbia was supposed to support Russia's efforts in the Balkans. In 2017, the Montenegrin High Court tried fourteen suspects, including two members of the GRU in absentia. Russia's heavy-handed approach failed to determine the outcome of the elections or inhibit Montenegro's application to NATO. Even though opposition parties sympathetic to Russia burned NATO flags, Montenegro's parliament voted 46–0 to become the twenty-ninth member of the alliance, joining officially in June 2017. Montenegro applied for membership in the European Union not long afterward. On May 9, 2019, the court found all fourteen co-conspirators, including the two GRU agents, guilty of plotting the coup.[37]

Russian interference continued to push west, as reports surfaced about Russian agents using social media to influence the June 2016 referendum on the United Kingdom's exit from the European Union. Most analysts concluded that Russia also tried to influence the outcome of the 2016 presidential election in the United States, but recognizing limitations on its ability to do so, it focused primarily on polarizing America's polity and undermining confidence in democratic principles, institutions, and processes. It succeeded.

Conditions in America were ripe for polarization. Many Americans were frustrated with politicians who did not seem to understand them and the problems they faced. Transformations in the global economy had left many without jobs. Though there were nearly nine million more jobs in the United States in 2016 than in 2007, the gains were not evenly distributed. Many white Americans, 700,000 of whom lost jobs over those nine years, were upset about their economic outlook. Furthermore, some remained impoverished in the wake of the housing and financial crises of 2008–2009. Others were burdened with student debt. Many voters were weary of protracted and indecisive wars overseas or disappointed that government remedies for problems, such as access to health care, had not delivered as advertised. People in neighborhoods depopulated after the departure of manufacturing jobs were vulnerable to crime and drug addiction, particularly a growing opioid crisis. Some neighborhoods became "food deserts" as businesses closed and people lost access to affordable and nutritious food.*[38] Those left behind were angry over increasing income disparity and other inequities in services such as education. There was a crisis of confidence in America, and Russia took full advantage.

Donald Trump in the Republican Party and Bernie Sanders in the Democratic Party appealed to a frustrated electorate looking for iconoclastic candidates. Sanders, a self-proclaimed Democratic Socialist, had nontraditional views. The most progressive candidate in the race, he advocated for universal health care, tax hikes for the rich, free college education, and climate change legislation. His views appealed to young voters disillusioned by their work prospects and mounting student debt. In the 2016 presidential primaries, Sanders captured 29 percent more votes among those under the age of thirty than either Trump or Hillary Clinton.[39]

Trump was a threat to the Republican Party establishment just as Sanders was to the Democratic Party establishment. While many were appalled by his rhetoric and personal behavior, other voters found in Trump what they thought the country needed: an antiestablishment candidate with wholly unconventional views

---

* In the 2010 Census, the Agriculture Department estimated that 11.5 million poor Americans (4.1 percent of the U.S. population) lived in food deserts.

who spoke his mind. They liked that he was "capable of saying anything to anyone at any time and anywhere."[40] Colleen Kelley, an associate professor of rhetoric, observed that a democratic socialist from Vermont and a billionaire from New York City both took advantage of disenfranchised voters who were tired of establishment politicians whom they regarded as corrupt, out of touch, and incompetent.

Democratic and Republican political party establishments opposed the two unconventional candidates. As the general election began, many Republicans, distraught over Trump's nomination as their presidential candidate, signed "Never Trump" letters. At the Republican convention, "Never Trump" delegates made a failed attempt to revise the convention's rules package to block Trump's nomination. Similarly, party leadership of the Democratic National Committee (DNC) undermined Sanders to smooth the path for former First Lady, senator, and secretary of state Hillary Rodham Clinton to receive the nomination. Members of the DNC highlighted less favorable aspects of Sanders, such as his religious beliefs, affinity for socialist policies, and soft spot for Communist dictators. Divisions were apparent not only between the Republican and Democratic Parties, but also within them, and the Kremlin was quick to exploit those divisions.

Two years earlier, the Kremlin had intensified its use of the internet to widen political and social discord in the United States. It had learned from previous efforts, and by 2016, conditions were ideal for its refined cyber-enabled influence campaigns. The Internet Research Agency (IRA), a front organization for the GRU, turned the social network ecosystem invented in the United States against the American people, their democratic system, and their common identity. The IRA used Facebook (including its Instagram), Twitter, Google (including its YouTube, G+, Gmail, and Google Voice), Reddit, Tumblr, Medium, Vine, and Meetup to post content or support false personas. Russian agents even used music apps and games like *Pokémon Go* to reinforce themes and messages. The IRA maximized the potential of Facebook's features, including Ads, Pages, Events, Messenger, and even Stickers.

Cumulatively, it reached 126 million people on Facebook, posted 10.4 million tweets on Twitter, uploaded more than 1,000 videos to YouTube, and connected with more than 20 million users on Instagram.[41]

The IRA was persistent and sophisticated. It used vulnerabilities in the U.S. information ecosystem to exploit fissures in society. Facebook and Instagram were perfect platforms for persistent messaging to amplify divisive issues. Agents cultivated strong ties within social media groups with content designed to gain approval. They also inserted content intended to drive groups to extreme positions, sometimes pitting members against one another or against members of other groups. Some IRA content cultivated support for U.S. policy favorable to Russia, such as the withdrawal of U.S. forces from Syria or Afghanistan. Russian agents recruited witting and unwitting American accomplices. The IRA and its abettors used Twitter to bend reactions to current events in divisive ways, using "click farms" to manufacture popularity and draw attention to extreme messages. On critical topics, the IRA created "media mirages," interlinked ecosystems that surround audiences with a cacophony of manipulated content. The GRU used digital marketing best practices, evolving the page logos and typography over time.[42]

Russian manipulation was effective because of social media companies' business models, their blind avarice, and their narrow focus on functionality without due consideration for how their platform could be used for nefarious purposes. Because tech companies prioritize holding users' attention to expose them to more ads, the companies' algorithms do not prioritize truth or accuracy, but instead help disseminate fake news and disinformation. The algorithms that determine the presentation of content in social media encourage further polarization and extreme views. For example, YouTube algorithms, in suggesting which videos to watch next, guide users toward more extreme and polarizing content. Those who interact on network platforms self-segregate into homogenous groups that share beliefs on contentious issues such as gun control, climate change, and immigration. Liberals interact with liberals and conservatives interact with conservatives. The most divisive and emotional topics amplify different rather than common views. The internet and social media thus provided the GRU with a low-cost, easy way to divide and weaken America from within.[43]

Russian efforts to discredit Clinton in favor of her Democratic rival Sanders and in favor of the candidacy of Donald Trump were connected to the overall objective of weakening American society through racial, religious, and political polarization. While Russian disinformation showed a clear preference for Trump and a concerted effort to discredit Hillary Clinton, the majority of the content focused on socially divisive issues such as immigration, gun control, and race. The IRA's main effort was to foment racial division, which it accomplished by amplifying white militia content while creating content for black audiences that highlighted the mistreatment of black Americans by police. The IRA began producing videos toward that end in September 2015. Of 1,107 videos across 17 channels, most content, an astounding 1,063 videos across 10 channels, was related to Black Lives Matter and police brutality.[44]

The IRA even tried to use race and anti-immigration sentiment to support fringe groups advocating for Texas and California to secede from the United States, using the same playbook Russian agents had used in support of the Brexit movement (which culminated in the 2020 departure of Britain from the European Union). It is likely that the GRU pulled that playbook out of KGB archives. In 1928, the Soviet-led Comintern (Communist International), an organization founded to spread Communist revolution globally, planned to recruit southern blacks to advocate for "self-determination in the Black Belt." By 1930, the Comintern had initiated an operation to encourage a separate black state in the South that would expand the Communist revolution to North America.[45]

To conceal itself, the IRA used false identities co-opted from existing organization names and set them up as offshoots of real groups. These included United Muslims of America and Black

Guns Matter. The Kremlin went to great lengths to achieve its goals, even recruiting Americans to propagate Russian-backed social media messages and participate in political rallies in the physical world. When the DNC announced that it had been hacked, the IRA created the online persona Guccifer 2.0, a "lone Romanian hacker," to publish the stolen documents on a WordPress blog. To conceal its identity, the IRA used a network of computers located outside Russia, including in the United States, and paid the bills using cryptocurrency. To make it difficult to identify hostile pages, it often used portions of existing memes or local stories and modified the content to suit its purposes.[46]

Diminishing trust in authoritative sources of information, such as American mainstream media, made cyber-enabled manipulation easier for the GRU. The IRA attacked the professionalism and integrity of journalists across all community groups while portraying WikiLeaks (an organization that through anonymous sources publishes news leaks) positively. While candidate Trump further diminished confidence in the mainstream media with cries of "fake news," the Russians set up actual fake news sites. For example, at least 109 IRA Twitter accounts masqueraded as news organizations. The IRA used conspiracy theories to diminish further confidence in information. Conspiracies on Facebook and Instagram covered topics such as Hillary Clinton's health and the DNC hacks. One extreme example, dubbed Pizzagate, was a fictitious story of a pedophile ring supposedly run out of a Washington, DC, pizzeria by Hillary Clinton and her campaign chairman John Podesta. Initiated and fueled by conspiracy theorists posting it on social media sites, the ridiculous story gained traction, forcing the *New York Times* and the *Washington Post* to debunk it. The GRU promoted other fantastical stories across both left- and right-leaning sites. Yet, the mainstream media often aided Russian disinformation even as reporters tried to debunk conspiracy theories: falsehoods reported tended to be falsehoods believed.[47]

Fomenting division and diminishing confidence in sources of information complemented the GRU attack on the integrity of the

U.S. electoral process. The systems that the IRA attacked proved as vulnerable to physical manipulation as the American polity was to emotional polarization. In March and April 2016, twelve GRU officers hacked the Clinton presidential campaign, the Democratic National Committee (DNC), and the Democratic Congressional Campaign Committee (DCCC). It was easy. The GRU monitored the computers of DCCC and DNC employees, implanted malicious computer code ("malware"), and extracted emails and other documents. The malware allowed Russian agents to track employees' computer activity, steal passwords, and maintain access to the network. When the DCCC and DNC identified the attack in May 2016, the GRU employed countermeasures to avoid detection and remain on the network.[48]

A week prior to the Democratic National Convention in July 2016, the GRU used two cutouts, DCLeaks and WikiLeaks, to release 19,254 emails and 8,034 attachments. Much of the content was embarrassing and exposed the DNC's effort to bolster Clinton's candidacy and suppress Sanders's popularity, forcing the resignation of DNC chair Debbie Wasserman Schultz and other top officials. IRA-generated Facebook and Instagram posts amplified fears of voter fraud, claiming that certain states were helping Secretary Clinton win. One site stated that civil war was preferable to an unfair election. Others falsely reported that militias were deploying to polling stations to prevent fraud and called on others to join them. False reports claimed that "illegals" were overrepresented in voter polls in Texas and elsewhere, or were voting multiple times with Democratic Party assistance. The page Being Patriotic posted a hotline for tips about possible cases of voter fraud. It seems that the GRU, like most Americans, did not expect Donald Trump to win; it was prepared to foment acrimony and division through claims that the election had been rigged in Hillary Clinton's favor. After the surprising result on Election Day, the right-targeted voter fraud narrative shifted to suggestions that President Trump would have won the popular vote had it not been for voter fraud. Trump boosted those claims, asserting that millions of people had voted illegally.[49]

After the election, Russian disinformation efforts intensified, with the IRA continuing to target right- and left-leaning groups. Posts and tweets advocated for the elimination of the Electoral College. Many called for in-the-streets action and marches to protest the election outcome. The marches allowed Russian agents to cross over from cyberspace to the physical world, as they had before the election in Houston, with dueling protests fielded by IRA's weaponization of social media. In the post-election period, agents attempted to discredit individuals and institutions that advocated for strong responses to Russian aggression in Europe and the Middle East and its waging of cyber-enabled information warfare in the United States. Their attacks against conservative think tanks and their personal attacks against me and the National Security Council staff heightened in the late summer of 2017. The post-election attacks employed many of the same tactics used against Hillary Clinton in the 2016 campaign. GRU efforts to discredit Special Counsel Robert Mueller spanned almost the entire time of his investigation into Russia's attack on the 2016 election.

Both campaigns exercised poor judgment that made it easier for the Kremlin to undermine confidence in the electoral process. For example, when Russian intelligence endeavored to compromise candidate Trump and those around him through potentially lucrative real estate deals, Trump's company attempted to negotiate for a Trump Tower Moscow, offering to give Putin a $50 million penthouse in the proposed tower. The Kremlin undoubtedly knew that the proposal put Trump at risk of breaking the Foreign Corrupt Practices Act. When Russian agents approached the Trump campaign claiming to have incriminating evidence on Hillary Clinton, Donald Trump Jr. responded, "I love it."[50] The agents never delivered the promised "dirt." And the Trump campaign's hiring of the political consulting firm Cambridge Analytica to influence potential voters with microtargeted messages based on the data of more than 87 million Facebook users created the perception that social media–based tools had skewed the election results.[51]

The Clinton campaign also reinforced the GRU's efforts, working through cutouts to obtain incriminating information on Trump from Russian intelligence agents. Clinton lawyer Marc Elias hired opposition research firm Fusion GPS to produce the "Steele dossier," named for the retired British MI6 officer Christopher Steele, whom Fusion GPS had hired to solicit allegations of Trump improprieties. Steele used undisclosed Russian intelligence sources to compile a fantastic report that was published ten days prior to President Trump's inauguration.

Thus, regardless of the outcome of the election, the Kremlin was well positioned to undermine not only Americans' faith in their country's democratic processes and institutions, but also public confidence in the man or woman who would win the election. Partisan politics magnified the effects of the Kremlin's campaign and perpetuated America's vulnerability to Putin's playbook. Yet the GRU gave equal opportunity to both parties to step into the trap of *kompromat*, the use of compromising material for negative publicity or extortion—or in this case, to diminish further Americans' confidence in their leaders and in democratic processes and institutions.

The U.S. government and both presidential campaigns were not prepared for Russian efforts to compromise the candidates and the democratic process. But they should have been. David Cohen, former deputy director of the U.S. Central Intelligence Agency, observed, "We've seen Russian interference in Europe for the past ten years. We saw identical techniques: stolen information, misinformation, all of that, in a variety of countries . . . and one of the things we did not do as well as we should have was sound the alarm. . . . We didn't do a good enough job of better preparing ourselves, of saying, 'The Russians did that there, so there is no reason to think they're not going to do the same thing here.'"[52]

★ ★ ★

DENIABILITY IS critical to Russia's disinformation and subversion. As evidence mounted of its attack on the U.S. election and its sustained effort to polarize American society, Putin stated that "the Russian state has never interfered and is not going to interfere into internal American affairs, including election process."[53] Russian denials often provide opportunities for willful ignorance among those disinclined to confront Russian actions. The Clinton and Trump campaigns found it convenient to turn a blind eye to Russian subversion. The DNC was embarrassed by the documents that revealed their effort to tip the balance in favor of Hillary Clinton during the primary and did not want to draw more attention to their actions. And candidate Trump exploited Secretary of State Hillary Clinton's use of a personal, unclassified email account to access and pass highly classified material as a focal point of criticism. The Trump campaign even encouraged more Russian hacking. At a rally in Doral, Florida, in July 2016, candidate Trump called on the Russians to find the thirty thousand missing personal emails that Clinton had deleted during her tenure as secretary of state. As for the Obama administration, after it became aware of the hacking, it did not respond in a concerted manner. Later, President Trump, because he conflated Russia's attacks on the election with the legitimacy of its outcome, seemed to take Vladimir Putin at his word when the former KGB case officer denied it. As late as July 2018, President Trump continued to cast doubt on the universal finding of all U.S. intelligence and law enforcement agencies that Russia had in fact interfered in American elections in 2016. On July 16, 2018, President Trump stated at a news conference with President Putin in Helsinki, Finland, "They said they think it's Russia; I have President Putin, he just said it's not Russia. I will say this: I don't see any reason why it would be."[54]

Disinformation enables Russian deniability. The Kremlin manipulates the news to confuse and sow doubt. Peter Pomerantsev, a former Russian television producer, identified the goal as the creation of an environment in which "people begin giving up on the facts."[55] A particularly egregious example came after British intelligence identified the two Russian agents who had attempted to murder Sergei Skripal and his daughter, Yulia, with a nerve agent in Salisbury, placing more than 140 people at risk, including children. Later, a British woman died after she handled the container the nerve agent was in. Russia claimed that the two agents were in Salisbury to visit cathedrals, and insinuated that they were a gay couple touring the United Kingdom. Russian officials propagated dozens of lies and conspiracy theories about the Skripal attempted murder through media and Twitter accounts. Indeed, the *Washington Post* likened the Kremlin's effort to "an elaborate fog machine to make the initial crime disappear." Stories blamed a toxic spill, Ukrainian activists, CIA agents, British prime minister Theresa May, and Skripal himself.[56]

President Trump authorized strong action in response to the Skripal poisoning, including the expulsion of sixty undeclared Russian intelligence officers and the closing of the Russian consulate in San Francisco, California.[57] He did so after a concerted effort across the U.S. government to understand the facts and coordinate a global response with the United Kingdom and other allies. I spent many hours on the phone trying to get some reluctant allies and partners to take resolute action. Although roughly twenty countries joined the United States and the United Kingdom in condemning the attack and expelling at least some Russian agents (153 were expelled worldwide), some allies' actions were disappointing. The Kremlin's disinformation efforts had been at least partially successful. The response to the attempted murder of Skripal revealed that the European consensus that emerged after the shock of Russia's annexation of Crimea and invasion of Ukraine was wearing off. Even in the face of an egregious attack using a banned nerve agent that could have caused hundreds of casualties in a NATO member country, some U.S. allies in Europe were reluctant to take strong action against Putin. They rationalized the attack as score settling with a defector. Putin's playbook had worn down their resolve. When I told President Trump about the small numbers of agents expelled from Germany and France, he was incensed; he felt that European nations should take more responsibility for their own security. He was right.

Denial enabled by disinformation allows Putin to get away with murder, literally. Russia has issued implausible denials to its role in crimes ranging from the attempted murder and murder of individuals to mass murder. In 2006, Russian agents murdered defector Alexander Litvinenko in London with polonium, a radioactive material that causes a slow and painful death. In 2009, Sergei Magnitsky, a Russian tax accountant who knew too much about President Putin and the oligarchs who surround him, was murdered in a Moscow prison. In 2015, Boris Nemtsov, an opposition politician, was murdered on a bridge just outside the Kremlin. And in 2017, Denis Voronenkov, an exiled critic of Mr. Putin's and a former member of the Duma, the lower house of the Federal Assembly of the Russian Federation, was gunned down in Kiev, Ukraine. The Kremlin murdered journalists as well as politicians. Notably, in 2006, Anna Politkovskaya, who was famous for her coverage of the Chechen wars.[58]

The Kremlin denied the Assad regime's undeniable use of chemical weapons and indiscriminate bombing to murder innocents in Syria. Yet, not only did it deny these events, but the Kremlin also produced and spread disinformation about them. When Assad's forces murdered more than seventy innocents with nerve agents on April 7, 2018, even before the attack occurred, Russia began to claim that there was intelligence of a potential chemical attack planned by Islamist militant groups. Later, the Kremlin claimed the attack was a false-flag event designed to blame Russia.[59]

Russia never took responsibility for murdering 298 people in the shootdown of Malaysia Airlines Flight 17 over Ukraine on July 17, 2014. Photos and video of the missile's route captured on social media, evidence of the missile launches on social media, and evidence at the wreckage site all proved incontrovertible, but when asked about Russia's role, Putin responded, "Which plane are you talking about?"[60]

Disinformation creates confusion about what to believe. Deniability, in the Kremlin case, fosters a sense of helplessness and incites fear concerning what Russia might do to target the United States and other free and open societies. As the philosopher and mathematician Bertrand Russell observed, those "who live in fear are already three parts dead." Putin's playbook generates a destructive cycle. Fear consumes compassion and contributes to the polarization and weakening of the targeted society.

* * *

RNGW IS designed to accomplish the Kremlin's objectives short of major armed conflict. But conventional military strength is important to intimidate weak neighbors and deter U.S. and NATO forces from responding to Russian aggression. Here Putin faces a challenge: Russia does not have the defense budget to compete with the United States and its NATO allies, either in advanced conventional weaponry or in the ability to conduct integrated land, aerospace, maritime, and cyberspace operations—what the U.S. military calls joint warfighting. But just as the internet and social media provided opportunities to revise *maskirovka* (tactical deception and disguise), Russia has integrated disruptive technologies into its military to exploit perceived U.S. and NATO vulnerabilities.

Since he took office, Putin led an ambitious program of military reform that integrated new technologies, improved discipline, and reorganized the force. There was much work to do. What was supposed to be a short, easy war to reestablish Russian control over Chechen territory lasted from December 1994 to August 1996, ending with the Russian Army's humiliating withdrawal. The Chechen War was a nightmare for poorly trained, underfed, ill-equipped, and undisciplined Russian soldiers, who sometimes surrendered to the enemy without a fight or sold arms to the Chechens for food or drugs and alcohol.[61]

As it undertook massive reforms in the 2000s, the Russian military did not try to match U.S. and NATO capabilities. Instead of exquisite systems, Russia invested in cheaper combinations of air defense, offensive cyber and electronic warfare, drones, long-range missiles, and massive artillery. This approach seemed to work. During the annexation of Crimea and the invasion of

Ukraine in 2014, the Russian military established air supremacy from the ground with sophisticated air defenses rather than expensive stealth fighter jets.

A reforming Russian military was emboldened under Putin. In the years following the invasion of Ukraine, Russia routinely held large military exercises in the Baltic Sea and on its most western border, staring down NATO, and joint exercises with the Chinese in far eastern Russia. Russian naval vessels and aircraft conducted dangerous intercepts of U.S., allied, and partner aircraft and vessels, including in the Nordic-Baltic region. An annual large-scale exercise named Zapad ("West") drew its name from strategic military exercises designed to demonstrate the military strength of the Warsaw Pact along the Soviet Union's "Western Front" during the Cold War. As with RNGW, Russia's conventional force prowess, shown through these maneuvers and the positioning of nuclear-capable missiles, was meant to have a psychological effect on NATO.

Still, given U.S. conventional military tactics and Russia's limited defense budget, traditional military reform was insufficient to incite fear and restore Russian national prestige. Putin was determined to expand Russia's nuclear arsenal and announce a nuclear doctrine designed to intimidate NATO countries and weaken the alliance. Its doctrine of "escalation control" called for the threat of early use of a nuclear strike on Europe to pose a dilemma for the United States: risk a nuclear holocaust or sue for peace on terms favorable to Russia. In developing the capability to enact its doctrine, the Kremlin violated the 1988 Intermediate-Range Nuclear Forces Treaty.

While threatening nuclear war, Russia demonstrated how its offensive cyber capabilities threatened the United States and its allies. Prior to its attack on the 2016 U.S. election, Russia conducted malicious cyber intrusions targeting U.S. critical infrastructure.[62] The Kremlin had already revealed its capabilities overseas. On Christmas Eve 2015, the lights went out in Eastern Ukraine, affecting more than two hundred thousand people. It was the first time that a cyber attack switched off a nation's power grid. As the

Russians were turning off the lights in Ukraine and hacking the Democratic National Committee, they inserted malicious code into American water and electric systems, as they had attempted to do earlier at U.S. nuclear power plants.[63] Like the nuclear strategy of escalation control, Russian cyber threats to infrastructure are meant to intimidate and deter the United States and other NATO allies from responding to Russian aggression against a member of the alliance.

* * *

ECONOMIC COERCION through dependence on Russian energy is another powerful tool of Putin's, as it augments the threats from Russia's conventional, nuclear, and cyber capabilities. The countries that gained independence from Soviet control after the demise of the Warsaw Pact and the dissolution of the Soviet Union are particularly vulnerable because they inherited a transportation and energy infrastructure that depends on Moscow. Moscow demonstrated in Belarus, Ukraine, Armenia, Tajikistan, and Kyrgyzstan that it will restrict access to energy supplies or use energy pricing tactics to coerce target countries. In 2010, Russia forced Ukraine to grant a twenty-five-year extension of the lease to its Black Sea Fleet's base in Crimea, one of the bases Russia used to annex Crimea by force years later. Moscow used economic coercion to convince Kyrgyzstan and Armenia to join the Eurasian Economic Union, an organization designed to compete with the European Union and extend Russia's influence over former Soviet territory.[64]

Even Germany made itself vulnerable through policy choices that eliminated alternative sources to Russian natural gas. Shameless corruption played a role. In 2005, during his final months as chancellor, Gerhard Schroeder gained approval for a multibillion-dollar Nord Stream pipeline project with the Russian state gas company, Gazprom, to transport gas from Russia to Germany. Soon after he left office, Schroeder became chairman of the pipeline shareholders' committee. In April 2017, Nord Stream II AG, the developer of the pipeline, signed a deal for a second pipeline under the Baltic Sea called Nord Stream 2 that will double the amount of gas being transported from Russia to Germany through the Baltic Sea. The pipeline not only deepens Germany's dependence, but also punishes Ukraine, which will lose up to $2 billion a year in transit fees, or 1.5 percent of its GDP. The pipeline also denies transit fees to other NATO and European Union members, including Poland, Slovakia, and Hungary. The Polish government deemed the pipeline a "hybrid weapon" born of Moscow and intended to divide the European Union and NATO.[65] In early 2020, the U.S. Congress placed sanctions on companies completing Nord Stream 2 in an effort to "stop construction." It was too little too late, however, as the pipeline was nearing completion. The sanctions succeeded mainly in souring relations between the United States and Germany.

* * *

GERMANY IS the most prosperous and powerful nation in Europe, which makes it a particularly attractive target for Putin—for weakening Europe is the priority in Putin's effort to break apart the post–Cold War order and reestablish Russia as a great power. Indeed, the continent remains the principal battleground for RNGW's combination of disinformation, denial, disruptive technology, and dependence to sow division and exhaust the will of European nations, the EU, and NATO.

Writing about the period between the world wars, the diplomat and historian George Kennan observed that Soviet diplomacy depended not on the "strength of their ideas," but on "the weakness of the Western community itself: from the spiritual exhaustion of the Western people."[66] Although the challenges Europe faced in the 2000s paled in comparison to the trauma of the First World War, they were sufficient to generate spiritual weariness if not spiritual exhaustion. At the end of the Cold War, Europe celebrated freedom from Soviet domination and Communist totalitarianism, but as newly freed peoples confronted practical problems such as inefficient agriculture and tired industries, solidarity based in newfound freedom slowly gave way to economic concerns and a tendency to take their rights for granted.[67]

To diminish confidence in the European Union, the Kremlin exploited events, magnifying crises such as the global financial crisis of 2008; the European financial crisis in 2015 that strained the euro; the refugee crisis from 2011 onward associated with the civil war in Syria and violence from Afghanistan to North Africa to the Maghreb; Britain's referendum to exit the European Union in 2016; the *gilets jaunes* ("yellow vest") movement in France in 2019, and the rise over several years of nativist, secessionist, and Euroskeptic political parties in Spain, Hungary, Italy, and Poland.[68] The weakening of Europe began well before the turn of this century, however. Economic internationalism of the 1990s affected Europe as well as the United States. Factories moved to cheap labor markets, and many citizens were left behind by the transforming global economy and growing income inequality.[69] As the promise of free-market capitalism at the end of the Cold War confronted the realities of lost jobs and income disparity, skepticism of the European Union grew. The Union expanded rapidly, growing from twelve countries to twenty-eight between 1995 and 2013. As the introduction of the euro moved monetary policy away from the states and EU bureaucracy and regulations grew, so did sentiment across the Continent that "faceless" men and women in Brussels, including the 732 elected members of the European Parliament, were usurping national sovereignty. Skepticism of the European Parliament and unbounded globalization gave rise to populist parties that polarized European politics and created opportunities for Russia to weaken NATO and the European Union.

Strained transatlantic relations between the United States and Europe created still more opportunities for Russia. The Atlantic alliance was adrift for much of the post–Cold War era. Although Europe supported the United States after the terrorist attacks of September 11, 2001, many Europeans opposed the 2003 invasion of Iraq, and opposition to that war grew in ensuing years. After President Barack Obama continued to withdraw the bulk of U.S. forces from Europe and violence spread across the greater Middle East, centered on the rise of ISIS and the civil war in Syria, some Europeans blamed U.S. precipitous disengagement for emboldening Russia and exacerbating crises. When President Obama declared his desire to pivot toward Asia, many Europeans concluded that he was turning his back on seven decades of transatlantic partnership. Then, in 2016, Republican candidate for president Donald Trump professed profound skepticism about NATO, suggesting that the alliance was "obsolete." He described the European Union as a competitor rather than a union of allies and like-minded nations that shared democratic principles. After President Trump took office in January 2017, his revival of the "America First" slogan seemed to herald disengagement from Europe and the abandonment of American leadership of the postwar international order. President Trump's sudden decision in October 2019 to withdraw U.S. special operations forces from northeastern Syria surprised NATO allies and put their forces in a precarious position. Those allies saw the lack of consultation as an example of diminished U.S. commitment to NATO and Europe. Soon thereafter, French president Emmanuel Macron described the European Union as on "the edge of a precipice," noting that the combination of Great Britain's impending departure from the Union and the EU's internal divisions could make it "disappear geopolitically." He also partly blamed Trump for the Union's struggles, saying that the U.S. president "doesn't share our idea of the European project." At the end of 2019, after starting a new diplomatic initiative with Russia, Macron stated that NATO was experiencing "brain death" and asked rhetorically how Turkey could remain a member of the alliance and still purchase sophisticated defense systems from Russia.[70]

Putin, of course, took advantage of tensions among European nations and between the United States and Europe. Indeed, in the first two decades of the twenty-first century, he has been undeterred due to a lack of unity among these allies, to diminished confidence among them, and to their failure to impose costs on the Kremlin sufficient to force Putin to abandon his playbook. Putin's perception of Europe as weak, combined with the unenforced red line in Syria in 2013, almost certainly contributed to his 2014 decision to annex Crimea and invade Eastern Ukraine. Perceived European impotence and American reluctance probably contributed to other Kremlin decisions, such as attacking elections in Europe and the United States, the Skripal poisoning in the United Kingdom, and support for Assad's use of chemical weapons to commit mass murder in Syria. Despite sanctions on Putin's regime and the Russian defense industry, Nord Stream 2 was a reminder of how Russia could extend its influence in Europe despite egregious violations of international law and infringements on European sovereignty. While subverting Europe politically, Russia was rewarded financially and gained coercive economic influence.

Russia's appearance of strength, however, belied significant weaknesses that cut across its economy, demographics, public health, and social services. As former U.S. secretary of state Madeleine Albright has observed, Putin's Russia played a poor hand well.[71] The United States and its allies, particularly in Europe, played a much better hand poorly. Or, as the Stanford professor of international relations Kathryn Stoner pointed out to me, understanding the game that is being played is more important than the face value of the cards a player holds. Understanding the Kremlin's strategy and the fear, sense of lost honor, and ambitions that drove its actions is the first step in parrying Putin's playbook and protecting our free and open societies.

CHAPTER 2

Parrying Putin's Playbook

A country that does not respect the rights of its own people will not respect the rights of its neighbors.

—ANDREI SAKHAROV

THE UNITED States and Europe were ill-prepared for Russia's toxic combination of disinformation, denial, dependence, and disruptive technologies. Responses to the Kremlin's sustained campaigns of subversion not only were slow and inadequate, but also tended to aid and abet those who sowed dissention and division. In her November 21 testimony before the House of Representatives Intelligence Committee, Dr. Fiona Hill observed that "when we are consumed by partisan rancor, we cannot combat these external forces as they seek to divide us against each other, to degrade our institutions and destroy the faith of the American people in our democracy."[1] When we are considering how best to counter Russian aggression, Hill's admonishment is the best starting point. Putin seeks to divide; Americans and Europeans should not divide themselves. Putin employs disinformation; Americans and Europeans should restore trusted sources of information. Putin cultivates dependencies; Americans and Europeans should depend more on each other and like-minded nations. Putin employs disruptive technologies to compensate for Russia's weaknesses; Americans and Europeans should counter those efforts and maintain their considerable competitive advantage. The United States and European nations should be confident.

When I met Patrushev in Geneva, the combination of fear and injured pride was palpable. He appeared tough, but his was the kind of toughness that comes from bitter disappointment, in his case, in a system he had spent his whole life defending. The Soviet Union was corrupt to its core, but Patrushev had been taught to look—from the inequality in a system that professed egalitarianism; from the brutality that belied the Soviet concept of social justice for workers as a fraud; and from the cynical patriarchy in the Stalinist order that during and after World War II killed six million of its people and put approximately one million others in barbaric prisons in which many more perished.[2] Patrushev's long face carried his disappointment and anger over the collapse of the corrupt system he had worked so hard to perpetuate. And since 2000, he had joined Putin in an effort to recreate that system—but not exactly. He, Putin, and the Siloviki dropped all pretense of egalitarianism, doubling down on nationalism and pride in Mother Russia. And they added a strong dose of greed, as both Putin and Patrushev became personally wealthy at the top of their corrupt system. Blaming the United States for their failings became a habit, and competition with the United States and Europe was necessary to distract the Russian people from those failings. It was also natural. Putin, Patrushev, and the Siloviki define themselves and their system, as they did during the Cold War, based on a perceived threat from the West.[3]

Because the Kremlin's base motivation is unlikely to change while Putin is in power, the United States, its allies, and like-minded partners must parry Putin's playbook and, in particular, its critical components of disinformation, denial, dependence, and the use of disruptive technology. Because the Kremlin's objective is to divide and weaken the United States and Europe from within, defeating Putin's sophisticated strategy will require strategic competence and a concerted effort to restore confidence in democratic principles, institutions, and processes.

Putin's aggressive behavior and Patrushev's tough demeanor mask Russia's underlying vulnerability and diminishing power relative to the United States and Europe. In 2019, Russia's GDP was roughly equivalent to that of the state of Texas and smaller than Italy's.[4] After Russia's annexation of Crimea and invasion of Ukraine, NATO countries finally increased defense spending. Excluding the United States, their combined $299 billion budget in 2019 compared favorably to Russia's 2018 defense budget of $61.4 billion. The U.S. defense budget alone was eleven times larger, at $685 billion in 2019.[5] But although Europe and the United States enjoy tremendous comparative advantages over Russia, parrying Putin's playbook requires mobilizing those advantages and remediating vulnerabilities that the Kremlin exploits.

* * *

AFTER TWO decades of Putin's rule, Russian aggression itself may be most effective at restoring trust in democratic principles and institutions. As the historian and author Timothy Garton Ash observed in 2019, Europe recognizes that it faces an existential threat of disintegration, "like the prospect of death, that concentrates minds."[6] Concentrating minds might lead to the abandonment of flawed assumptions that have, in the past, masked the growing threat. Because of flawed assumptions, the United States and its European allies, in spite of all good intentions, have allowed and, at times, encouraged Russian aggression.

I return here to the idea of strategic narcissism, for America's failure to develop an effective response to Russian aggression was based in it and, in particular, in a kind of wishful thinking: that after the collapse of the Soviet Union, Russian leaders would accept the status quo. Multiple U.S. administrations neglected the emotional drivers behind Putin's actions. Even after the pattern of Russian attacks and subversion of European nations was undeniable, over-optimism about prospects for change in Russian policy delayed effective responses.

Nearly eight years before I met Nikolai Patrushev in Geneva, Secretary of State Hillary Clinton met her counterpart, Foreign Minister Sergey Lavrov, in that same city. Only seven months had passed since Russia's invasion of Georgia, the first war in history in which cyber attacks were used in combination with a military offensive and a sustained disinformation campaign. Clinton presented Mr. Lavrov with a "reset button" meant to symbolize a fresh start in the relationship. She described her reset attempt as "a very effective meeting of the minds" that she hoped might lead to "more trust, predictability, and progress."[7] Optimism about the reset policy grew as work progressed on the New Strategic Arms Reduction Treaty (New START), which reduced the number of strategic nuclear missile launchers by half and limited the number of deployed strategic nuclear warheads. Also positive was Russian support for the expansion of a Northern Distribution Network to supply U.S. troops in Afghanistan and new sanctions on Iran. In March 2012, President Obama was caught on an open microphone whispering to Russian president Dmitry Medvedev—Putin would return to the presidency from the position of prime minister two months later—that he would have "more flexibility" after the U.S. presidential election in November of that year. Obama was referring to the potential for a new arms agreement, but his comment also communicated to Medvedev a willingness to overlook Russia's transgressions in the interest of making progress on that and other priorities.[8] Seven months later, during his reelection campaign for president, Obama mocked his opponent, Senator Mitt Romney, for describing Russia as a geopolitical foe: "The 1980s are now calling to ask for their foreign policy back. Because the Cold War's been over for 20 years."[9]

Over-optimism led to complacency as the Obama administration pursued a Russia policy based on its hopes to work with the Kremlin rather than the needs to deter and defend against Russian aggression. Those hopes soon vanished when Russia annexed Crimea, invaded Ukraine, intervened in Syria, hacked the Clinton campaign and the DNC, and attacked the 2016 presidential election. In the 2000s, as the Russian threat grew more complex and sophisticated, the United States wrongly assumed that Russia's goals aligned with those of the United States. It believed that diplomatic efforts could bring the Kremlin in from the cold to join the community of responsible nations and abandon its disruptive behavior. Psychologists define optimism bias as the tendency of those beginning a treatment to believe in the success of the treatment even if the result is uncertain. President Obama and Secretary Clinton were not the first to succumb to optimism bias and wishful thinking while pursuing improved relations with Russia, nor would they be the last.

In the summer of 2001, President George W. Bush met with President Vladimir Putin and reported that "I looked the man in the eye. I found him to be very straightforward and trustworthy. We had a very good dialogue. I was able to get a sense of his soul; a man deeply committed to his country and the best interests of his country. And I appreciate so very much the frank dialogue."[10] Putin's talent for deception and manipulation was on full display as he told President Bush a fabricated story about how he had saved from a fire in a *dacha* a cherished Russian Orthodox cross given to him by his mother and worn around his neck. As he would later do with the Clinton Foundation and the Trump organizations, he tried to do a favor for President Bush by arranging a lucrative job in a Russian oil company for one of Bush's friends.[11] By the end of his second term, Bush was forced to revise his assessment of Putin. In August 2008, as both presidents were in the receiving line to greet Chairman Xi Jinping at the opening of the Beijing Olympics, Russian forces were invading Georgia.

President Donald Trump continued this trend of U.S. presidents believing that they could appeal to mutual interests, build personal rapport with Putin, improve the relationship between Washington and Moscow, and change Russian strategic behavior. Trump often stated that improved relations with Russia "would be a good thing, not a bad thing." The candidate was appreciative of Vladimir Putin's flattery, stating in December 2015, "When people call you brilliant, it's always good, especially when the person heads up Russia."[12] Trump treated some of Putin's most brazen criminal actions with dismissiveness and moral equivalency. For example, in 2017, when asked by Bill O'Reilly in a Fox News interview if he respected Vladimir Putin even though, as O'Reilly stated, "he's a killer," the president responded, "There are a lot of killers. You think our country's so innocent?"[13] President Trump, in public statements made before and after his election, appeared to waver in his determination to hold Russia accountable despite policy decisions that strengthened Europe's defenses and imposed significant costs on Putin and those around him, mainly in the form of sanctions. Indeed, the president seemed at times to abet Russian disinformation and denial. For example, after U.S. intelligence agencies determined unequivocally that there had been a Russian attack on the 2016 U.S. presidential election, President Trump described his conversation with Putin: "He said he didn't meddle. I asked him again. You can only ask so many times. But I just asked him again, and he said he absolutely did not meddle in our election. He did not do what they're saying he did." Commenting further in Helsinki, Finland, in July 2018, after a one-on-one meeting with Mr. Putin, the president stated that "I have great confidence in my intelligence people, but I will tell you that President Putin was extremely strong and powerful in his denial today."[14]

While some speculated that President Trump sometimes appeared to be an apologist for Russia and Mr. Putin because the Kremlin was extorting him with damaging evidence of business improprieties or embarrassing personal conduct, Trump's over-optimism about improving Russian relations fit a pattern of optimism bias and wishful thinking across two previous administrations.[15] And the unreciprocated efforts to improve relations with Putin left U.S. presidents vulnerable to the KGB case officer's subterfuge. At the 2018 press conference in Helsinki, when asked directly by a reporter if he had "compromising material" on President Trump, Putin did not give a direct answer. He responded, "Well, distinguished colleague, let me tell you this, when President Trump was in Moscow back then, I didn't even know that he was in Moscow. I treat President Trump with utmost respect, but back then when he was a private individual, a businessman, nobody informed me that he was in Moscow. . . . Do you think that we try to collect compromising material on each and every single one of them? Well, it's difficult to imagine utter nonsense on a bigger scale than this. Please disregard these issues and don't think about this anymore again."[16] Putin was never going to be Donald Trump's friend. He used the Helsinki summit to undermine the U.S. president and keep alive speculation about the slanderous contents of the Steele dossier.

The basis for Trump's persistent optimism bias, even as Putin undermined him publicly, had an added dimension. For some of the self-proclaimed strategists around President Trump, the pursuit of improved U.S.- Russian relations despite continued Russian aggression was based mainly on two rationalizations: first, a misunderstanding of history and an associated nostalgia for the alliance with the Soviet Union during World War II; and second, a peculiar sense of kinship with and affinity for Russian nationalists. This latter rationalization is based on a perceived commonality of interest in confronting Islamist terrorism and protecting what these Trump strategists regarded as wholesome and predominantly Western, Caucasian, and Christian cultures from dilution through multicultural, multi-ethnic, and multi-religious immigration.[17] In a July 2018 interview with Tucker Carlson of Fox News, President Trump said that the characterization of Russia as an adversary was "incredible" because of the country's tremendous sacrifices during World War II. "Russia lost 50 million people and helped us win the war," President Trump said. Some Americans and Europeans view Russia as the repository of a purer version of Christianity and, under Putin, a bastion of conservatism that is protecting Western civilization from postmodern ideas that are anathema to some conservatives.[18]

But both rationalizations are fundamentally flawed. The alliance with the Soviet Union in World War II was an "alliance of necessity." In the midst of that war, Russia had initially tried to stay out of the conflict by signing the cynical Molotov-Ribbentrop Pact, which resulted in the brutal dismemberment of Poland and the inevitable annexation by the Soviet Union of the three Baltic states. It was only when Nazi Germany turned on its accomplices that the Soviets found themselves unexpectedly fighting on the side of the Western Allies and (after the Pearl Harbor attacks of December 1941) the United States. It was an alliance that Soviet dictator Joseph Stalin had tried his best to avoid; he had been hostile to the governments and people of the West.[19] The only factor that held the unlikely allies together was Adolf Hitler. And while it is true that the Soviet Union bore the largest sacrifice of fighting in terms of lives lost, once the war ended, the alliance of necessity dissolved and gave way to a cold war between the two powers.

Despite the U.S. desire to regard Russia as an erstwhile ally grateful for American bloodshed for a common cause and the $11.3 billion in U.S. assistance under the Lend-Lease policy, Russia's memory of the alliance in World War II does not evoke warm feelings among Kremlin leaders.[20] Some Russians view U.S. and U.K. delays in opening a second front in France as an intentional effort to allow the Soviets and the Germans to bleed each other to death on the Eastern Front. And they believe their exclusion from the joint American-British effort to build an atomic bomb was part of a plan to dominate the Soviet Union and the postwar world. If the prospect of improved relations with Putin relied on a natural confluence of interests with respect to Europe or to Russian nostalgia for the World War II alliance against Nazi Germany, that prospect was dim.

Ignorance of history combined with bigotry to generate another source of delusional thinking about Putin's Russia. Some Americans were easy targets for Russian disinformation because they felt a kinship with and a cultural affinity for Russia as a defender of social conservatism and Christianity. That basis for optimism about improved relations with the Kremlin was not confined to the United States; it was even more prevalent in parts of Europe. For example, Hungarian prime minister Viktor Orban expressed alignment with Russia, declaring that Hungary would be "breaking with the dogmas and ideologies that have been adopted by the West" in order to build a "new Hungarian state." Some saw Putin as a modern-day crusader who was protecting

Christianity from Islamist terrorists after U.S. interventions in the Middle East made the world less secure. The Russian Orthodox Church, which acts as an arm of the Kremlin and Russian intelligence services, praised Putin's intervention in Syria as part of the "fight with terrorism" and a "holy battle." Russia actively cultivates these feelings of racial and religious kinship to further polarize and weaken Western resolve to confront the Kremlin's aggression.[21]

* * *

ONCE SPECIOUS rationalizations for seeking improved relations as an end in themselves are rejected, we can develop a consistent strategy designed not only to defend against Russia's ongoing campaign and deter further aggression, but also to set conditions for a post-Putin era in which Russian leaders recognize that they can best advance their interests through cooperation rather than confrontation with the West. The public and private sectors both have an important role to play. Because Putin's playbook depends so heavily on disinformation and denial, defense begins with exposing the Kremlin's efforts to sow dissension within and between nations.

Governments have powerful tools available to identify malicious cyber actors and act against them. Law enforcement and sanctions against individuals and organizations engaged in political subversion have proven effective. Because most of the evidence that underpins indictments and sanctions is public, the results of law enforcement investigations, such as the Mueller Report in the United States, are particularly valuable in pulling back the curtain on Russian cyber-enabled information warfare. Named after Special Counsel Robert Mueller, the report on the two-year investigation into Russia's attack on the 2016 presidential election exposed the level of Russian interference in the 2016 presidential campaign and the overall effort to divide Americans through cyber-enabled information warfare.[22] Armed with the Mueller investigation and other sources of information, on March 15, 2018, the Trump administration placed sanctions on Russian individuals and companies, including those associated with the IRA and GRU.[23] The U.S. Department of Justice also announced criminal counts against twenty-six Russian nationals and three Russian companies.[24]

In the fight against Putin's playbook, citizens and their representatives in government have an important role to play. As Fiona Hill suggested, they might resolve, at the very least, not to be their own worst enemies. The reaction to the Mueller Report among President Trump, his supporters, and his opponents demonstrated how political divisiveness can mask what all sides should have agreed upon: that Russia attacked the U.S. election and that the attacks continued beyond the election to divide Americans and reduce confidence in democratic principles, institutions, and processes. Some glossed over that point of agreement to either echo President Trump's description of the investigation as a "witch hunt" or to claim that the report did not go far enough to reveal either "collusion" between the Trump campaign and the Russians or obstruction of justice on the part of the president. As Hill observed in her testimony before the impeachment inquiry committee in November 2019, Russia's goal was to put the U.S. president, no matter who won the election, "under a cloud." She warned that those who support fictional narratives reinforce the Kremlin's campaign.

Sadly, Hill's observation seemed to fall on deaf ears. February 2020 revelations that Russia was using disinformation to bolster the candidacies of Bernie Sanders and Donald Trump spurred President Trump to fire Acting Director of National Intelligence Joseph Maguire as the president dismissed Russia's continued subversion of America's democratic process as a "hoax." Meanwhile, some Democrats resurrected the already investigated allegations that Trump was somehow in collusion with the Kremlin. Putin could not have written a better script. Social media remained the Kremlin's weapon of choice.[25]

Deterring Russian aggression in cyberspace requires more than a purely government response. While the U.S. National Security Agency (NSA) has exquisite capabilities to attribute actions in cyberspace, it is often reluctant to do so because attribution might reveal its tools and methods. The scale of the problem alone demands efforts across the public and private sectors. Social media and internet companies must continue the work they began after the 2016 election to expose and counter disinformation and propaganda. Facebook, which took the most blame for the vulnerabilities its system created, identified and deleted Russian bots on both Facebook and Instagram. Facebook also had Cambridge Analytica (the UK-based firm that harvested data from millions of people's Facebook accounts without consent) delete Facebook data and improve users' awareness of how to strengthen their privacy features. Twitter identified and deleted bots as well. But Russian bots and trolls adapted, trying to stay ahead of those protecting infrastructure, exposing disinformation, or countering denial. Moreover, these defensive actions did not adequately address the safeguarding of personal data that Russian or other malign actors might use in cyber-enabled information warfare or the economic incentives that drive users toward extreme content.

Because social media companies have economic incentives to gather and use personal data to generate revenue (mainly through advertising), regulation may be necessary. A combination of removing the cloak of anonymity for some users (such as advertisers), protecting individuals' personal data, and requiring internet and social media companies to be held liable for damaging compromises of data or blatant abuse of their platforms are all actions that could shift industry incentives in favor of protecting against disinformation and denial. Regulation may also help ensure that those companies do not become the arbiters of freedom of speech in democratic societies.

Private-sector efforts can be particularly valuable in countering Russia because they are unclassified and can be released to the media, the public, and law enforcement. For example, a private company helped attribute responsibility for the attempted murder of Sergei and Yulia Skripal in Salisbury, United Kingdom. Bellingcat, an international research and investigation collective, conducted an open-source investigation to identify the attackers and connect them to their GRU offices. Combinations of government and private efforts counter Russian denial in the physical as well as the cyber world. Private-sector efforts can create a firehose of truth to counter the Kremlin's firehose of falsehoods. Still, no combination of private- and public-sector efforts to expose and defend against RNGW activities will solve the problem once and for all. As attacks on the 2020 presidential election made clear, Russian agents will adapt continuously to avoid detection, circumvent defenses, and launch new offensives.

Countering cyber attacks such as data theft or damage to systems must go beyond the "perimeter" defense or even so-called defense in depth, in which multiple layers of security controls are placed throughout an information technology system. Because capable state actors such as Russia can penetrate elaborate defenses given adequate time and resources, defense requires a good offense. Organizations like the NSA in the United States conduct continuous reconnaissance in cyberspace to identify and preempt attacks before they can penetrate the system perimeter. Furthermore, the creation of U.S. Cyber Command in 2010 marked crucial efforts to "direct, synchronize, and coordinate cyberspace planning and operations to defend and advance national interests in collaboration with domestic and international partners."[26] The command is tackling the problems of integrating and scaling capabilities to the magnitude of the threat under the concept of "increasing resiliency, defending forward, and constantly engaging."[27] This form of active defense appeared successful during the U.S. midterm elections in 2018. In keeping with U.S. Cyber Command's doctrine of persistent engagement and causing problems for adversaries before they penetrate our systems, cyber operators reportedly blocked internet access to the IRA on the day of the elections.[28] In the future, private-sector firms are likely to participate in active defense. Attribution and punishment after attacks will remain important, but these have proven inadequate to deter or prevent attacks designed to cripple critical systems or infrastructure, extort victims, or wage cyber-enabled information warfare. Amplified sharing of information and expertise within government and between government and industry should help provide protection to the dot-com, dot-gov, and dot-mil internet domains.

Just as the GRU and SVR learned from early attempts to undermine Western democracies, the United States and other nations might learn from countries on the Russian frontier that were on the receiving end of Putin's playbook. In 2007, Estonia came under a sustained cyber offensive that included disinformation, denial, and cyber attacks on infrastructure. Its disagreement with Russia over the relocation of a World War II statue sparked an offensive that began with distributed denial-of-service attacks, a flood of internet traffic that overwhelms servers and shuts down websites. Estonians lost access to news outlets, government websites, and bank accounts. Russian media stoked the crisis with disinformation, such as false reports that Soviet war graves were being destroyed. Russia has continued this sustained campaign of disinformation and propaganda for over a decade.[29] Recognizing that it needed improved and sustained defenses, Estonia, under the direction of President Toomas Henrik Ilves, formed a civilian cyber defense reserve unit, attained a high level of security through end-to-end encryption and two-factor authentication, and constantly monitored systems for potential threats. Estonia demonstrated that a clear understanding of the problem, determined leadership, a comprehensive strategy, and close cooperation across public and private sectors can defend successfully against malicious actors. Estonia's cybersecurity now includes high-functioning e-government infrastructure, digital identity, mandatory security baselines, and a central system for identifying and responding to attacks. Private-sector service providers are reviewed to assess and reduce risk. President Ilves recalled that citizen and private-sector involvement is essential to effective defense and that the keys to success were incentivizing online security measures, implementing cybersecurity public education programs, and constantly monitoring cyberspace and power grids.[30]

In Finland, the government sought the participation of its citizens in an effort to track, fight, and prevent cyber attacks. Mandatory cyber education is meant to "bolster Finland as an information society" and contribute to cyber research. The National Cyber Security Centre is accessible to all. It provides information, posts vulnerabilities, and runs "exercises" that include simulated cyber attacks to increase awareness of vulnerabilities and motivate organizations to protect themselves. The goal is to achieve comprehensive security.[31]

Although it is challenging to scale up efforts of small countries like Estonia and Finland to a country the size of the United States, their examples demonstrate the potential for collaboration not only between government and industry, but also with academia and civil society.

Karen Edwards, a Silicon Valley executive who helped build the Yahoo brand, one of the prime internet companies, knew that defensive measures were inadequate to counter Russian disinformation. She was angry about what was happening to her country as Russians and extremist groups polarized society and created a crisis of confidence in democratic processes and institutions. Present at the creation of the first Silicon Valley boom in the 1990s, Edwards was, even then, both enthusiastic about the promise of the internet and wary about how it might be abused by nefarious actors. Two decades later, the Stanford University and Harvard Business School graduate had an idea of how a good offense in the presentation of information might extend beyond defense and preempt elements of cyber-enabled information warfare. Edwards and her business partner, Raj Narayan, started a company called Soap AI in Palo Alto, California. The company is based on an innovative solution that combines the positive potential of the internet with emerging artificial intelligence technologies to defeat those using disinformation and propaganda to polarize American society. After diagnosing the problem as information overload, mistrust of media, and a lack of diverse perspectives,

Edwards, Narayan, and their team designed a machine-learning platform that allows users to understand better what is happening in the world by accessing verified sources of information, reducing the clutter associated with clickbait, and ensuring access to a range of perspectives.[32] Soap AI uses a "scrub cycle" in which artificial intelligence sorts stories from verified sources across a range of perspectives. Soap presents multiple opinions on an event or story so readers can make their own judgments based on correct information.[33] A dynamic combination of offensive, defensive, and preemptive solutions to the disinformation dimension of Putin's playbook could turn what the Kremlin and other authoritarian governments consider a vulnerability—that is, America's inherent decentralization and resistance to authoritative direction from the center—into a strength.[34]

Education is vitally important, not only to alert citizens to the dangers of Russia's disinformation campaigns but also to restore confidence in democratic principles, institutions, and processes. A Russian proverb describes education as light and ignorance as darkness. A public informed about challenges to national security and to issues that adversaries use to sow dissension, such as race, gun control, and immigration, will prove less vulnerable to manipulation. Education inoculates society against efforts to foment hatred and incite violence on the basis of race, religion, politics, sexual orientation or any other sub-identity.

Finally, education combined with the restoration of civility in public discourse can reduce the vitriol that widens the fissures in society that Russia and others exploit. A renewed focus on civics education in the United States and other Western societies is important to deter and defeat Russia's campaigns of disinformation and denial. While keeping in mind the importance of self-criticism, civics curricula in Western nations might emphasize the virtues of those nations' free, open, and democratic societies. For example, while acknowledging that the American experiment is flawed and incomplete, curricula in the United States might ensure that citizens appreciate the nobility of an unprecedented multi-century effort to ensure democracy, individual rights, equal opportunity, and liberty for all. Political leaders and the media have a vital role in that connection; as do citizens, who might resolve, whenever they discuss points of disagreement, to give at least equal time to points of agreement. Cindy McCain, the widow of Senator John McCain, initiated a program to promote civility in public discourse. As in cyber defense, improving education and restoring civil discourse will take broad public involvement.

All these efforts are especially critical on the European continent, which remains on the front lines of Putin's aggressive ambitions. The United States and European and other democratic nations must recognize that parrying Putin's playbook requires strong collective action. The first step is to regain self-respect not only within individual nations, but also among them as free and open societies.

Europe needs to regain psychological as well as physical strength. The European Union will be only as strong as its members. Europe's larger states could and should lead; in the next decades, Germany and France are particularly critical to the strength of Europe and the strength of the transatlantic alliance. So are the states that won freedom after the Iron Curtain fell. Will European culture sustain the common identity essential to generating the will to defend itself? The Kremlin is betting that the answer is no. In 2019, Putin declared that liberalism had "become obsolete."[35] It will depend on European leaders and their citizens to prove him wrong.

* * *

AS THE United States, NATO, and others counter Russian efforts that fall below the threshold of a military response, they should not underestimate the danger of Russian conventional military and nuclear capabilities. NATO conventional and nuclear strength remains an important deterrent of further aggression, as an emboldened Kremlin could miscalculate and take actions that spark a disastrous military confrontation. It is for this reason that NATO member states should fulfill the pledge made at the Wales Summit in 2014 to invest the equivalent of 2 percent of their GDP in defense. After the end of the Cold War, Europe's military power atrophied because of the belief that great power competition was a relic of the past. Because Putin is trying to collapse the alliance, it is possible that he could, based on an assessment of weak resolve within NATO nations, precipitate a crisis in the Baltics or elsewhere. He might actually want a target nation to invoke Article 5 of the NATO Charter, which states that "an armed attack against one or more of them in Europe or North America shall be considered an attack against them all," and then attempt to dissuade other NATO nations from recognizing the attack. That failure would deliver a hard psychological blow to the alliance.

The United States and its NATO allies must also develop and field capabilities that counter Russia's disruptive military technologies, including its new nuclear weapons. The United States and NATO are behind in countering Russian capabilities in electronic warfare, layered air defense, and a range of other disruptive capabilities designed to close the gap in advanced military technologies.[36] Important conventional capabilities include missile defense, long-range precision fires, and air defense against drones. The U.S. withdrawal from the 1988 Intermediate-Range Nuclear Forces Treaty in response to Russia's violation was necessary to maintain deterrence in Europe and make clear that Russia's irresponsible doctrine of escalation control could only lead to catastrophe for all parties. And should Russia and other nations, such as China, agree to negotiate a treaty limiting or eliminating types of intermediate-range weapons, as the INF Treaty did in 1988, or to preserve New START, the treaty signed in April 2010 (and the follow-up to START I, which reduced the number of strategic nuclear missile launchers by half and established an inspection and verification regime), the United States should remain ready to enter into verifiable agreements that limit the scale and scope of the most destructive weapons on earth. As the United States and NATO invest in future military systems, they should keep

Russian countermeasures in mind and design simple, less expensive systems that degrade gracefully, rather than complex, expensive systems susceptible to catastrophic failure.

The combination of actions, initiatives, and capabilities to parry Putin's playbook should aim to deter Russia by denial—that is, by convincing the Kremlin that it cannot accomplish its objectives through its pernicious form of aggression, the use of military force, or nuclear extortion under its doctrine of escalation control. Should Russian aggression continue or expand, however, the United States and like-minded nations should be prepared to exploit the Kremlin's many vulnerabilities. Those include Putin's and the Siloviki's personal vulnerability to public scrutiny, the Russian people's growing desire for a say in how they are governed, and the frailty of an economy overburdened by corruption, self-imposed isolation, and a demographic time bomb.

* * *

TRUTH AND transparency are important offensive as well as defensive weapons to defeat the Kremlin's use of lies and obfuscation. Putin's brash actions internationally belie both his weakness and his fear of losing power. His rule, extended by sham elections, has grown old. When he returned to the presidency after a break as prime minister that was meant to give the illusion that he respected the Constitution, he encountered massive protests. People were angry in part because Putin raised the retirement age even as more Russians became aware that he had become a billionaire many times over only by looting the country.[37] By 2019, protests were a regular occurrence in Russian cities as Putin's popularity dropped. In regional elections in September of that year, Alexei Navalny, an anticorruption activist and lawyer, developed what is known as a "smart-voting" strategy. Creating a list of candidates across the country who he believed could defeat those backed by Putin's United Russia party, he urged opposition-minded citizens to vote for those on that list. The strategy resulted in victories for a record 160 candidates across the country[38] and showed that, even as Putin controls the media and restricts the opposition, elections still matter in Russia.[39] Exposing Putin's personal finances and the finances of the Siloviki who surround him may further embolden a Russian opposition movement that has survived despite brutal Kremlin repression and absolute control of the media, parliament, courts, and security services. Support for opposition groups, anticorruption organizations, and surviving investigative journalists in Russia is an appropriate counter to Putin's playbook and a way to communicate support for the Russian people while countering Kremlin aggression.

It is likely that opposition to Putin will grow with time as he enters his third decade in power. In January 2020, Putin proposed changes in the Constitution designed to extend his rule beyond 2024. The changes stripped power away from the presidency and empowered the Parliament and the State Council. Prime minister Dmitry Medvedev resigned. Medvedev was given a new position that had not existed previously, deputy chairman of the Security Council. Putin's choice for Medvedev's successor as prime minister, Mikhail Mishustin, is a technocrat previously tasked with modernizing the Russian Federal Tax Service. Without his own political base, Mishustin seems unlikely to impede Putin's authority. Indeed, the choice is broadly seen as a way to allow Putin to retain de facto power, perhaps by making policy though the newly empowered State Council.

Putin is very sensitive to the truth, especially about his personal life and his finances. When a small Moscow paper reported that he had divorced and was engaged to a famous gymnast, he had the paper shut down. When the vast data leak known as the Panama Papers revealed in 2016 how he had secured lucrative deals for his friends in exchange for a cut of the profits, he deflected attention from this by accusing Hillary Clinton of inciting protests in his country.[40]

* * *

RUSSIAN ECONOMIC dependence on its energy sector is another key vulnerability, and the European nations should take advantage of it. Reducing dependence on Russian oil and gas presents an opportunity to impose costs greater than Putin and the stagnant Russian economy can bear. And because it is possible that, faced with continued economic stagnation, Putin will precipitate another crisis to stoke Russian nationalism and distract the Russian people from their discontent, dependence on Russian energy could prove to be a security as well as an economic liability for Germany and other European states.

Finally, despite Putin's 2020 proposal to change the Constitution and create a position from which he can continue to wield power, the United States, Europe, and other like-minded nations should think comprehensively about a post-Putin Russia. It is important to recognize that Russian society is still emerging from a traumatic period of transition that witnessed the collapse of the Soviet Union, the birth of a new Russian state, changed borders, and transformation of its economic and political systems. It was, as former U.S. secretary of state Condoleezza Rice observed, simply "too much to overcome."[41] Although the West will have limited influence over how the transition from Putin to a new order occurs and how that new order addresses the challenges and opportunities facing Russia, the United States and other nations might prepare now to play a supportive role.

Lessons from Russia's failed transition should inform that support. A post-Putin government would have three options: repression, serious reform, or an incompetent execution of the first two options. The West should approach a post-Putin Russia with the goal of welcoming it into a Euro-Atlantic security system that aims to preserve peace and promote prosperity. If the "power vertical" (i.e., the recentralization of the power of the presidency and federal center) that Putin helped to create collapses, America and other nations, informed by failed efforts in the 1990s, should support democratic and institutional development in Russia.[42] Preparation might begin with an expansion of grassroots programs, exchanges, and education programs (such as the Fulbright scholarship) that circumvent Putin's repression of civil society to reach the Russian people directly. And the failed effort to promote reform in Russia in the 1990s should also lead to the recognition that reform in Russia will depend on the Russian people. Still, as Secretary Rice observed, "Russia is not Mars and the Russians are not endowed with some unique, antidemocratic DNA."[43]

The West must remain open to the possibility, however, that a new government may not abandon the Kremlin's aggressive policies and may instead perpetuate Putin's playbook. Under Putin, the Kremlin's fears are about losing power *internally*; a rush to allay imagined Russian fears of the West would be foolhardy. Deterrence should remain a top priority for the United States and NATO, so that it is clear to whoever follows Putin that further expansion and continued subversion would be too costly.

★ ★ ★

IN THE near term, Putin and Russia have forged a "comprehensive strategic partnership" with Chairman Xi Jinping and the Chinese Communist Party. President Putin described China as "our strategic partner." Chairman Xi reciprocated with "We've managed to take our relationship to the highest level in our history." Additionally, he referred to President Putin as his "best friend and colleague."[44] The two authoritarian regimes aid and abet each other in their mutual effort to collapse the postwar political, economic, and security order.[45] In 2017, joint military exercises in the Baltic Sea signaled to the world the start of this new partnership. In 2018, China joined Russia's annual military exercise in Siberia for the first time. The following year, India and Pakistan were also invited to join the exercise otherwise known as Vostok, meaning "East." Also, in 2019, hundreds of Russian and Chinese military flights violated U.S. allied airspace from the Baltic Sea to the Sea of Japan. On July 23, 2019, a joint Russian-Chinese flight of bomber aircraft entered the air defense identification zone of South Korea and Japan, triggering intercepts by fighter jets from both countries.[46] In December 2019, Russian and Chinese ships joined the Iranian navy for an exercise in the Indian Ocean and the Gulf of Oman. Also, trade between Russia and China increased significantly, growing from $69.6 billion in 2016 to $107.1 billion in 2018, with the increased trade even conducted in the nations' own currencies as a step toward reducing reliance on the U.S. dollar.[47]

The two countries' warming relationship has resurrected discussions of the Nixon administration's triangular diplomacy with the Soviet Union and China. Under triangular diplomacy, the United States endeavored to have a closer relationship with each nation than each nation had with the other. Under Putin and Xi, however, the prospects for improved relations are dim. A Russian grand alliance with China is unnatural because Russia would be a minor and weaker party. Putin's playbook has made clear that great power competition is not a relic of the past. He and Xi are drawn together, in part, because they are both authoritarian leaders who are determined to undermine free and open societies. Despite Putin's dangerous aggression against the United States, Europe, and the rest of the free world, the danger from Xi Jinping's Chinese Communist Party is greater based on the scale of the challenge and the pernicious nature of China's strategy. For all Putin's brazen attacks against our country, it is China that in many ways presents the larger, and more complicated, threat to the United States.

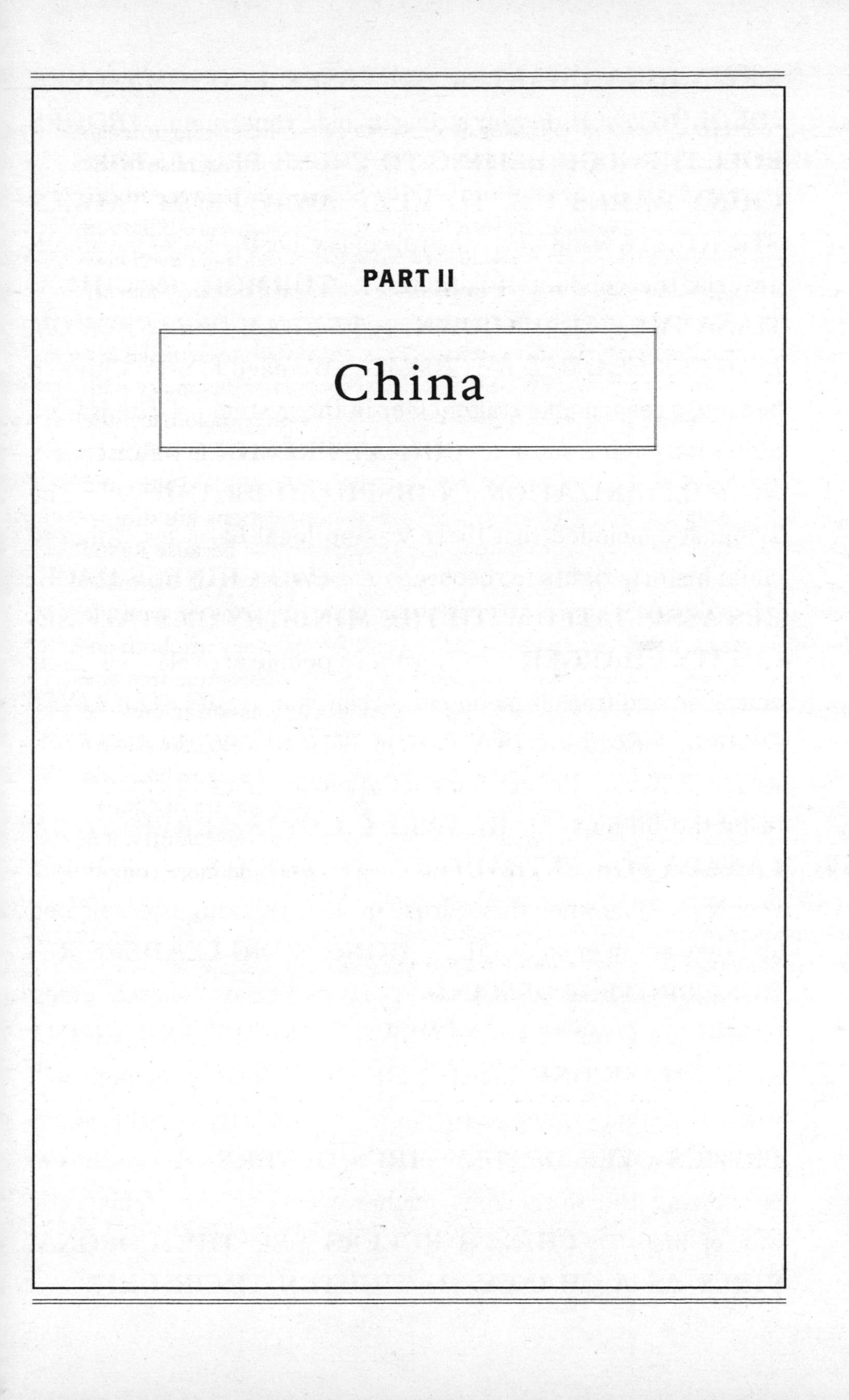

PART II

China

NEW CHINA CHARTER STRESSES ECONOMY OVER IDEOLOGY . . . Hide your strength, bide your time . . . **TROOPS ROLL THROUGH BEIJING TO CRUSH PROTESTORS** . . . **CHINA WARNS U.S. TO KEEP AWAY FROM TAIWAN STRAIT** . . . A world where the rule of law, not the law of the jungle, governs the conduct of nations . . . **TURMOIL IN CHINA: TIANANMEN CRACKDOWN** . . . *VIETNAM DRILLS FOR OIL IN BLOCK CLAIMED BY CHINA* . . . We need to urge China to become a responsible stakeholder in the system . . . The United States is a Pacific power . . . **CHINA'S PRESIDENT PLEDGES NO MILITARIZATION IN DISPUTED ISLANDS** . . . The Tribunal concluded that there was no legal basis for China to claim historic rights to resources . . . **TWO CHINESE HACKERS ASSOCIATED WITH THE MINISTRY OF STATE SECURITY CHARGED** . . . After these people are released, their education and transformation must continue . . . *BATTLE LINES DRAWN: A FULL-BLOWN TRADE WAR BETWEEN AMERICA AND CHINA* . . . I'm getting a lot of money in tariffs its coming in by the billions . . . **HUAWEI C.F.O. IS ARRESTED IN CANADA FOR EXTRADITION TO THE U.S.** . . . Freedom is only possible when this "virus" in their thinking is eradicated and they are in good health . . . **HONG KONG LEADERS REBUFF PROTEST DEMAND** . . . If one of these sides is going to lose, it's going to be the NBA . . . **TRUMP AND CHINA SIGN PHASE ONE TRADE DEAL** . . . The agreement will work if China wants it to work . . . **CHINA SILENCES CRITICS OVER DEADLY VIRUS OUTBREAK** . . . Taiwan is showing the world how much we cherish our democratic way of life . . . **CHINA'S RULERS SEE THE CORONAVIRUS AS A CHANCE TO TIGHTEN THEIR GRIP** . . .

RUSSIA
Nur-Sultan
KAZAKHSTAN
Ulaanbaatar
MONGOLIA
Harbin
KURIL IS.
Bishkek
KYRGYZSTAN
Ürümqi
TAJIKISTAN
Vladivostok
Sapporo
SEA OF JAPAN
Beijing
N. KOREA
Pyongyang
Seoul
S. KOREA
JAPAN
Tokyo
Ōsaka
Islamabad
PAKISTAN
YELLOW SEA
CHINA
Chengdu
Chongqing
Wuhan
Shanghai
EAST CHINA SEA
RYŪKYŪ ISANDS
New Delhi
NEPAL
Kathmandu
BHUTAN
Thimphu
TAIWAN STRAIT
Taipei
TAIWAN
BANGLADESH
Dhaka
INDIA
Guangzhou
Macau
Hong Kong
NORTH PACIFIC OCEAN
MYANMAR
Naypyidaw
LAOS
Hanoi
HAINAN
Vientiane
VIETNAM
Yangon
THAILAND
Bangkok
CAMBODIA
Phnom Penh
Ho Chi Minh City
BAY OF BENGAL
SOUTH CHINA SEA
Manila
PHILIPPINE SEA
PHILIPPINES
SPRATLY ISLANDS
SRI LANKA
Sri Jayewardenepura-Kotte
Melekeok
PALAU
BRUNEI
Bandar Seri Begawan
Kuala Lumpur
MALAYSIA
SUMATRA
Singapore
BORNEO
INDONESIA
0 500 1000 1500 km
0 500 1000 mi

CHAPTER 3

An Obsession with Control: The Chinese Communist Party's Threat to Freedom and Security

He who controls the past controls the future. He who controls the present controls the past.

—GEORGE ORWELL, *1984*

I HAD never been to China when I arrived with the president and his traveling party on November 8, 2017. From my first day on the job in the White House almost nine months earlier, China was a top priority. U.S.-China policy was a prominent feature of the Trump presidential campaign. The Chinese were eager to arrange a summit between the two presidents at Mar-a-Lago after the warm and successful visit of Prime Minister Abe Shinzo of Japan just three weeks after the president's inauguration. China figured prominently in what President Obama had identified to President Trump as the biggest immediate problem his administration would face: what to do about North Korea's nuclear and missile programs.[1] I thought it was vital to frame a long-term strategy for China prior to the Mar-a-Lago Summit, scheduled for April 2017, so that the initial discussions between the two leaders and the two main working groups that Secretary of State Rex Tillerson was eager to initiate (one on security, the other on trade and economic relations) would be informed by policy goals and more specific objectives.

In March, the Principals Committee of the National Security

Council convened to recommend the agenda and objectives for the Mar-a-Lago Summit. At the outset of the meeting, I highlighted the fundamental assumption underpinning U.S. relations with China since paramount leader Deng Xiaoping initiated market reforms and the opening of China in 1978: after being welcomed into the international political and economic order, China would play by the rules, open its markets, and privatize its economy.[2] And as the country became more prosperous, the Chinese government would respect the rights of its people and liberalize. I observed that the intentions, policies, and actions of the Chinese Communist Party (CCP) had rendered those assumptions demonstrably false. The party has no intention of playing by the rules associated with international law, trade, or commerce. China is a threat to free and open societies because its policies actively promote a closed, authoritarian model as an alternative to the rules-based order. This matters because the CCP aims to accomplish its objectives at other nations' expense. In particular, its strategy intends to shift global economic leadership and geopolitical alignment toward China and away from the United States. It was past time, I believed, to effect one of the most significant shifts in foreign policy and national security strategy in recent American history.

I spent most of my career in Europe and the Middle East. I knew a little Chinese history but had a lot to learn. My "professor" was the NSC senior director for Asia, Matt Pottinger. Matt learned Chinese in high school in Boston and then studied for two years overseas in Beijing and Taiwan. He was fluent not only in Mandarin, but also in Chinese history and literature. He covered China as a journalist for eight years, including as a reporter for the *Wall Street Journal*. When covering the devastating 2004 tsunami in Southeast Asia, Matt was impressed by the effectiveness, discipline, and compassion of the U.S. Marines who provided humanitarian relief. He decided to seek an age waiver and join the Marine Corps as an officer. After nine months of intense physical preparation, he reported, at the age of thirty-two, to Officer Candidate School in Quantico, Virginia. Subsequent assignments in Iraq and Afghanistan as an intelligence officer convinced him that the United States had tremendous capabilities but did not always employ those capabilities well. American leaders often failed to understand the nature of complex competitions with enemies and adversaries. Matt came to the attention of then–Major General Michael Flynn. In 2010, the two coauthored, with Paul D. Batchelor, a monograph entitled *Fixing Intel*. In 2015, when Donald Trump declared his candidacy for president, Flynn had retired from the army as a lieutenant general, and Pottinger had left the Marine Corps and was working at an investment firm in New York City. After Trump was elected, Flynn, who was named national security advisor, asked Pottinger to join the NSC. I was grateful for the opportunity to serve with Matt. He worked tirelessly to get sound policies and strategies in place not only for China and North Korea, but also for the entire Indo-Pacific region. And his sense of humor brought much-needed levity to the hard work and long hours on the NSC staff.

The discussions at Mar-a-Lago were meant, in part, to convey a significant change in U.S. policy. We communicated to our counterparts that we were particularly concerned about China's aggressive actions in the South China Sea, where their People's Liberation Army was building islands in order to lay claim to 1.4 million square miles of water through which approximately three trillion dollars of trade flows each year. But the main theme of the summit was China's unfair trade and economic practices, which we described as a form of economic aggression that the United States could no longer tolerate. The discussions were cordial. I got the impression from our guests that they had heard these points before, believed that time was on their side, and doubted our will to back any of our concerns with action.

But when Air Force One touched down in Beijing nearly seven months later, a new China policy was largely in place. Government bureaucracies were shifting away from an approach that regarded China's growing power at the West's expense as an inevitable phenomenon that was best accommodated rather than challenged. The new policy acknowledged that we were in a competition with China—a competition that the United States was losing because of a failure to understand the emotions, ideology, and aspirations that motivated Chinese Communist Party policy. Since the 1990s, U.S. policy toward China betrayed all the elements of strategic narcissism: wishful thinking, mirror imaging, confirmation bias, and the belief that others will conform to a U.S.-developed "script." China aided in that self-delusion as the CCP used co-option and coercion to tighten its control internally and extend its influence internationally while concealing its true intentions. Our two days in Beijing heightened my sense of urgency to infuse our approach to China with a strong dose of strategic empathy.

The first step was to appreciate the influence of historical memory on Chinese Communist leaders. John Fairbank, the godfather of American sinology, noted in 1948, in the first edition of his seminal book *The United States and China*, that to understand the policies and actions of Chinese leaders, "historical perspective is not a luxury, but a necessity."[3] During the visit, Chairman Xi Jinping and his advisors used history to convey messages to President Trump, to the Chinese people, and to the world. The selective use of history—both the history they invoked and the history they averted—revealed the emotions and worldview that drive Chinese Communist Party goals. State Councilor Yang Jiechi, my Chinese counterpart, had decided on a "state visit plus" of tremendous grandeur that would take us to three sites adjacent to one another at the center of Beijing: the Forbidden City (the seat of Chinese emperors across five centuries), the Great Hall of the People (a vast building completed in 1959 as part of the tenth anniversary of the founding of Communist China), and Tiananmen Square (the site of Mao's mausoleum and the focal point of the massive protests against Chinese Communist Party rule in 1989 that the People's Liberation Army brutally suppressed).

* * *

OUR HOSTS were State Councilor Yang, Foreign Minister Wang Yi, Chinese ambassador to the United States Cui Tiankai, and Vice Foreign Minister Zheng Zeguang. Our party included U.S. secretary of state Rex Tillerson, White House senior aide Jared Kushner, U.S. ambassador to China Terry Branstad, White House press secretary Sarah Huckabee Sanders, U.S. trade representative Robert Lighthizer, and director of social media and assistant to the president Dan Scavino. Chairman Xi and his wife, the famous singer Peng Liyuan, greeted President and Mrs. Trump at the gates of the Forbidden City. The leaders and their wives moved ahead of the rest of our party. As we walked through the West Glorious Gate, I looked for Matt Pottinger, who had been walking behind us. I discovered later that the guards had denied him access. Matt knew too much. Our hosts were gracious, but clearly intended to use the visit to the Forbidden City to convey messages without the encumbrance of someone who might subject those messages to skepticism and keen appraisal.

The main message was consistent with a speech Xi Jinping had delivered just two weeks earlier, at the Nineteenth National Congress in the Great Hall of the People: the Chinese Communist Party was relentlessly pursuing the "great rejuvenation of the Chinese nation." The Forbidden City was the perfect backdrop for Chairman Xi to place in historical perspective his determination to "take center stage in the world and to make a greater contribution to humankind."[4] The visit portrayed this great rejuvenation as an inevitable return to an earlier era during which Imperial China was a powerful "Middle Kingdom (中国)." The Forbidden City was built during the Ming dynasty, which ruled China for 276 years (from 1368 to 1644), a period considered one of China's golden ages, in which its economy, territorial control, and culture reached unprecedented efflorescence. It was during this dynasty that Zheng He, an admiral in the Ming fleet, embarked on seven voyages around the Western Pacific and Indian Oceans, more than half a century before Columbus. Zheng He's "treasure ships," among the largest wooden ships ever built, brought back tribute from all parts of their known world. But despite the success of the seven voyages, the emperor concluded that the world had nothing to offer China. Citing the expense of the fleet, he ordered the treasure ships scuttled and Chinese ports closed. Xi viewed the nineteenth and twentieth centuries as an aberrational period during which European nations and their colonies, and then, later, the United States, achieved economic and military dominance.[5] The 2017 visit to the Forbidden City was meant to depict China's increasingly active foreign policy as a return to a natural order. The Forbidden City was the destination for foreigners to bow before the emperor's authority, pay tribute, and to supplicate for privileges that the emperor might bestow upon them.

Xi wanted his visitors to recognize as inevitable that Chinese power would once again underpin an international system in which Chinese leaders granted privileges in exchange for recognition of China's superiority. The visit by the U.S. president and First Lady was, in part, a continuation of the "coming-out party" to announce China's return to power that began with the spectacular opening ceremony at the Beijing Olympics in 2008. Like that show and the closing ceremony that placed modern technological innovation in the context of five thousand years of Chinese history, the tour of the city and the remarkable performances based on three Chinese operas before dinner were reminders that Chinese dynasties stood at the center of the earth and that Chinese emperors were the guarantors of harmony on earth and the arbiters between earth and heaven.

The chairman's message was meant for the Chinese people as well as for President Trump and our party. We walked in the footsteps of countless foreign delegations, such as the British mission, led by Lord George Macartney, that visited the Qianlong emperor of the Qing dynasty in the 1790s. Traveling from the port to the Forbidden City, the British found the route lined with banners emblazoned with large characters announcing that the Europeans had come to "pay tribute to the Great Emperor." During President Trump's visit, state television's live coverage served the same purpose: to show the Chinese people a foreign power acknowledging China's and Chairman Xi's power. Qianlong and Chairman Xi both saw the narrative of national greatness as necessary for maintaining domestic order.[6]

While the images broadcast to China and the world from the Forbidden City were meant to project confidence in the Chinese Communist Party, they belied profound insecurity. Like Chairman Xi, the emperors who occupied the Forbidden City practiced a remote and autocratic style of rule vulnerable to corruption and internal threats. Since the end of the Han dynasty, in AD 220, China's core provinces were ruled only half the time by a strong central authority. Even when China was ruled by powerful governments such as that of the Han, Tang, Song, Ming, or Qing dynasties, China was subject to domestic turmoil and foreign invasion. As with those who went before him, Xi's outer confidence masked a sense of foreboding that he might suffer a fate similar to that of previous rulers.[7] A few months later, in 2018, Chairman Xi did away with term limits and extended his rule indefinitely.

In its very design, the Forbidden City seemed to reflect the contrast between leaders' outward confidence and inner apprehension. Our guide walked us through the three great halls at the city's center: the Taihedian, the Hall of Supreme Harmony (where the emperor presided over ceremonies), the Zhonghedian, the Hall of Central Harmony (where officials kowtowed to the emperor before ceremonies in the Taihedian), and the Baohedian, the Hall of Preserving Harmony (where emperors held banquets to entertain heads of state, kinsmen, and government ministers). Those grand structures were meant not only to impress, but also to defend from threats that might come from inside or outside the city's walls. The emperor was housed at the center of the walled complex for protection and was constantly surrounded by his guards. The emperors who sat on the elaborate throne in the Hall of Central Harmony made decisions based largely on fear and anxiety.

For example, the Yongle emperor Zhu Dai, who built the Forbidden City, having overthrown his nephew to take power, was even more concerned about internal dangers than he was about the possibility of another Mongol invasion. To identify and eliminate opponents, the emperor set up an elaborate spy network. To preempt opposition from scholars and bureaucrats, he directed the executions not only of those suspected of disloyalty, but also their entire families, including women and children. Among the victims were four scholars who became known as the Four Martyrs. One of these scholars, Fang Xiaoru, when threatened with the elimination of nine kinship lines, replied defiantly that he was "fine with ten." All his blood relations were murdered along with all his students and peers—a total of 873 people. The Chinese Communist Party used similar tactics several centuries later. [8]

I could not help but think of the contrast between the city's grandeur and its occupants' insecurity. The art and architectural style, however, reflected the basis for Confucius's social creed that hierarchy and harmony fit together and are interdependent. Our guide explained that architectural styles conformed to the *Treatise on Architectural Methods*, an eleventh-century manual that specified particular designs for ranks in Chinese social structure. As we entered the Hall of Supreme Harmony, the largest building in the Forbidden City, he pointed out that its double-layered-roof design was reserved for the emperor only. The grand throne is surrounded by six immense golden pillars engraved with dragons to evoke the supreme power of the emperor. Behind the throne, a carved gilt screen and incense burners in the shape of unicorns signify the submission of all other kingdoms. After the fall of the Ming dynasty in 1644, the new rulers of the Qing dynasty preserved the Ming architecture, but changed the names of some of the buildings. The Hall of Imperial Supremacy became the Hall of Supreme Harmony. Consistent with the Confucian teaching that the fundamental duty is to "know thy place," the emperors promoted deference to hierarchy (both by the Chinese people and vassal states) as the path to harmony. But their effort to preserve hierarchical order and control was anything but harmonious for those subjected to their brutality. As the Manchu forces advanced into China proper, Chongzhen, the last Ming emperor, left his throne carved with images of dragons and unicorns, and hanged himself from a tree on Meishan, a hill overlooking the palace.

As our guide described the construction of the Forbidden City, it was clear that the rulers' determination to preserve the hierarchical order and guard against internal threats meant hardship rather than harmony for the Chinese people. Zhu Di mobilized one hundred thousand artisans and one million laborers to build the city; they did so in just fourteen years. Peasants dragged large carved stones seventy kilometers from the quarry to the construction site. Working in below-freezing temperatures, fifty laborers per stone sloshed water in front of the sled that bore the stone to create an icy surface before dragging the sled over rough ground. The most powerful symbol of hierarchy and the deference that Chinese rulers expected lay not in the architectural style of the city, but in the sacrifice the Chinese people made to construct it. Laborers in the Forbidden City, like those who worked eighty kilometers away to strengthen the Great Wall and protect against another Mongol invasion, were on the unfortunate end of superior-inferior relations that underpinned the hierarchical order.

As we walked through the city, it was easy to view Chairman Xi as supremely confident. He wanted to be seen as the unchallenged ruler of an increasingly powerful and apparently harmonious country. Yet for Xi and his predecessors, the pomp of the office masked deep insecurity, and harmony concealed brutal repression. After the tour of the Forbidden City, I thought of the relationship between the emperor and the Chinese people during previous eras as analogous to the dominance that the Chinese Communist Party seeks over all aspects of Chinese political and social life today. Xi and Communist Party leaders expect the same degree of deference to hierarchy and collective effort to achieve the China Dream, what Xi has described as Chinese prosperity, collective effort, socialism, and national glory. As China's power increased, so did leaders' uncertainty and fear. Danger came in many forms. The sweep of Chinese dynastic history reveals cycles of prosperity followed by increased population; the growth of corruption; some combination of natural disaster, famine, rebellion, and civil war; political and economic decline; and, finally, collapse. The *Romance of the Three Kingdoms*, one of China's classic texts, warns, "After a long split, a union will occur; after a long union, a split will occur (分久必合, 合久必分)." Communist Party leaders and their dynastic progenitors viewed control through autocratic hierarchy as the best guarantor of harmony and protection against chaos. Control then and today required protecting China not only from internal influences that might challenge the hierarchical order, but also from threats along China's vast frontier. (China today shares its 13,743 miles of border with fourteen countries, including Russia, India, Vietnam, and North Korea).

Our guide showed us where the last royal occupant of the Forbidden City, Emperor Puyi, was stripped of power in 1911, at the age of five, during China's Republican Revolution. He remained in the old Imperial apartments at the back of the palace until 1924. Puyi abdicated in the midst of the "century of humiliation," a period of Chinese history that Chairman Xi had described to President Trump and those who joined the two leaders for dinner at Mar-a-Lago six months earlier. The century of humiliation was the unhappy era during which China suffered major internal fragmentation, lost wars, made major concessions to foreign powers, and endured brutal occupation. Humiliation began with Great Britain's defeat of China in the First Opium War from 1839 to 1842. It ended with Allied and Chinese defeat of Imperial Japan in 1945 and Communist victory in the Chinese Civil War in 1949.[9]

As the tour ended, I was even more convinced that our dramatic shift in policy was needed and long overdue. The Forbidden City was supposed to convey confidence in China's national rejuvenation and return as the Middle Kingdom. But for me, it exposed the fears as well as the grand ambitions that drive the Chinese Communist Party's efforts to extend China's influence along its frontiers and beyond and regain the honor lost during the "century of humiliation." The party was obsessed with control because control was necessary to allay its fears and fulfill its ambitions.

* * *

THE HISTORY that our Chinese hosts omitted was as revealing as the history they promoted. The two leaders and their wives preceded us into the Hospital for Cultural Relics at the National Palace Museum. As we observed craftspeople restoring artifacts, it was clear that Xi was resurrecting what Mao had tried to destroy: historical memory of China's Imperial past. Mao was an iconoclast; Xi was a nostalgic. Mao destroyed order and invited the chaos of continuous revolution; Xi evoked Confucian moral order to maintain control and encourage conformity.

The 1.5-ton portrait of Chairman Mao Zedong that hung over the Gate of Heavenly Peace, facing Tiananmen Square, was impossible to miss. But our guide did not mention it; nor did he make any mention of Mao, even though the square occupies the space between that great portrait and the mausoleum that holds Mao's crystal coffin and his embalmed body. It was at the Gate of Heavenly Peace that, on October 1, 1949, Mao announced the founding of the People's Republic of China (PRC). He and his fellow revolutionaries believed that the state had to be torn down to save it. The Bolshevik Revolution in Russia seemed a workable model. By the time Mao gave his speech in 1949, the Chinese people had endured fourteen years of brutal Japanese occupation after the invasion of Manchuria in 1931 and a costly civil war that followed from 1945 to 1949.

Xi repeatedly spoke of Japan's brutal occupation of China and portrayed the Chinese Communist Party as a savior that had liberated the Chinese people from Japanese oppressors. Even as we looked out at Tiananmen Square, our hosts cast their efforts to achieve "national rejuvenation" as the party's triumph over the century of humiliation. But as I looked upon the square's gray vastness, my mind could not help but replay images of fanatical Red Guards from Mao's Cultural Revolution in 1966 or the tanks that brutally repressed the student demonstrations of 1989.

Chairman Xi dwells on the century of humiliation for another reason: to gloss over the first decades of party rule that followed, which were even worse. The application of Maoist theory between the end of the Civil War in 1949 and Mao's demise nearly three decades later killed tens of millions of Chinese through misrule, policy-induced famine, and political purges, to say nothing of the disastrous Maoist-inspired revolutions in other parts of the world.

Six years after National Security Advisor Henry Kissinger orchestrated President Nixon's opening to China in 1972, Deng Xiaoping, who was purged during the Cultural Revolution and forced to work in a tractor factory, succeeded Mao as paramount leader. Deng gradually dismantled Maoist policies. From 1978 to 1989, he focused on economic growth, stability, educational progress, and a pragmatic foreign policy. In 1981, five years after Mao's death, the party declared that the Cultural Revolution was "responsible for the most severe setback and the heaviest losses suffered by the Party, the state, and the people since the founding of the People's Republic."[10] Under Deng and his successors, such as Jiang Zemin (1989–2002) and Hu Jintao (2002–2012), the rejection of Maoist economic policies and political excesses was explicit. But that changed when Xi Jinping assumed the premiership in 2012.

To master the past as a means of securing his future, Xi cultivated a more benign interpretation of Chinese Communist Party history, one based on three phases of progress. First, Mao Zedong ended the century of humiliation. Second, Deng Xiaoping and his successors generated wealth. Third, Xi Jinping returned China to greatness. Xi's portrayal of Mao as savior rather than tyrant represented more than a manipulation of history; it required the suppression of personal trauma. Xi and his family suffered physical and psychological abuse during the Cultural Revolution. His father, Xi Zhongxun, a senior party official and veteran of the revolution, was imprisoned and tortured. The Red Guards ransacked his childhood home and forced his family to flee. One of Xi's sisters died from the hardship. Xi was brought before a jeering crowd during a "struggle session," a humiliation ritual used during the Cultural Revolution, where his own mother denounced him. Like many of his contemporaries who are now at the top of the party, Xi was forced to work in the countryside. He rarely speaks of the horrors inflicted on his family at the outset of the Cultural Revolution from 1966 to 1968. Instead, his propaganda apparatus portrays his seven years of hard labor as an uplifting coming-of-age story that explains his resilience as well as his empathy for the hardships suffered by the less fortunate people of rural China. Xi's reluctance to criticize Mao and the Cultural Revolution is more than a form of Stockholm syndrome, in which victims develop positive feelings toward their captors and sympathy for their causes. Xi is unwilling to renounce the Mao era mainly because he understands that any historical questioning of the Communist Party past could morph into skepticism of and opposition to the Communist Party present. Contemplation of the Maoist period's failures might raise doubts about the party's ability to deliver on the China Dream through absolute control. Losing control of the past is, for autocrats, the first step toward losing control of the future.[11]

* * *

MANIPULATING THE collective memory of the Chinese people requires ever greater feats of censorship and nationalistic education under Xi. Deng Xiaoping's reforms generated prosperity, but they also caused ideological incoherence. The contradictions between Communist orthodoxy and a highly globalized economy were starker than ever thirty years after Deng said, "Let some people get rich first" and "Getting rich is glorious."[12] Authoritarian capitalism created ample opportunities for corruption and produced a bourgeois class larger than any other self-proclaimed Communist country has ever seen. When he assumed leadership of the party in 2012, Xi was determined to reemphasize the ideological underpinning of CCP rule but to couch it in a rhetoric of Chinese chauvinism and national destiny. In speeches, he revived Mao's claim in his *Little Red Book* (a collection of 267 of the dictator's aphorisms) that it was "an objective law independent of man's will [that] the socialist system will eventually replace the capitalist system." In tandem, he has promoted a "community of common destiny for mankind," a bid for global leadership that strongly echoes his dynastic forebears, based on the idea that China reigns supreme over *tianxia*—"everything beneath heaven."

Xi was the consensus pick to lead the party in 2012, a bona fide member of what Australian journalist John Garnaut has labeled the "princeling cohort." The princelings, direct descendants of the party elders who fought and won the revolution in 1949, share existential angst that they may succumb to the historical cycle that destroyed every dynasty that came before them. For Xi and his contemporaries at the top of the CCP, maintaining control and achieving national rejuvenation can be matters of life or death.[13]

This attitude was further reinforced by the Tiananmen Square protest. In May 1989, hundreds of thousands of protesters gathered at Tiananmen Square to demand democratic governance, free speech, and a free press. Within a week, many of the protesters began hunger strikes. The Chinese government declared martial law and dispatched mechanized PLA units to the capital. On the night of June 3, the PLA closed in on the center of Beijing, firing live ammunition into crowds of people on the streets. The army stormed the square at 1 a.m. on June 4. Estimates of civilian deaths ranged from several hundred to ten thousand. The massacre generated global outrage. As I glimpsed Tiananmen Square, I remembered that history and thought once more about the paradox of China's growing power and fragility.

For the CCP's leaders, the lesson of Tiananmen was never to loosen its grip on power. Xi and the party see 1989 as a period during which the Chinese Communist Party might have joined the Soviet Union in collapse. As with Putin and the Russian Siloviki, party leadership viewed Mikhail Gorbachev as weak. Gorbachev, who visited Beijing amid the Tiananmen Square protests to celebrate the fortieth anniversary of Soviet-Communist China relations, lost faith in the primacy of the Soviet party elites and compromised. Xi and his cohort believe that Gorbachev's effort to make the Communist Party of the Soviet Union a "party of the whole people" was misguided and led to the Soviet Union's demise.

While obsessed with party purity and order at home, the CCP is determined to advance its system of authoritarian capitalism abroad to expand Chinese power and influence at the expense of countries that adhere to democratic principles and free-market economic practices. On the morning of the second day of our visit, we stood on the steps of the Great Hall of the People as the American and Chinese leaders, clad in overcoats against the brisk autumn air, reviewed a People's Liberation Army honor guard. Off to one side was a throng of Chinese and American elementary school children leaping up and down and enthusiastically waving the flags of the two nations. The kids, who had been cued to begin their cheering too soon, were visibly exhausted by the time the two leaders passed in front of them to enter the Great Hall and begin the day's talks. Pottinger (who wouldn't be stopped from attending the second day's meetings) leaned over and deadpanned into my ear, "The children are getting an early start on their social credit scores." Such bits of humor were what made our intense schedule feel tolerable—and would have landed a Chinese blogger in jail.

The ceremony and the tour of the Forbidden City left me with the impression that the party's leaders believe that they have a fleeting window of strategic opportunity to strengthen their rule and revise the international order in their favor. To seize upon this opportunity, the CCP integrates internal and external efforts to expand its comprehensive national power. Internally, realizing the so-called China Dream requires unprecedented economic growth, popular support for national rejuvenation, and tight control of the population. Externally, satisfying the narrative of national rejuvenation renders a dramatic expansion of Chinese economic, political, and military influence indispensable. The CCP's strategy relies on co-option and coercion to influence China's population, other nations, and international organizations to act in the party's interests. The party also attempts to conceal its intentions and its actions to preclude competition. This strategy of co-option, coercion, and concealment integrates a range of cultural, economic, technological, and military efforts. What makes this strategy potent and dangerous, not only to the United States and the free world but also to China's citizens deemed a threat to the party's ambitions, is the integrated nature of the party's effort across government, industry, academia, and the military.

* * *

TO MAINTAIN its exclusive grip on power in the post-Mao period, the party strove to meet the population's rising expectations mainly through increased economic opportunity. Since Deng's reforms, the Chinese people achieved an astonishing rate of growth, which pulled more than 800 million people out of poverty. In the first ten years of the twenty-first century, China's middle class grew by 203 million people. China became the world's second-largest economy and the largest exporter. Infrastructure and construction projects transformed China's harbors, airports, railways, and roads, connecting the Chinese people to one another and the world to an unprecedented degree. By the early 2000s, half the world's cranes were building gleaming skyscrapers in China's rapidly expanding cities. The party's goal was to double income levels between 2010 and 2020. This proved unsustainable. Since 2015, China has marked less than 7 percent growth every year, and by 2020, China's leaders saw this key pillar of their legitimacy, economic growth, fracturing. Policies designed to maintain high rates of growth generated long-term frailties in the economy.[14] Vast debt fueled inefficient growth but did not produce profitable returns. Overinvestment in particular sectors led to overcapacity and losses. By early 2020, economic growth had decelerated to the lowest rate in twenty-nine years as capital investment by Chinese firms dropped. To boost the decelerating economy, China cut banks' reserve rates to free up $126 billion for loans. But then the outbreak of the novel coronavirus in early 2020 and the associated quarantine and travel restrictions affected nearly half of China's population, slowing China's economy further. It seemed possible that China's economic policies designed to keep the party's exclusive grip on power and allow China to sprint to catch up to the United States might risk what party leaders feared most: internal dissent based on a failure to meet the peoples' rising expectations.

The logical way to continue the economic growth that began with Deng's reforms in the 1980s would have been to reform markets even further, unleashing free enterprise and deemphasizing large, inefficient state-owned companies that lacked incentives to increase productivity and pursue innovative technologies. Instead, under Xi, the party strengthened the primacy of state-owned enterprises (SOE). Although SOEs are inefficient and major sources of waste and corruption, they are critical to maintaining the party's control over the economy and co-opting the population. SOEs are also foundational to the party's plans to shift the economy toward high-end manufacturing and dominate critical sectors of the emerging global economy. Xi moved to "strengthen, optimize, and enlarge" state companies, directing more than $1 billion in mergers to create national champion industries such as railway, metals, mining, shipping, and nuclear energy.[15]

* * *

UNDER THE party's strategy of co-option, coercion, and concealment, China's authoritarian system has become ubiquitous. To ensure their grip on power even if they fail to meet their goals for improved standards of living, party leaders emphasized propaganda and accelerated the construction of an unprecedented surveillance state that is more intrusive than that imagined by George Orwell in his novel *1984*. Indeed, the party invented the term *brainwashing*, and today's efforts have their roots in the thought reform movement that Zhou Enlai initiated in 1951 and the CCP perfected during the Cultural Revolution. Twenty-first-century brainwashing has been upgraded with new technology. For 1.4 billion Chinese people, government propaganda is a seamless part of everyday life. Chinese television news follows a regular agenda: ten to fifteen minutes on Chairman Xi and other CCP leaders, five to ten minutes on Chinese economic achievements, and five to seven minutes on the failures of the rest of the world. There is also routine coverage of the theme that the United States wants to keep China down. Students in universities and high schools must take lessons in "Xi Jinping Thought on Socialism with Chinese Characteristics for a New Era," the chairman's fourteen-point philosophy that emphasizes the party's "people-centric" approach to governance and the many benefits of the CCP's supreme leadership over everything. Xi Jinping Thought is the subject of the most popular app in China. The app, whose name translates to "Study Strong Country," requires users to sign in with their mobile numbers and real names before earning study points through reading articles, commenting daily, and taking multiple-choice tests about the party's virtues and wise policies.[16]

The social credit score is one of the party's many tools for co-opting the population into conformity and coercing recalcitrant individuals. The party uses its control of the internet and all forms of communication in combination with artificial intelligence technologies to monitor activities and conversations. The resulting social credit score is meant to determine eligibility for almost all social services, such as loans, internet access, government employment, education, insurance, and transportation.

Like its dynastic predecessors, the party leadership is particularly concerned about dissent in China's border regions. The party has acted most aggressively toward the ethnic minority populations and in territories annexed in recent history. In western

Xinjiang, for example, where the ethnic-majority Uighurs mainly practice Islam, the party has engaged in systematic repression designed to coerce the population into forswearing their religious and cultural identity in favor of the party's nationalist ideology. By 2019, the party had detained at least a million Uighurs in concentration camps where they are subjected to systematic brainwashing. Uighur families are forced to house loyal party members so their progress in reeducation can be monitored. The CCP has demolished historic mosques. Ethnic Han have been forcefully resettled into Xinjiang to dilute Uighur culture. Xinjiang has become a testing ground for maintaining ideological purity and psychological as well as cultural control. In Xinjiang's concentration camps, prisoners begin the day with a flag-raising ceremony; they pass time singing Communist Party songs, praising the party and Xi Jinping, and studying Chinese language, history, and law. The CCP responded to international criticism of these repressive tactics with denial, but evidence mounted. In November 2019, the *New York Times* uncovered a startling cache of documents allegedly leaked by a member of the CCP. The more than four hundred pages of records revealed party orders to crush all minorities' opposition, imprison more than one million people in concentration camps, and carry out systematic brainwashing and cultural control. Included in the documents were internal speeches by Chairman Xi directing officials to show "absolutely no mercy" as they crack down on minority populations. He also directed follow-up efforts to extend restrictions on Islam to other parts of China. Local officials who resisted the party's orders were purged; a county head in southern Xinjiang was jailed for quietly releasing more than seven thousand inmates.[17] The CCP is also repressive, albeit more subtly, in Tibet, and it has continuously chipped away at local autonomy and individual rights in the former colonial territories of Hong Kong and Macao.

In Tibet, where the Buddhist majority regards the Dalai Lama as their spiritual leader, the party blended co-option with coercion under a campaign of "stability maintenance." To appear as a benefactor, the party refurbished rather than razed temples and historical sites. As in Xinjiang, however, Communist Party cadres monitor every village, oversee political education, and manage every monastery and religious institution. Enabled by new technology, the party intends to scrutinize daily behavior so it can identify and swiftly punish dissent. The party also claimed the right to approve "high reincarnations" that select future Dalai Lamas, the foremost leader of the "Yellow Hat" school of Tibetan Buddhism.

In June 2019, the party's effort to tighten its control over Hong Kong's population sparked sustained protests that continued into 2020. The protests were initially in response to a law that would allow local authorities to extradite criminal fugitives wanted on the mainland. The demonstrators demanded suspension not only of the bill, but of other means of eroding Hong Kong's democratic autonomy. The party responded by waging a sustained campaign of propaganda to discredit the protestors and by carrying out coercive measures against companies and individuals that supported them. A landslide victory for pro-democracy candidates in the November 2019 election indicated widespread support for the protest movement and Hong Kong's semiautonomous status. Demonizing dissent and blaming foreign forces were the same tactics employed after the Tiananmen Square massacre thirty years earlier. After President Trump signed a bill expressing support for the protestors and authorizing sanctions against individuals and entities that used force against them, thousands of people assembled in front of the Hong Kong City Hall for a pro-American and pro-democracy rally.[18] The party conducted a global propaganda campaign to portray the protests as a foreign-backed color revolution designed to destabilize China.

Efforts to prevent dissent and maintain control through co-option and coercion span the entire country. Religion is one of the party's perceived threats because it encroaches on the void left after the collapse of Maoism. Mao attacked religion as "vulgar superstition," but his effort to replace spirituality with Communist ideology and his own cult of personality failed. The Catholic Church and fast-growing Protestant religions concerned Xi and the party, although their campaign against Christianity was less brazen than the campaign against Islam. In 2018, for example, the party attempted to co-opt the Catholic Church by ceding veto power to the Vatican over bishop nominees in return for Rome's recognition of party-appointed bishops. Despite its effort, about half the country's ten million Catholics continued to worship underground and reject churches run by the state. When Protestant congregations proved difficult to control because of their diversity, the party forcefully removed crosses from the tops of churches and even demolished some churches to make an example of those that had failed to register with the government. To provide an alternative to Christianity and Islam, Xi and the party resurrected the Confucian moral code, with its emphasis on deference to hierarchy and preservation of harmony, as a form of folk religion intended to strengthen the CCP's grip on power. The party has also significantly boosted patronage of Buddhism and Daoism as "Chinese" alternatives to what it regards as foreign belief systems.[19]

Suppression of religion extended to suppression of ideas associated with Western liberalism. Any principles or values that might challenge the absolute control of the party had to be eliminated. Particularly dangerous were materials that extolled individual rights, including freedom of expression, representative government, and rule of law. In 2019, for example, the Ministry of Education ordered a nationwide check on all university constitutional law textbooks. Within weeks, a popular textbook written by Beijing University law professor Zhang Qianfan was pulled from bookstores throughout the country. Zhang noted in an interview that "constitutional law, as an academic discipline, should not be politicized." Not long after it was posted on a social media platform, the interview also disappeared.[20]

* * *

THE PARTY'S effort to stifle human freedom and extend authoritarian control does not stop at China's borders. China uses a combination of co-option and propaganda to promote its policies and its worldview. China's expanding influence in the world, what scholars and policy makers call *tianxia* (天下, meaning "everything under heaven"), goes beyond the peaceful development of a new international order sympathetic to Chinese interests. Chinese leaders aim to put in place a modern-day version of the tributary system that Chinese emperors used to establish authority over vassal states. Under that Imperial system, kingdoms could trade and enjoy peace with the Chinese Empire in return for submission.[21] If the Chinese Communist Party succeeds in creating a twenty-first-century version of the tributary system, the world will be less free, less prosperous, and less safe. China intends to establish the new tributary system through a massive effort organized under three overlapping policies: Made in China 2025, One Belt One Road (OBOR), and Military-Civil Fusion.

Made in China 2025 is designed to make China a largely independent science and technology innovation power. To achieve that goal, the party is creating high-tech monopolies inside China and stripping foreign companies of their intellectual property through theft and forced technology transfer. SOEs and private companies work in concert to achieve the party's objectives. In some cases, foreign companies are required or coerced to enter into joint ventures with Chinese companies to sell their products in China. These Chinese companies mostly have close ties to the party, making routine the transfer of intellectual property and manufacturing techniques to their partners and, by extension, to the Chinese government. Thus, foreign companies entering into the Chinese market often make huge profits in the short term, but after transferring their intellectual property and manufacturing know-how, they see their market share diminish as Chinese companies, advantaged by state support and cheap labor, produce goods at a low price and dump those goods into the global market. As a result, many international companies lose market share and even go out of business. Made in China 2025 aims to fuel China's economic growth with a vast amount of transferred technology and eventually dominate sectors of the emerging global economy that will give it military as well as economic advantages.

The party's international efforts to achieve national rejuvenation and realize the China Dream come together under the One Belt One Road (OBOR) initiative, later labeled the Belt and Road Initiative (BRI) for foreign audiences, to mask its China-centric nature. OBOR calls for more than one trillion dollars in new infrastructure investments across the Indo-Pacific and Eurasian continents and beyond. While the initiative initially received an enthusiastic reception from nations that saw an opportunity both for economic growth and to satisfy their need for improved infrastructure, by 2018 it had become clear to many of those nations that CCP investment came with many strings attached, most prominently unsustainable debt and widespread corruption. Under the CCP's integrated strategy, economic motives are inseparable from strategic designs. OBOR projects are meant to gain influence over targeted governments and place the "Middle Kingdom" at the hub of routes and communications networks. New or expanded transportation and shipping routes will ease the flow of energy and raw materials into China and Chinese products out. More routes would significantly reduce the risk that the United States or other nations could interdict those flows at critical maritime chokepoints, such as the Strait of Malacca (the main shipping channel between the Indian Ocean and the Pacific).[22] To ensure control at key geographic points, the CCP uses investment and indebtedness as the basis for servile relationships between the Middle Kingdom and modern-day vassal states. OBOR is, in large measure, a colonial-style campaign of co-option and coercion.

Belying the party's description of OBOR as development of a "community for shared future for mankind," the initiative has instead created a common pattern of economic clientelism that the Chinese Communist Party eagerly exploits.[23] The CCP first co-opts countries with large, high-interest loans from Chinese banks. Once they are indebted, the party coerces that country's leaders to align with the party's foreign policy agenda and goal of displacing the influence of the United States and its key partners (e.g., Japan, Australia, India, and European nations). Although Chinese leaders often depict these deals as "win-win," many OBOR projects have proven to be a one-way toll road that ensures China's access to a client country's energy and raw materials, creates artificial demand for Chinese products and a Chinese labor force, and allows China to control critical physical and communications infrastructure. These deals, rather, fit the description of "triple wins" solely for China: Chinese companies and workers abroad cycle money back into the Chinese economy, Chinese banks enjoy high-interest payments, and the Chinese government gains powerful influence over the target country's economic and diplomatic relations.

For developing countries with fragile economies, the OBOR sets a ruthless debt trap. When countries are unable to service loans, China sometimes trades debt for equity to gain control of the debtor country's ports, airports, dams, power plants, or communications networks. The list of countries for whom China set the debt trap reveals a shrewd strategy to control routes vital to commerce and freedom of navigation. By 2020, the risk of debt distress was growing in thirty-three countries with OBOR financing, and eight poor countries (Pakistan, Djibouti, the Maldives, Laos, Mongolia, Montenegro, Tajikistan, and Kyrgyzstan) already had unsustainable levels of debt.[24]

China's tactics vary based on the relative strength or weakness of the target states' leaders and institutions. When faced with large-scale investment projects, countries with weak political institutions often succumb to corruption, which makes them even more vulnerable to China's strategy. In Sri Lanka, for example, then-president Mahinda Rajapaksa incurred debts far beyond what his nation could bear. He agreed to a high-interest loan to finance Chinese construction of a port, despite no immediate or apparent need for a new harbor on the small island nation. The prime minister was later defeated electorally, but the Sri Lankan government remained severely indebted. Following the commercial failure of the port, Sri Lanka was forced to sign a ninety-nine-year lease to a Chinese state-owned enterprise. Although Chinese officials announced that the port would not be used for military purposes, two Chinese submarines docked there in advance of Japanese prime minister Abe Shinzo's visit to the country in 2014.

The Maldives, a small island nation of four hundred thousand people off the coast of India, was another attractive target because it controls a maritime territory of high strategic importance, one more than three times larger than the United Kingdom. When China approached the country and struck a deal with President Abdulla Yameen (who, along with other officials, greatly profited from the inflated value of contracts), the Maldives incurred a combination of debt and guaranteed loans of more than $1.5 billion, more than 30 percent of their GDP. (Unreported guarantees could make the total loans as high as $3 billion.) In 2018, Chinese efforts to influence the presidential election in the Maldives failed due to a backlash against corruption, indebtedness, and the associated loss of sovereignty.[25]

Malaysia, the second-largest recipient of OBOR funding after Pakistan, was another important target for the CCP due to the country's strategic location at the heart of Asia, with 4,500 kilometers of coastline and a shamelessly corrupt government. After Prime Minister Najib Razak and his co-conspirators embezzled $4.5 billion from the country's sovereign wealth fund—$681 million went into Najib's personal bank account—China arrived to bail him out. China also financed a Malaysian rail project for $16 billion, over twice the actual cost, a scheme that Chairman Xi personally approved. Five months later, Najib flew to Beijing to sign the deal. Beginning in mid-2017, the much-needed cash flow from Chinese state-owned banks was initiated, helping the president of the third-wealthiest nation in Southeast Asia cover for his embezzled funds.[26]

In Kenya, the railway project to connect the port city of Mombasa with Nairobi significantly underdelivered in revenue and increased public debt to unsustainable levels. Kenyan economist David Ndii described the railway in harsh terms as heralding "a new age of Oriental colonialism." Seeing Kenyan government officials "groveling" to the Chinese and making excuses for excesses, such as the maltreatment of Kenyan workers, Ndii was reminded of "the chiefs who sold their people into slavery . . . and signed away their lands to European imperialists for blankets and booze."[27]

The new vanguard of the CCP is a delegation of bankers and party officials armed with duffel bags full of cash. Corruption enables a new form of colonial-like control that extends far beyond the strategic shipping routes in the Indian Ocean and South China Sea. In Ecuador, China financed a great dam in the jungle at the base of an active volcano. The $19 billion agreement allowed China to receive 80 percent of Ecuador's oil exports at a discount; China sells the oil at a markup for profit. Two years after the dam opened in 2016, thousands of cracks appeared in its machinery, and the reservoir was clogged with silt and trees. The first time the turbines were activated, the power surge shorted out the national electrical grid.[28]

In Venezuela, China profited from the corrupt authoritarian regime of Nicolás Maduro even as the dictator destroyed the country's economy. China kept the dictator's regime on life support with a $5 billion credit line in 2018, and in return secured oil at a discount and resold it at a markup, profiting as the Venezuelan people became destitute.[29] The CCP also supports other dictatorships with new technical means of co-option and coercion, such as surveillance technologies, facial recognition, and restricted internet.

The Military-Civil Fusion policy is the most totalitarian of the three prongs; it reveals starkly how Xi has moved away from the market-reform trajectory of Deng Xiaoping. Under Xi's rule, SOEs and private companies alike must act at the direction of the party. First in 2015 and then again in June 2017, the party declared that all Chinese companies must collaborate in gathering intelligence. "All organizations and citizens," reads Article 7 of China's National Intelligence Law, "must support, assist with, and collaborate in national intelligence work, and guard the national intelligence work secrets they are privy to." Chinese companies work alongside universities and research arms of the People's Liberation Army not only to achieve its economic goals but also to extend China's influence internationally. Chinese companies have become arms of the party as it dominates key sectors of the global economy, leads in the development of dual-use technologies, and modernizes the PLA. Capturing private companies under its systemic efforts through Made in China 2025 allows the party to conceal its intention to move ahead of the forerunning nations (e.g., the United States) that lead in technologies with both economic and defense applications. Chinese companies steal or force the transfer of intellectual property; abet the party's bribery and compromising of foreign political and business leaders; and create financial and infrastructural vulnerabilities to allow espionage or intelligence operations.[30]

But Military-Civil Fusion extends beyond the use of Chinese companies to include efforts that are varied, comprehensive, and unconventional. In addition to espionage through traditional channels such as cyber-theft by the Ministry of State Security or undeclared intelligence personnel at Chinese diplomatic missions, the party tasks some Chinese students and scholars in U.S. and other foreign universities and research labs to extract technology. Many of the returning scholars and scientists are then received in one of more than 150 "Overseas Chinese Scholar Pioneering Parks," located in high-technology development zones, for what is essentially an intelligence debriefing.[31] Chinese entities marking themselves as nongovernmental science and technology organizations and advocacy groups are particularly effective. Founded in 2015, the Shenzhen-based China Radical Innovation 100 (CRI 100) is a self-described nonprofit development platform that targets innovation hubs overseas such as Silicon Valley and Boston in the United States and Tel Aviv in Israel. CRI 100 boasts a "new international cooperative innovation model" that, in fact, consists of extracting and sending back the results of cutting-edge research in U.S. universities and labs via the centers it has established overseas, such as the Radical Boston Innovation Center. In

2019, the CRI listed the Massachusetts Institute of Technology, the University of Michigan, Carnegie Mellon University, and Oxford University as partners. At the Boston-based North America Chinese Association of Science and Technology, over 85 percent of the members have doctoral degrees from top U.S. universities and work in the top research labs in corporate America. Also affiliated with CRI is the six-thousand-member Silicon Valley Chinese Engineers Association, which provides "channels to allow members to engage in China's rapid economic development."[32]

Military-Civil Fusion fast-tracks transferred and stolen technologies to the People's Liberation Army in such areas as maritime, space, cyberspace, biology, artificial intelligence, and energy. Military-Civil Fusion also encourages state-owned enterprises and private companies to acquire companies or a strong minority stake in companies with advanced technologies so they can be applied not only for economic, but also military and intelligence, advantage. China is known to have more than a dozen organizations that direct a nationwide effort to access foreign technologies and recruit the scientists and engineers to work in China. An example is the Thousand Talents program that targets for recruitment professors and researchers who have access to cutting-edge technology. In January 2020, Professor Charlie Lieber, the chair of Harvard University's Department of Chemistry, who had received a $50,000 monthly salary, $150,000 in annual living expenses, and more than $1.5 million under the program for a special laboratory in Wuhan University of Technology, was arrested for lying to FBI investigators. Sometimes, U.S. national defense funding supports the CCP's highly organized technology transfer activities.[33] Scientists affiliated with the People's Liberation Army have worked on projects funded by the U.S. Department of Defense and the Department of Energy. One of many examples is the Shenzhen-based Kuang-Chi Group, described in the PRC media as "a military-civilian enterprise." The Kuang-Chi Group was founded largely on U.S. Air Force–funded research on metamaterials at Duke University. The group is applying research as it works with the People's Liberation Army's space-based reconnaissance platforms.[34]

Under Military-Civil Fusion, so-called "private" companies in telecommunications are the CCP's most valuable tool for industrial espionage. China's use of communications infrastructure for espionage is not a theoretical possibility; it is ongoing and gaining in scale. In 2018, for instance, African Union officials accused China of having spied on its headquarters' network system in Addis Ababa over a period of five years. The headquarters building had been built by a Chinese SOE as China's gift to the African Union.

Chinese cyber espionage is responsible for what former NSA director Gen. Keith Alexander described as the "greatest transfer of wealth in history." In fact, a U.S. Council of Economic Advisers study estimated that the loss from malicious cyber activities to the U.S. economy in 2016 alone could have been as high as $109 billion.[35] A grim revelation of China's cyber capabilities came in December 2018, when American and U.K. law enforcement exposed a sustained Chinese hacking operation of tremendous scale. The Chinese Ministry of State Security used a hacking squad known as APT10 to target U.S. companies in the finance, telecommunications, consumer electronics, and medical industries as well as NASA and Department of Defense research laboratories. APT10 hacked into internet service providers in the United States, Britain, Japan, Canada, Australia, Brazil, France, Switzerland, South Korea, and other countries, extracting clients' intellectual property and sensitive data. For example, the hackers obtained personal information including Social Security numbers for more than one hundred thousand U.S. naval personnel.[36]

China's theft and development of cutting-edge technologies is fundamental to its military modernization program. The party's narrative of national rejuvenation prioritizes the development of a powerful military capable of defending expanding global interests. The PLA has used stolen technologies to pursue advanced military capabilities such as hypersonic missiles, anti-satellite weapons, laser weapons, modern ships, stealth fighter aircraft, electromagnetic railguns, and unmanned systems.[37] Their vision is of a force capable of overmatching the United States in future war. Much of the technology needed is dual use. U.S. venture capital and private equity companies help the PLA achieve its vision by providing much of the capital to Chinese companies engaged in research and development of quantum computing, artificial intelligence, and other technologies.

Some transfers from U.S. companies have permitted the Chinese defense industry not only to develop its own capabilities, but also to drive the U.S. defense industry out of the international arms market with inexpensive knockoffs. For example, the Chinese drone company DJI controlled over 70 percent of the global market share in 2018 thanks to its unmatched prices. The company's unmanned systems even became the most frequently flown commercial drones in the U.S. Army until they were banned for security concerns.[38]

Chinese espionage is successful in part because the CCP co-opts individuals, companies, and political leaders to collect intelligence and turns a blind eye to their activities. The party cultivates sympathetic agents in the United States and other nations with the lure of short-term profits. Companies from the United States and other free-market economies often do not report theft of their technology because they are afraid of losing access to the Chinese market, hurting their stock prices, harming relationships with customers, experiencing professional embarrassment, or prompting federal investigations.

Co-option crosses over to coercion when the CCP demands that companies adhere to the party's worldview and forgo criticism of its repressive and aggressive policies. For example, when a Marriott social media account manager "liked" a pro-Tibet tweet in 2018, the hotel company's website and app were blocked in China for a week, and the manager was fired after pressure from the Chinese government. In the same year, Mercedes-Benz, the German car manufacturer, was forced to apologize to the Chinese people for quoting the Dalai Lama in an Instagram post.

In late July 2019, a young pilot of Cathay Pacific Airways was arrested by Hong Kong police during a protest. After being chastised by the CCP for not acting fast enough against employees involved in the protests, Cathay's CEO and chief commercial officer resigned. Other leading Hong Kong companies rushed to condemn the protests.[39]

A glaring example of China's coercion of American companies came in October 2019, when Daryl Morey, the general manager of the Houston Rockets basketball team, tweeted his support for the Hong Kong protesters. Intensely sensitive to any hint of intrusion into its internal affairs, Chinese state-run television canceled its broadcast of Rockets games. The NBA lost contracts totaling approximately $100 million, and Chinese officials demanded that the league issue an apology. The party also threatened to cut off the extremely lucrative revenue stream from China if Americans did not silence their opinions about Hong Kong. Immediately, leading NBA figures, including star players James Harden and LeBron James and coach of the Golden State Warriors Steve Kerr, obliged, chastising the Rockets general manager for expressing his opinion. This external application of the methods used for internal control, such as the social credit score system, succeeded with Orwellian efficiency.

The brilliance of the social credit score is that it coerces by coopting people's social networks. If, for example, a Chinese citizen protests against the government in a way deemed threatening, not only will the protester's score fall (preventing him from purchasing train tickets, renting apartments, obtaining loans, etc.), but so will the scores of all his friends and family because of their association with him. This then prompts his friends and family to "unfriend" or ostracize the protester, and possibly speak out against his antigovernment "unsocial" behavior. The Chinese Communist Party thus has co-opted its citizens to enforce and reinforce the state's coercive measures. The party also co-opted the NBA in an eerily similar fashion, using the mere threat of losing access to the profitable China market.

The muted international response to China's oppression of its people and coercion of its neighbors attests to the effectiveness of the party's campaign of political and economic co-option. China uses both foreign investment and access to its market to incentivize countries and foreign companies to conform to its interests and position on sensitive issues. The party develops a broad range of incentives and influence efforts to manipulate political processes in target nations and engineer policies favorable to Chinese interests. In 2018 and 2019, Australia, New Zealand, and the United States uncovered sophisticated CCP influence campaigns that augmented China's considerable economic leverage with the purchase of influence within universities, the bribery of politicians, and the harassment of the Chinese diaspora community to become advocates for CCP policy.[40]

China's influence efforts are also designed to promote a "China model" as an alternative to liberal democratic governance and free-market economics. China exploits the openness of free societies to promote views and policies sympathetic to the CCP. China's influence campaign is well organized and sophisticated. Consistent with its long-standing Maoist motto "We have friends all over the world," the Chinese People's Association for Friendship with Foreign Countries cultivates relationships with local officials through sister city and state-to-province relationships. Those relationships are based on "principles" consistent with China's foreign policy. While some relationships are positive financially and culturally, many aim to co-opt officials and influence them to take positions that strengthen the CCP initiatives and undermine U.S. policy.

As Xi intensified efforts to control his own population and co-opt or coerce foreign entities to support his policies, the CCP curtailed or stopped many long-standing dialogues and projects between Chinese organizations and U.S. think tanks and universities. Chinese archives and libraries closed. Yet, the CCP maintained access to programs with U.S. institutions that advanced its interests and gave it access to key technology while constraining or shutting out U.S. researchers. Chinese think tanks and universities mandated centralized approval of topics and foreign participants for conferences. Think tanks work as arms of the CCP and follow Xi Jinping's directive to "go global" and advance the Chinese narrative. A scholar with long experience in China observed that Chinese interlocutors no longer had "much to say beyond the Xi Catechism." Instead of demanding reciprocity, some U.S. think tanks and universities chose self-censorship because they feared losing access to China completely if they criticized CCP practices, such as the establishment of a surveillance state, the incarceration of political prisoners, the internment of Xinjiang's Muslim population in reeducation camps, or the party's intensifying campaign of intimidation of Taiwan to achieve unification.[41]

* * *

THE CHINESE Communist Party's obsession with control and its drive to achieve national rejuvenation converge on Taiwan, the island territory that gained autonomy as the last bastion of the Republic of China after the ROC's defeat at the hands of the Communists during the 1945–1949 Civil War. Taiwan constitutes a particular danger to the mainland's autocracy and authoritarian capitalist economic system because it presents a democratic, free-market alternative. Taiwan liberalized its economy in the 1960s and undertook political reforms in the 1980s that transitioned governance from one-party rule to a multiparty democracy. Taiwan's tremendous success confronts the Chinese with a thriving system that implicitly debunks the CCP's oft-cited contention that Chinese people are poorly suited for representative government and individual rights.

Understanding that Taiwan's success heightens the party's most fundamental insecurities and undercuts the promise of the China Dream, the party under Xi intensified rhetoric concerning the final annexation of Taiwan. The People's Republic of China dwarfs Taiwan, geographically as well as economically, but some numbers paint a vastly different picture. For example, compare Taiwan's per capita income, in purchasing power terms, of $57,000 (higher than that of Germany, the United Kingdom, or Japan) with China's $21,000 (below that of Kazakhstan, Mexico, or Thailand). Taiwan is a threat because it provides a small-scale, yet powerful example of a successful political and economic system that is free and open rather than autocratic and closed.[42]

Taiwan has been the object of a relentless CCP campaign of co-option and coercion. Although the establishment of U.S. diplomatic relations with China in 1979 was accompanied by the Taiwan Relations Act, which stipulated the peaceful resolution of Taiwan's future status, the CCP immediately began a campaign to bolster political sympathies in Taiwan for reunification. Cooption efforts included expanding investment and trade to make the island more dependent on the mainland. By 2000, China became Taiwan's biggest export market, accounting for more than a quarter of its exports.[43] Expanded trade with China made Taipei more vulnerable to Beijing's economic leverage.[44] To punish unfriendly Taiwanese political leaders, for example, the CCP temporarily stopped issuing individual travel permits just prior to Taiwan's 2019 parliamentary elections.[45] Other coercive measures against Taiwan included diplomatic isolation as China made renouncing Taiwan a condition for foreign investment and access to its market. Between Xi Jinping's ascension to power in 2012 and the year 2018, The Gambia, São Tomé and Príncipe, Panama, the Dominican Republic, El Salvador, and Burkina Faso ended diplomatic relations with Taiwan in exchange for Chinese investment and access to the Chinese market for their exports.

In the run-up to the 2020 presidential election in Taiwan, the party did its best to ensure defeat of the incumbent, Tsai Ing-wen, and the Democratic Progressive Party due to her party's position that Taiwan is an independent country. Those efforts, based in part on the CCP's erosion of citizens' rights in Hong Kong and its heavy-handed tactics, backfired. Tsai secured over 57 percent of the ballot and a record 8.2 million votes, well ahead of her rival, Han Kuo-yu, and his Kuomintang, or Chinese Nationalist Party, which favors closer ties with China. The setback is likely to heighten Xi's desire to push for unification. Xi's foreign minister, Wang Yi, commented after the election that "Taiwan is an inalienable part of China's territory" and that "those who split the country will be doomed to leave a stink for 10,000 years."[46]

A cause for greater concern is the possibility that the PLA will intensify preparations for a cross-strait invasion of Taiwan. After Xi removed term limits on the presidency, a move that allowed him to rule indefinitely, some speculated that he did so to force unification on Taiwan during his tenure. Statements of CCP officials under Xi were aggressive; many implied military action. In 2019, Chairman Xi said in a speech that Taiwan "must and will be" reunited with the mainland. China's preparations for a cross-strait assault include rapid modernization and expansion of its navy and air force and increased patrols around Taiwan of bomber, fighter, and surveillance aircraft.[47]

The sustained campaign of co-option and coercion aimed at Taiwan may represent the most dangerous flashpoint for war, but it is only the first priority in a much larger CCP campaign designed to achieve hegemony in the Asia Pacific region. Since 1995, when China occupied Mischief Reef, a low-tide elevation within the Philippines' exclusive economic zone (EEZ), China has become increasingly assertive in the South China Sea. Occupying an area twice the size of the state of Alaska, the South China Sea lies east of Vietnam, west of the Philippines, and north of Brunei. It is a vital waterway through which one third of the world's cargo transport flows. While the PRC made expansive territorial claims in the South China Sea since its establishment, it has since 2012 acted to seize control of the area through quasi-military and military deployments and the construction of military bases on reefs and artificial islands. When its neighbors protested China's attempted "land grab," the PLA conducted belligerent shows of force. In 2016, the Permanent Court of Arbitration tribunal in The Hague ruled the Chinese claim as having no legal basis. Yet, in encounters that followed, heavily armed China Coast Guard patrol vessels threatened to shoot foreign fisherman in the disputed waterway. Economic coercion was even more effective than military intimidation. After The Hague ruling, Philippine president Rodrigo Duterte said he would ignore the Permanent Court of

Arbitration's decision and instead advance partnership with China in oil exploration plans. In return, he received $24 billion in pledges of investment and credit lines. Later, that number grew to $45 billion, including railways, bridges, and industrial hubs.[48]

China continues to expand its military systems around the South China Sea, Taiwan, and in the East China Sea, near Japan's Senkaku Islands. China's military strategy in the region is often dubbed "anti-access and area denial," or A2/AD. Aimed at establishing exclusionary control, the strategy integrates cruise and ballistic missiles and air defenses. The PLA has modernized its land, maritime, and air systems to extend military power out to the "second island chain," comprising the Ogasawara and Volcano Islands of Japan and the Mariana Islands of the United States. It is also demonstrating the ability to impose costs on American air and naval forces should they attempt to intervene during a conflict. China hopes to gain coercive power over nations and territories in the region through not only demonstrated military prowess, but also economic coercion and the use of information warfare and maritime militias. Its efforts to create exclusive areas of primacy across the Indo-Pacific region are particularly challenging because they are integrated as components of a "total competition" that is "the peaceful equivalent of total warfare."[49]

* * *

WHAT CHINA'S campaign of co-option, coercion, and concealment has in common with Putin's playbook is the objective of collapsing the free, open, and rules-based order that the United States and its allies established after World War II, the order that some believed, after the collapse of the Soviet Union and the end of the Cold War in the 1990s, was no longer contested. What Russia's annexation of Crimea and China's imposition in the South China Sea have in common is the strategic behavior of probing. Historians A. Wess Mitchell and Jakub Grygiel describe probing as the use of aggressive diplomacy, economic measures, and military actions to test the willingness of the United States and its allies to contest efforts to displace U.S. influence and to replace the free and open order with a closed, authoritarian system sympathetic to Russian and Chinese interests. The revisionist powers of Russia and China increasingly coordinate their actions under what they deemed in June 2019 a "comprehensive strategic partnership of a new era."

During my second full day in the White House in 2017, I hosted an "all hands" meeting with the NSC staff during which I shared my assessment that China and Russia were emboldened by what they perceived as American retrenchment and disengagement from arenas of competition. The trip to Beijing convinced me even further that it was past time to reenter those arenas and compete to counter China's campaign of co-option, coercion, and concealment.

CHAPTER 4

Turning Weakness into Strength

If names cannot be correct, then language is not in accordance with the truth of things. And if language is not in accordance with the truth of things, affairs cannot be carried on to success. (名不正, 则言不顺; 言不顺, 则事不成 *míng bùzhèng, zé yán bù shùn; yán bù shùn, zé shì bùchéng)*

—CONFUCIUS

OUR LAST meeting in the Great Hall of the People was with Li Keqiang, the premier of the State Council, the titular head of government. If anyone in our party, including President Trump, had any doubts about China's view of its relationship with the United States, Premier Li's long monologue should have removed those doubts. He began with the observation that China, having already developed its industrial and technology base, no longer needed the United States. He dismissed U.S. concerns over unfair trade and economic practices, indicating that the U.S. role in the future global economy would be to provide China with raw materials, agricultural products, and energy to fuel its production of the world's cutting-edge industrial and consumer products. President Trump listened for as long as he could and then interrupted the premier, thanked him, and stood up to end the meeting.

As we drove to the hotel to prepare for the state dinner back at the Great Hall, Matt Pottinger and I discussed how starkly Premier Li's monologue had revealed the CCP's break from Deng Xiaoping's guidance during China's opening and reform period in the 1990s: "hide our capabilities and bide our time, never try to take the lead, and be able to accomplish something."[1] After the 2008 financial crisis, Chinese leaders gained confidence in their economic and financial model as Western economies lost confidence in theirs. Many in China believed that the United States had caused the crisis due to the subprime mortgage problem. The U.S. inability to regulate its own banks had led to a loss of faith in the Western capitalist model and the search for a new one. Chinese leaders aggressively marketed their statist economic model across the Indo-Pacific region and globally brandished their growing power. They also made clear that they expected servile relationships with their neighbors. In 2010, Yang Jiechi, who was then China's foreign minister, told his counterparts at a meeting of the Association of Southeast Asian Nations (ASEAN) in Hanoi, Vietnam, "China is a big country and you are small countries."[2]

The next day, as we departed Beijing on the way to the Asia-Pacific Economic Cooperation (APEC) conference in Danang, Vietnam, I was grateful for the extraordinary hospitality and the grandeur of the "state plus" visit. Pottinger and I knew, however, that our Chinese counterparts would be disappointed in the result. The United States and like-minded nations were undergoing a fundamental shift from strategic engagement with China to competitive engagement. It was an inevitable course correction as China's aggressive foreign and economic policies, having gone unchecked for so long, could no longer be ignored. The range of CCP actions under China's strategy of co-option, coercion, and concealment cast doubt on the assumption that positive engagement with CCP leaders would convince them to become responsible stakeholders in the rules-based international order. We were entering a new era, one that required us to use new tactics to convince party leaders that it was in their interest to play by the rules internationally, relinquish a degree of control, and return to the path of reform and openness.

U.S. policy toward China has suffered from strategic narcissism since the American Revolution, as businessmen, missionaries, and diplomats over two centuries have tended to define

China based on economic, religious, and political hopes rather than realities.[3] After the break in relations that followed the Chinese Civil War and the Korean War, China was receptive to President Nixon's overtures in the 1970s because of the two countries' shared enemy. During the Cold War, Nixon and his national security advisor, Henry Kissinger, pursued "triangular diplomacy," which took advantage of the PRC's and Soviet Union's mutual wariness by forging a closer relationship with each of the Communist powers than they had with each other. Even Mao became an enthusiastic partner under the construct of "the enemy of my enemy is my friend." "We were enemies in the past, but now we are friends," Mao told Kissinger in 1973. "A horizontal line—the U.S.—Japan—China—Pakistan—Iran—Turkey, and Europe" could "together deal with the bastard" (i.e. the Soviet Union).[4] But after the collapse of the Soviet Union, the U.S. relationship with the PRC returned to the hope that the United States could change China.

U.S. leaders and policymakers from the George H. W. Bush administration through the Obama administration believed that economic, political, and cultural engagement would lead to the liberalization of China's economy and, eventually, its authoritarian political structure.[5] Hopeful aspirations for reform overwhelmed any desire to confront China's unfair economic practices, technology theft, abysmal human rights record, and increasingly aggressive military posture. Only one year after the Tiananmen Square Massacre, President George H. W. Bush maintained, "As people have commercial incentive, whether it's in China or in other totalitarian countries, the move to democracy becomes inexorable."[6] President Bill Clinton argued for China to join the World Trade Organization, despite the risk that its state-directed economy could distort global markets to its advantage. To sell China's membership, Clinton claimed that "By joining the W.T.O., China is not simply agreeing to import more of our products; it is agreeing to import one of democracy's most cherished values: economic freedom. The more China liberalizes its economy, the more fully it will liberate the potential of its people."[7]

Though President Obama announced a "pivot" or "rebalance" to Asia, the policy rested on vestiges of hope that a cooperative relationship with China would finally emerge. In April 2012, National Security Advisor Tom Donilon cut language from a speech referencing human rights and U.S. military presence and instead added the phrase "pursuing a stable and constructive relationship with China."[8] In November 2013, Susan Rice, who had replaced Donilon as national security advisor, announced that the United States would "seek to operationalize a new model of major power relations."[9] It did not take long for Chairman Xi to embrace that language while taking a series of actions to undermine U.S. interests. First, the CCP began to build islands in the South China Sea and directly challenge territorial claims of East Asian nations. Next, China declared unilaterally an air defense identification zone above a large area of the East China Sea, including the Senkaku/Diaoyu Islands, which are Japanese territory. Soon thereafter, news broke that China was building multiple military bases on the islands in the South China Sea. The PLA Navy and maritime militias intruded into other nations' territorial waters. In 2015, when the United States and other nations objected to the reclamation efforts, Chairman Xi promised that the islands were only for maritime safety and natural disaster support.

In 2015, President Obama demanded that China halt its campaign of economic cyber espionage. At a joint press conference in the Rose Garden of the White House later that year, Chairman Xi and President Obama announced that they had reached a "common understanding" that neither government would knowingly support cyber theft of corporate secrets or business information. But the Chinese attempted a large-scale cyber attack the next day.[10] In an ineffective effort to cover their tracks, the CCP shifted the lead of its cyber offensive from the People's Liberation Army to the Ministry of State Security and began to use more sophisticated techniques.[11] Some thought that because America's decline relative to China's rise was inevitable, that China should be accommodated to avoid competitions that might lead to confrontation. "We have more to fear from a weakened, threatened

China than a successful, rising China," Obama once stated.[12] But avoiding competition only emboldened the CCP. China became more aggressive in cyberspace as well as in the South China Sea.

When, in July 2016, the Permanent Court of Arbitration ruled against China's specious claims of control and its unlawful building of islands in the South China Sea, the PLA Navy maneuvered warships into those waters, rammed fishing vessels, and sailed recklessly near U.S. Navy ships on the grounds these activities violated the United Nations Convention on the Law of the Sea (UNCLOS). And by 2018, it was clear that Xi had lied, as satellite imagery showed construction of missile shelters and radar facilities. Later, the PLA added air defense and anti-ship missiles to those facilities.[13] False hope allowed the CCP to conceal its actions and intentions while developing the ability to coerce other nations into recognizing its claims.

The Obama administration was not the first to base its China policy on the belief that engagement would foster cooperation, but Pottinger and I believed that it should be the last. We set out to build bipartisan support for the most significant shift in U.S. policy since the end of the Cold War.

By the time Pottinger and I worked on the China strategy, evidence overwhelmingly proved that the CCP was neither playing by the rules economically nor going along the expected path of reform. Moreover, its policies were actively undermining U.S. interests. By incentivizing China's transition to a free-market economy and a more liberalized government with favorable trade terms, access to advanced technology, investment, and membership in international organizations, the United States had actually enabled the growth in power of a nation whose leaders were determined not only to displace the United States in Asia, but also to promote a rival economic and governance model globally.[14]

★ ★ ★

OUR NATIONAL Security Council staff's assessment of China policy in 2017 began with an emphasis on strategic empathy. We needed to ground our approach to China in a better understanding of the motivations, emotions, cultural biases, and aspirations that drive and constrain the CCP's actions. The recognition that the CCP was obsessed with control and determined to achieve national rejuvenation at the expense of U.S. interests and the liberal international order led to the adoption of new assumptions.[15] First, China would not liberalize its economy nor its form of government. Second, China would not play by international rules and would instead try to undermine and eventually replace them with new ones more sympathetic to its interests. Third, China would continue to combine its form of economic aggression, including unfair trade practices, with a sustained campaign of industrial espionage to dominate key sectors of the global economy and lead in the development and application of disruptive technologies. Fourth, China's aggressive posture was designed to gain control of strategic locations and infrastructure to establish exclusionary areas of primacy. Finally, absent more effective competition from the United States and like-minded nations, China would become more aggressive in promoting its statist economy and authoritarian political model as an alternative to free-market economics and democratic governance. For me, the trip to Beijing confirmed those new assumptions and reinforced my belief that the United States and other nations with a stake in this competition could not remain passive in countering the CCP's strategy. We could no longer adhere to the narcissistic view that defined China in aspirational terms: what the West hoped China might become.

Any strategy to reduce the threat of the CCP's aggressive policies must be based in a realistic appraisal of how much influence the United States and other outside powers have on the evolution of China. There are structural limits on influence because the party will not abandon practices it deems critical to maintaining control. Despite the CCP's best efforts, however, China is not and will never be monolithic. There are opportunities to expand engagement with entities that are not dominated by the party, such as true commercial, academic, religious, and civil society enterprises. And while there are historical, cultural, and structural limits on U.S. and other foreign influence in China, those limits should not be an argument for passivity in confronting the CCP's oppression of the Chinese people domestically or its economic and military aggression internationally.

We concluded, as we crafted a new strategy, that there was reason for optimism. The 1989 Tiananmen Square uprising, the protests in Hong Kong thirty years later, and the thriving democracy in Taiwan demonstrate that the Chinese people are neither culturally predisposed toward dictatorship nor happy to surrender fundamental rights, including having a say in how they are governed. Despite our limitations, the United States and like-minded partners possessed tremendous latent potential for influencing the party's behavior because we had been largely absent from arenas of competition. Those who worked on the strategy felt that the shift from engagement to competition had initiated a multigenerational effort. The need to compete enjoyed growing support across the political spectrum in the United States and from other nations, international corporations, and academic institutions. Awareness of the threat that the policies of the CCP posed to freedom and prosperity was growing.

Before we departed Beijing, at the press conference held at the end of a long day of meetings, President Trump summarized China's unfair trade and economic practices. He then looked over to Chairman Xi and said, "I don't blame you. I blame us." The message was that it would be unnatural for the United States and its partners to remain passive as the CCP undermined democracy, liberal values, and free-market economic practices abroad while repressing its people at home. But competition should not be seen as leading to confrontation. Pottinger and I believed that if the United States and our allies and partners began to compete effectively, it would be possible to turn what the CCP saw as weakness into strength. Competing might also generate confidence in the principles that distinguished our free and open societies from the closed, authoritarian system China was promoting. We saw competitive advantage in freedom of expression, of assembly, and of the press; freedom of religion and freedom from persecution based on religion, race, gender, or sexual orientation; the freedom to prosper in our free-market economic system; rule of law and the protections it affords to life and liberty; and democratic governance that recognizes that government serves the people rather than the other way around.

* * *

THE CCP views freedom of expression as a weakness to be suppressed at home and exploited abroad. The free exchange of information and ideas, however, may be the greatest competitive advantage of our societies. We have to defend against Chinese agencies that coordinate influence operations abroad—such as the Ministry of State Security, the United Front Work Department, and the Chinese Students and Scholars Association—but we should also try to maximize positive interactions and experiences with the Chinese people. Those who visit and interact with citizens of free countries are most likely to go home and question the party's policies, especially those that stifle freedom of expression. So, the people who direct academic exchanges or are responsible for Chinese student experiences should ensure that those students enjoy the same freedom of thought and expression as other students. That means adopting a zero-tolerance attitude for CCP agents who monitor and intimidate students and their families back home.

Foreign students at universities abroad, regardless of their country of origin, should gain an appreciation for the host nation's history and form of governance. When universities and other hosting bodies protect the freedoms that these students should enjoy, it serves to counter the propaganda and censorship to which the students are subjected in their home country. Perhaps most important, Chinese and other foreign students should be fully integrated into student bodies, to ensure they have the most positive academic and social experience.

The protection of students' ability to express themselves freely should extend to expatriate communities. The U.S. and other free nations should view their Chinese expatriate communities as a strength. Chinese abroad, if protected from the meddling and espionage of the CCP, are capable of making their own judgments about the party's activities. As the party becomes more aggressive in controlling its population to maintain its exclusive grip on power, Chinese expatriates may further appreciate the benefits of living in societies that permit freedom of expression. It is appropriate, for example, for free and open societies not only to disabuse their Chinese visitors of the party's anti-Western propaganda, but also to create safe environments for Chinese expatriates to question the CCP's policies and actions. Investigations and expulsions of Ministry of State Security and other agents should be oriented toward protecting not only the targeted country, but also the Chinese expatriates within it.

Expatriates also have the potential to counter the party's predatory actions under Made in China 2025, One Belt One Road, and Military-Civilian Fusion. As the U.S. Senate Committee on Homeland Security concluded in November 2018, the United States should not only block efforts to recruit Chinese expatriates for espionage, but also provide "more incentives for highly educated Chinese talents to participate in the U.S. economy."[16]

Freedom of expression and freedom of the press also play a key role in promoting good governance to inoculate countries from bad deals under One Belt One Road (OBOR). Uganda provides an example of how the combination of law enforcement and investigative journalism countered China's predatory economic behavior to U.S. advantage. In 2015, the Ugandan government agreed to borrow $1.9 billion from a Chinese bank to build two dams. An investigation in 2018 revealed shoddy construction in the unfinished dams, later that year, a New York court convicted a Chinese energy company representative of paying bribes to African officials. Ugandan leaders then asked a U.S. consortium to bid on a new oil refinery project, a bid it won.[17] Uganda demonstrates the potential associated with freedom of the press enabling public accountability under the rule of law. Over time, the combination of exposing China as an untrustworthy partner and providing alternatives to its predatory behavior will give the CCP incentives to alter its behavior.

As with freedom of expression, the CCP views tolerance of diversity as a threat. It is in this area that the United States and other countries might draw a strong contrast. Although some might see expanded immigration from an authoritarian state as a danger, I believe the United States and other free and open societies should consider issuing more visas and providing paths to citizenship for more Chinese, especially those who have been oppressed at home. Immigrants who have experienced an authoritarian system are often most committed to and appreciative of democratic principles, institutions, and processes. They also make tremendous contributions to our economy. Should the CCP intensify the coercion of its own people in Hong Kong and elsewhere as it did in Xinjiang or engage in brutality reminiscent of the Tiananmen Square Massacre, the United States and other nations should consider offering visas or granting refugee status to those able to escape the repression. Following the bloody crackdown in Tiananmen Square, President George H. W. Bush issued an executive order that granted Chinese students in the United States the right to stay and work. In the following decade, more than three quarters of the highly educated mainland Chinese students stayed after graduation. Many Chinese Americans who remained in the United States after Tiananmen to become U.S. citizens were at the forefront of innovation in Silicon Valley. The Chinese diaspora could, through its familial connections, provide a significant counter to the CCP's propaganda and disinformation.[18]

* * *

THE CCP views its centralized, statist economic system as bestowing advantages, especially the ability to successfully coordinate efforts across government, business, academia, and the military. And it views America's and other nations' decentralized, free-market economic systems as rendering them unable to compete with China's centrally-directed strategies, such as Made in China 2025, OBOR, and Military-Civilian Fusion. That is why the United States and other free-market economies need to demonstrate the competitive advantages of decentralization and unconstrained entrepreneurialism while defending themselves from Chinese predation. Here, the private sector plays a vital role. Companies and academic institutions at the forefront of developing and applying new technologies must recognize that China is breaking the rules to take advantage of our open societies and free-market economies. A first step toward preserving competitive advantage is to crack down on Chinese theft of our technologies. Although there have been significant reforms in national security reviews of foreign investments, another effective defense would be to enforce requirements that U.S. companies report investment by China-related entities, technology transfer requests, and participation in the CCP's core technology development or PLA modernization programs.[19]

There is much room for improvement in the effort to prevent China from using the open nature of the U.S. economy to promote not only its state capitalist model, but also to perfect its surveillance police state. Many universities, research labs, and companies in countries that value the rule of law and individual rights are witting or unwitting accomplices in the CCP's use of technology to repress its people and improve PLA capabilities. For dual-use technologies, the private sector should seek new partnerships with those who share commitments to free-market economies, representative government, and the rule of law. Many companies are engaged in joint ventures or partnerships that help the CCP develop technologies suited for internal security, such as surveillance, artificial intelligence, and biogenetics. Others accede to Chinese investments that give the CCP access to such technology. In one of many examples, a Massachusetts-based company provided DNA sampling equipment that helped the CCP track Uighurs in the Xinjiang region.[20] Google has been hacked by China, used by the CCP to shut off the Chinese people's access to information, and refused to work with the U.S. Department of Defense on artificial intelligence. Companies that knowingly collaborate with CCP efforts to repress the Chinese people or to build military capabilities that might one day be used against those companies' fellow citizens should be penalized.

Tougher screening for U.S., European, and Japanese capital markets would also help restrict firms' complicity in helping the CCP's authoritarian agenda. Many Chinese companies directly or indirectly involved in domestic human rights abuses and violation of international treaties are listed on American stock exchanges. Those companies benefit from U.S. and other Western investors. There are more than seven hundred Chinese companies listed on the New York Stock Exchange, about sixty-two on the NASDAQ Composite index, and more than five hundred in the poorly regulated over-the-counter market.[21] One company that is a candidate for delisting is Hikvision, a company responsible for facial recognition technology that identifies and monitors the movement of ethnic Uighurs. Hikvision produces surveillance cameras that line the walls of Chinese concentration camps in Xinjiang. Together with its parent company, the state-owned China Electronics Technology Group, Hikvision is on the U.S. Commerce Department Entity List (what many call "the Blacklist"). Free-market economies like ours have far more leverage than they are using because they control the vast majority of the world's capital.

Defensive measures, however, are inadequate. Free and open societies need to become more competitive through reform and investments. China here has a clear advantage in the adoption of new technologies. Its centralized decision-making system, government subsidies, underwriting of risk, the relative lack of the kinds of regulations and bureaucratic hurdles typical in the United States and other democratic nations, and the lack of ethical impediments (e.g., in the areas of biogenetics and autonomous weapons) all foster fast application of technologies in the civil sector and the PLA. Although the United States and other nations should not compromise their ethics, many of the weaknesses relative to China are self-imposed. For example, the U.S. national security institutions suffer from chronic bureaucratic inertia. The slow, inflexible nature of defense budgeting and procurement in the United States has long been studied, with little effective change. But the stakes are now too high to tolerate the lack of predictable multi-year procurement budgets, convoluted procurement systems, and deferred defense modernization. The sheer difficulty of doing business with the Department of Defense discourages the most innovative small companies from contributing to defense capabilities and makes it difficult to innovate within the life cycle of emerging technologies. The old model of multi-year research and development to design and test a capability is no longer valid. The U.S. Department of Defense and military services risk exquisite irrelevance as the PLA develops new capabilities and countermeasures that vitiate long-standing American military advantages. Reducing barriers to collaboration between the private sector and national security and defense-related industries could release the potential of free-market innovation in this critical area.

But even streamlining bureaucracy will prove insufficient to compete with the vast investments China is making in emerging dual-use technologies that will advantage its data economy and its military capabilities. That is why government and private-sector investment in technologies in the areas of artificial intelligence, robotics, augmented and virtual reality, and materials science will prove crucial for the United States to maintaining differential advantages over an increasingly capable and aggressive PLA.[22] Defense cooperation across the Indo-Pacific region should extend to multinational development of future defense capabilities, with the ultimate goal of convincing the CCP that it cannot accomplish objectives through the use of force. Multinational cooperation in the development of space and cyberspace capabilities could also deter Chinese aggression in these contested domains. And Taiwan's defense capabilities must be sufficient to ward off China's designs for what would be a costly war with the potential of expanding across large portions of East Asia.

* * *

JUST AS the CCP views our free markets as a disadvantage, so, too, they see the rule of law in the United States and other democratic nations as a relative weakness. The CCP considers the supremacy of the law as an unacceptable encumbrance; so too, the requirement to treat all people equally before the law and the standards of fairness in the application of the law.[23] Here, again, what the Chinese see as weakness is in fact a foundational advantage of free and open societies that we must apply to competition with the CCP. It is the rule of law and, in particular, investigations conducted under due process of the law (the results of which are then made public) that give the people, companies, and governments the information they need to counter Chinese espionage. For example, in 2019, when it became obvious that Chinese communications infrastructure combined with a sustained cyber-espionage campaign posed a severe threat to economic security and national security, the United States, Australia, New Zealand, Japan, and Taiwan banned the Chinese telecommunications company Huawei from their networks and urged others to follow suit.[24] In February 2020, the U.S. Department of Justice charged Huawei and its subsidiaries with racketeering and conspiracy to steal trade secrets.[25] Law enforcement investigations will continue to play an important role, but CCP infiltration of universities, research labs, and corporations is so pervasive that others, including investigative journalists, are needed to expose the full range of Chinese industrial espionage.

Freedom of expression, entrepreneurial freedom, and protection under the law are interdependent. Together, they bestow competitive advantages useful not only in countering Chinese industrial espionage and other forms of economic aggression, but also in defeating CCP influence campaigns designed to mute criticism and generate support for CCP policies. From 2018 to 2020, studies of Chinese influence in mature democratic countries, including Australia, Germany, Japan, and the United States, exposed methods the CCP uses to cultivate witting and unwitting agents across national and local governments, industry, academia, think tanks, and civil society organizations.[26] The free press in those countries exposed the CCP methods, and CCP agents were prosecuted under due process of law. International cooperation among like-minded nations magnifies competitive advantage. For example, in December 2018, the United States and its closest allies revealed China's twelve-year-long cyber offensive in twelve countries and applied a complementary range of sanctions and indictments.[27]

★ ★ ★

STRENGTHENING DEMOCRATIC governance at home and abroad could be the best means of inoculating free and open societies against the CCP's campaign of co-option, coercion, and concealment. The party sees its exclusive and permanent grip on power as a strength relative to pluralistic democratic systems. There is growing evidence, however, that citizens' participation in the democratic process in countries targeted by the CCP has been effective in countering predatory policies under One Belt One Road. From 2018 to 2020, Chinese "investment" was no longer playing well with populations who were the real victims of the CCP's "debt traps." In 2019, the new (and former) prime minister of Malaysia, Mahathir Mohamad, promised to renegotiate or terminate the "unequal treaties" with Beijing—a term designed to evoke Chinese memories of the century of humiliation. New governments in small countries such as Sri Lanka, the Maldives, and Ecuador exposed the degree to which Chinese-funded and -constructed infrastructure projects had indebted them, violating their sovereignty.[28]

Strengthening democratic institutions and processes in target nations may be the strongest remedy to China's aggression. After all, citizens want a say in how they are governed and want to protect their nation's sovereignty. Wang An migrated to the United States from China in the 1950s and founded the groundbreaking computer company Wang Laboratories. Of his adopted country, the United States, he observed, "As a nation we do not always live up to our ideals, [. . .] but we have structures that allow us to correct our wrongs by means short of revolution." That is why support for democratic institutions and processes is not just an exercise in altruism. Democracy is a practical means of competing effectively with China and other adversaries who attempt to promote their interests at the expense of other nations through corrupt practices. If functioning democracies identify predatory actions by the CCP and act to hold leaders accountable for defending against them, the party will have coercive influence only on authoritarian regimes who prioritize their leaders' affluence and exclusive grip on power over the welfare of their citizens.

* * *

A GOOD offense based on the competitive advantages of our free and open societies requires a strong defense against the CCP's sophisticated strategies. The case of Huawei, the Chinese telecommunications giant at the forefront of the CCP's effort to control global communications infrastructure and the data it carries, provides an example, showing the effectiveness of bold, aggressive measures from the United States and the continuing importance of working with our allies. From the outside, Huawei looks like a highly successful company. Founded by former PLA military technologist Ren Zhengfei in 1987, Huawei surpassed Cisco as the world's most valuable telecommunications company after it stole the latter's source code. By 2020, Huawei controlled approximately 30 percent of the global market share in telecommunications equipment. It had also made tremendous progress toward its goal to dominate the emerging market in fifth-generation communications networks, or 5G. Throughout its expansion, Huawei benefited from a comprehensive campaign of cyber espionage and subsidies from the CCP.[29] It and other telecommunications companies, such as ZTE, are indispensable to Made in China 2025 and Military-Civil Fusion because they are on the cutting edge of acquiring the technologies (e.g., microchips and energy storage) and manufacturing know-how critical to Chinese self-sufficiency in high-end manufacturing. They also provide a communications backbone that allows the exfiltration of data critical to the future global economy. The "data economy," as it is known, is an emerging global digital ecosystem consisting of data suppliers and users. Whoever controls data, the protocols associated with it, and analytical tools powered by quantum computers will be in a position of tremendous competitive advantage. The CCP stands to gain significant intelligence, military, and economic advantages if a company it created and a company that must by law act as an extension of the Chinese government captures global data flows.[30] Moreover, control of communications infrastructure gives Huawei and, by extension, the CCP the ability to cripple communications and data flows vital to national defense and routine economic and financial activity.

The United States took defensive measures when, in 2019, it imposed tariffs on Chinese imports and banned U.S. companies from using networking equipment from Huawei. Earlier that year, Huawei's chief financial officer, Meng Wanzhou, was indicted by the U.S. Department of Justice and arrested in Canada on charges involving circumvention of sanctions on Iran.[31] In response, the CCP engaged in hostage taking, arresting two Canadian citizens without cause. Then, they sentenced to death another Canadian citizen for drug offenses after retrying him in one day. The CCP reactions demonstrated clearly that China cannot be regarded as a trusted partner and also exposed the lie that Huawei is a private company not associated with CCP policies.

While many countries joined the United States in restricting Huawei from its communications infrastructure, France decided to allow the company to build two of its three 5G networks. France, in effect, gave China's Ministry of State Security easy access to 67 percent of its telecom networks as well as the internal computer network of France-based companies. But it is actually worse than that because 5G, a system up to one hundred times faster than the 4G network, will permeate every aspect of citizens' personal lives, corporate world, national infrastructure, transportation, health, and defense. U.S. policy makers made a similar mistake two decades earlier, when they allowed China Telecom to establish a network in California. The rationalization was that the network's limited geographic scope would mitigate any security risks. It took the United States a decade and a half to understand the immensity of the Trojan horse it had let in. Terabytes of the most sensitive corporate, personal, and government data were redirected from all over North America to Beijing via China Telecom points of presence in the United States and Canada.[32]

There should no longer be any dispute concerning the need to defend against Huawei and its role in China's security apparatus. In 2019, a series of investigations revealed incontrovertible evidence of the grave national security danger associated with security vulnerabilities in a wide array of Huawei's telecommunications gear. An independent researcher found that many Huawei employees are simultaneously employed by China's Ministry of State Security and the PLA's intelligence arm. Furthermore, Huawei technicians have used intercepted cell data to help autocratic leaders in Africa spy on, locate, and silence political opposition.[33] China's use of major telecommunications companies to control communications networks and the internet overseas is a one-way street: American and other Western companies have little to no presence in the Chinese market. On the global scale, with tremendous subsidies, illicit financing techniques, and industrial espionage, Chinese companies are attaining monopolistic control of the industry, in another example of how a failure to defend against Chinese economic aggression turned a strength of free-market economies into a weakness.[34] A priority area for multinational cooperation should be the development of infrastructure broadly and, in particular, 5G communications to develop trusted networks that protect sensitive and proprietary data.

The United States and other free and open societies must work together to defend against the broad range of Chinese economic aggression to include unfair trade and economic practices. The Obama administration attempted to counter China's unfair trade practices with a painstakingly negotiated multilateral trade agreement with eleven nations, including seven from the Asia-Pacific region. The Trans-Pacific Partnership, however, would have met fatal opposition in either a Hillary Clinton or a Donald Trump administration after 2017. In 2018, the Trump administration imposed the first in a series of escalating tariffs on Chinese imports as a defensive measure against Chinese industrial overcapacity, overproduction, and dumping of products on the international market. Although the national security justification for initial tariffs on steel and aluminum (and simultaneous tariffs on those products imported from other countries, including close allies) was questionable, the subsequent "trade war" that followed marked a return of the United States to the arena of economic competition with China. Subsequent tariffs lent a sense of urgency to the extended negotiations, mainly between the exceptionally knowledgeable and determined U.S. trade representative Robert Lighthizer and China's vice premier, Liu He. The initial result was a phase-one trade deal that President Trump and the vice premier signed in January 2020. More important than Chinese promises to buy more U.S. goods were pledges to reduce barriers to entry to the Chinese market, to avoid currency manipulation, and to implement a new Chinese law to protect intellectual property and sensitive technology. The trade deal, however, marked only the end of the beginning of what will be a protracted competition with the United States and its fellow free-market economies on one side versus China and countries that opt into its statist economic model on the other. The grievances concerning China's unfair economic practices (e.g., government subsidies to state-owned enterprises) are likely to prove intractable because the party cannot address those grievances without loosening its grip on power.

* * *

AS CHINA protects and promotes its statist economic model, it will be important for the United States and like-minded countries to demonstrate collective resolve. Otherwise, China will take a divide-and-conquer approach. Multinational cooperation will also prove vital to protecting the sanctity and usefulness of international organizations such as the WTO, which the CCP seeks to subvert and bend toward its interests.

The need to compete within international organizations extends beyond trade and economic practices. China has systematically embedded officials in key high-ranking management positions in major global organizations. In 2016, for example, it used the secretary generalship position of the International Civil Aviation Organization (ICAO) to aid in its campaign of diplomatic isolation toward Taiwan. And it has used the United Nations Human Rights Council to advance CCP norms that allow states to justify abuses in pursuit of national interests.

China's most egregious abuse of an international organization centers on its membership in the World Trade Organization. When it signed the WTO accession agreement in 2001, Beijing made many commitments that remained unfulfilled nearly two decades later. These included refusing to report state subsidies enjoyed by Chinese firms and continuing the practice of forcing foreign firms to transfer proprietary core technologies as the price for access to the Chinese market. Because China threatens economic retaliation, few firms bring complaints to the WTO. In addition to co-option and coercion, the CCP added concealment, changing its rules to make transfers "voluntary," although they are still mandatory to gain access to the market. China continues to claim special status as a developing market and argues, essentially, that it should enjoy global market access while failing to adhere to global rules and standards. The United States and other nations committed to free, fair, and reciprocal trade may at some point have to consider threatening China's removal from the WTO if it does not reform its practices and adhere to the standards met by that body's other members.

A natural economic "decoupling" is occurring between China and free-market economies based not only on unfair economic practices, but also on the increasing risk of doing business with an authoritarian regime. The United States might help organize that decoupling in a way that protects against slowing economic growth and disruptions to global supply chains. The most effective countermeasures to CCP policies lie in the private sector. China's dishonest tactics and abuses have been uncovered, and companies are questioning whether access to the Chinese market is worth the cost. The reduction of investment in the Chinese market and the withdrawal of manufacturing and other industries from China may prove to be the only way to convince CCP leaders that true economic reform is in their best interest.

★ ★ ★

THE UNITED States and other free and open societies should be confident. There are opportunities to compete effectively, counter CCP aggression, and encourage internal change. China's behavior is galvanizing opposition among countries that do not want to be vassal states to the Middle Kingdom. Meanwhile, inside China, the tightening of control is also eliciting opposition among those who did sense the prospect of liberalization during the reform period. Despite the outward confidence of Li Keqiang and other officials, many Chinese intellectuals, businesspeople, and policymakers are increasingly aware that they have failed to solve fundamental problems in their society and economy. Many feel as if they are sitting on a powder keg of instability. The 2019–2020 protests in Hong Kong accentuated that reality, as did slowing economic growth and public anger over deficiencies and dishonesty in the government handling of the novel coronavirus. Even in the area of technology development and application under programs like Made in China 2025, it is not clear that the party's bold attempt to create an autarchic economic powerhouse will succeed. The CCP's obsession with control is not compatible with the academic and entrepreneurial freedom foundational to innovation and competition in the global market. Moreover, the party's attempt at social engineering under the one-child policy in place between 1979 and 2015 resulted in a rapidly aging population with vast gender imbalances. The implications of that demographic distortion are unclear but are certain to be profoundly negative.

Still, more important than our recognition of the relative strengths of our system to China's is our determination to protect those strengths. We might learn from China's accomplishments in pursuit of "comprehensive national strength." In particular, the United States and other nations might use the competition with China to galvanize improvements in areas where they are lagging. Those initiatives might include educational reform, improvements to infrastructure, and a sound approach to economic statecraft that better integrates public and private investment consistent with free-market principles.

SOME HAVE argued that competition with China is dangerous because it is tantamount to a Thucydides Trap. This was coined to express the high likelihood of a military conflict between a rising power (China) and a declining power (the United States), that emerges from a long-term structural change in global power.[35] The way to avoid the trap is to neither gravitate toward war nor toward passive accommodation, the most extreme of the potential options, but instead to find a middle way. When engaging with our Chinese counterparts, I explained our need to compete fairly as the best means of avoiding confrontation. Had the United States remained complacent about China's violations of international law and national sovereignty in the South China Sea (e.g., China's continued reclamation and militarization efforts), conflict would have become more likely. Had we remained inactive as China used state actors to steal key U.S. technology, their clandestine campaigns would have grown more aggressive rather than have decreased in scale. Transparent competition can prevent unnecessary escalation between the two countries and enable cooperation on pressing challenges where interests overlap. Competition need not foreclose on cooperation on problems such as climate change, environmental protection, food and water security, pandemic prevention and response, and even North Korea's nuclear and missile programs.

But it seems likely that an economic slowdown will heighten the CCP's fears and encourage more draconian means of ensuring its exclusive grip on power and greater efforts to blame the United States and others for China's problems. The unsound economic policies that the CCP pursued to catch up to and surpass the United States and the free world may, paradoxically, prevent its leaders from delivering on the triumphant narrative encapsulated in the China Dream.[36] Possessing neither democracy's ability for the people to administer correctives from within the system nor the tolerance for peaceful expression of discontent, China could see the possibility of active opposition to the party. The CCP's slow response to the coronavirus outbreak in early 2020, as local officials initially tried to cover it up and then used ham-handed censorship to stifle criticism of the party, were indicators of the system's weakness. In anticipation of potential opposition, the party is racing to perfect its technology-enabled police state. It will likely intensify these efforts. And as economic growth slows and party leaders' anxiety grows, China's foreign policy and military strategy could lead to dangerous confrontation in flashpoints such as the South China Sea, the Taiwan Strait, and the Senkaku Islands. The Chinese have a saying for what could occur: "to shoot accidentally while polishing a gun" (cā qiāng zǒu huǒ (擦枪走火). That is why the United States and its allies must possess the will and military capabilities to convince the CCP that it cannot accomplish objectives through the use of force.

The United States and other nations must counter the party's narrative that any accusations are meant to "keep China down" by containing it. Competing effectively with China diplomatically and economically as well as militarily should be understood as the best way to avoid confrontation. In 2019, at an event in the Chinese embassy, Chinese ambassador Cui Tiankai delivered a speech during which he portrayed the new U.S. approach to China as an effort to arrest China's rise and deny its people the promise of the China Dream. Matt Pottinger responded in Mandarin with an explanation of the shift in language from cooperation and engagement to competition. He quoted Confucius's doctrine of the rectification of names: "If names cannot be correct, then language is not in accordance with the truth of things. And if language is not in accordance with the truth of things, affairs cannot be carried on to success" (míng bùzhèng, zé yán bù shùn; yán bù shùn, zé shì bùchéng 名不正，则言不；言不顺，则事不成).[33] Competition should aim to convince Chairman Xi and party leaders that they can achieve enough of their dream without doing so at the expense of their peoples' rights or the security, sovereignty, and prosperity of other nations' citizens.

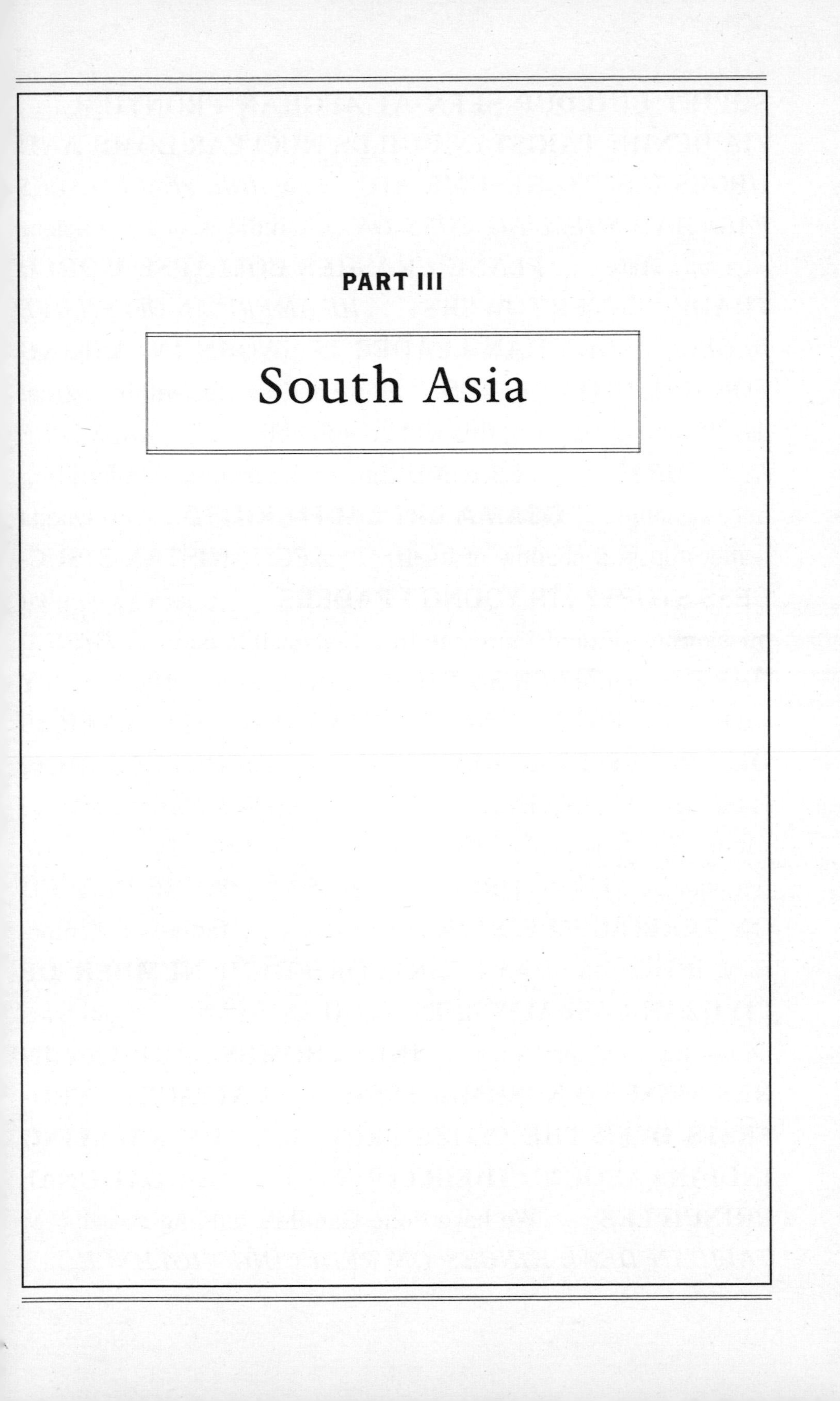

PART III

South Asia

SOVIET BUILDUP SEEN AT AFGHAN FRONTIER . . . **ZIA DENIES PAKISTAN BUILDS NUCLEAR BOMB AND URGES U.S. TO RESUME AID** . . . *KABUL PEACE TALKS FAIL, AND SHELLING GOES ON* . . . India is now a nuclear weapons state . . . **PLANE CRASHES COLLAPSE WORLD TRADE CENTER TOWERS** . . . *THE AMERICAN OFFENSIVE BEGINS* . . . **AFGHAN LEADER IS SWORN IN, ASKING FOR HELP TO REBUILD** . . . If we deliver, this will be a great day. If we don't deliver, this will go into oblivion . . . *MASSACRE IN MUMBAI* . . . America, it is time to focus on nation-building here at home . . . **OSAMA BIN LADEN KILLED** . . . Al Qaeda leadership is a shadow of itself . . . **AFGHANISTAN'S SUCCESS STORY? ITS YOUNG LEADERS** . . . America's combat mission in Afghanistan came to a responsible end . . . *PESHAWAR SCHOOL ATTACK LEAVES 114 DEAD* . . . **"PROBABLY THE LARGEST" AL QAEDA TRAINING CAMP EVER IS DESTROYED IN AFGHANISTAN** . . . *AYMAN AL ZAWAHIRI PLEDGES ALLEGIANCE TO THE TALIBAN'S NEW EMIR* . . . I immediately cancelled the meeting and called off peace negotiations . . . **AT U.S. URGING, PAKISTAN TO BE PLACED ON TERRORISM-FINANCING LIST** . . . India is a democracy; it is in our DNA . . . **RECORD-HIGH NUMBER OF CIVILIAN CASUALTIES IN AFGHANISTAN** . . . Countries do not fight endless wars . . . **INDIA BOMBS PAKISTAN IN RESPONSE TO KASHMIR TERRORIST ATTACK** . . . **PROTESTS OVER THE CITIZENSHIP ACT ARE RALLYING INDIANS AROUND THEIR COUNTRY'S FOUNDATIONAL PRINCIPLES** . . . We have done Gandhi's bidding . . . *A U.S. TALIBAN DEAL HINGES ON REDUCING VIOLENCE* . . .

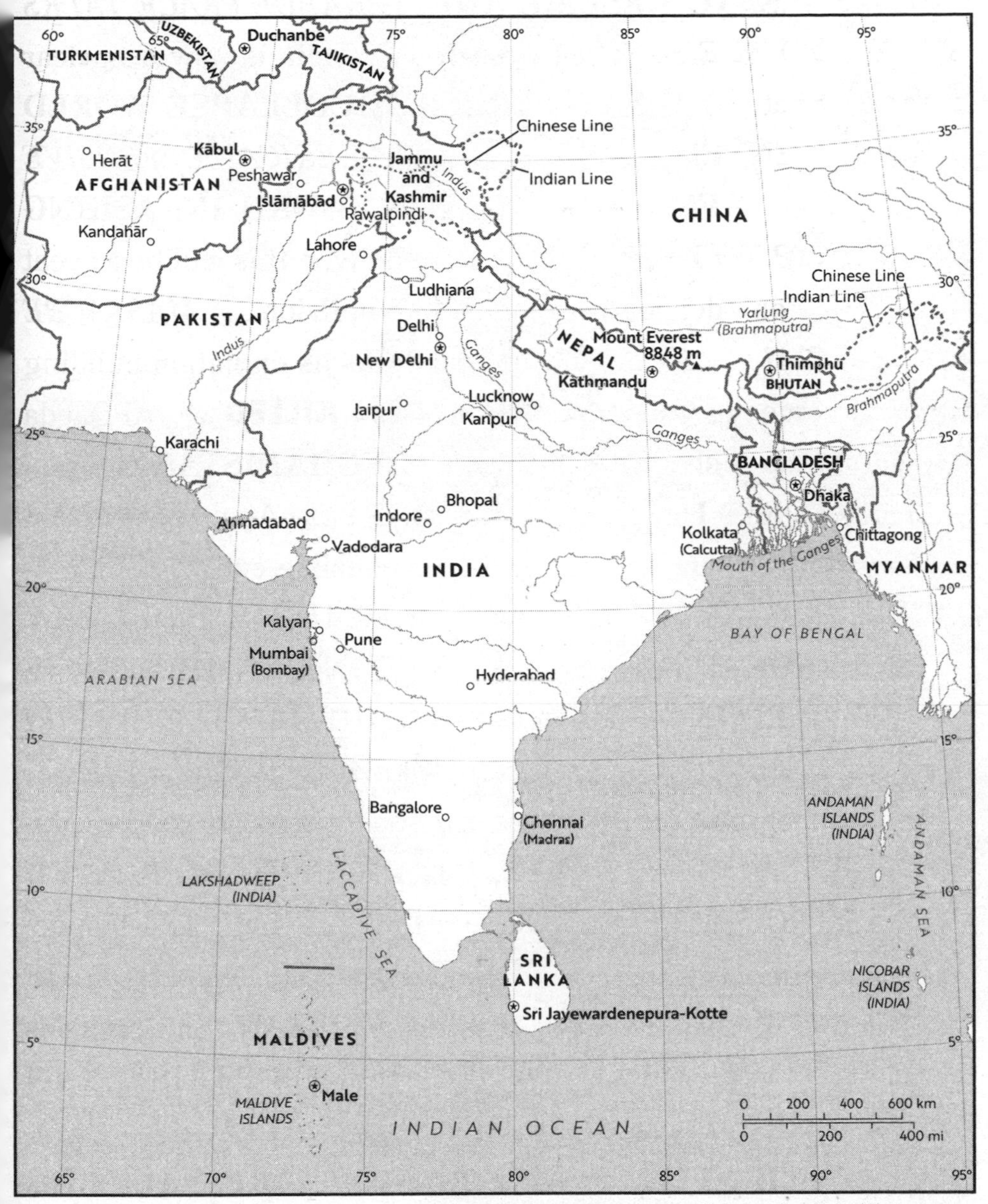
TURKMENISTAN
UZBEKISTAN
Duchanbe
TAJIKISTAN
AFGHANISTAN
Herāt
Kābul
Peshawar
Islāmābād
Rawalpindi
Kandahār
Jammu and Kashmir
Chinese Line
Indian Line
Indus
CHINA
Lahore
Ludhiana
PAKISTAN
Delhi
New Delhi
Ganges
NEPAL
Mount Everest 8848 m
Kathmandu
Chinese Line
Indian Line
Yarlung (Brahmaputra)
Thimphu
BHUTAN
Brahmaputra
Jaipur
Lucknow
Kanpur
Ganges
Karachi
BANGLADESH
Dhaka
Ahmadabad
Indore
Bhopal
Vadodara
INDIA
Kolkata (Calcutta)
Chittagong
Mouth of the Ganges
MYANMAR
BAY OF BENGAL
Kalyan
Mumbai (Bombay)
Pune
Hyderabad
ARABIAN SEA
Bangalore
Chennai (Madras)
ANDAMAN ISLANDS (INDIA)
ANDAMAN SEA
LAKSHADWEEP (INDIA)
LACCADIVE SEA
SRI LANKA
Sri Jayewardenepura-Kotte
NICOBAR ISLANDS (INDIA)
MALDIVES
Male
MALDIVE ISLANDS
INDIAN OCEAN
0 200 400 600 km
0 200 400 mi
60° 65° 70° 75° 80° 85° 90° 95°
35° 30° 25° 20° 15° 10° 5°

CHAPTER 5

A One-Year War Twenty Times Over: America's South Asian Fantasy

Every principle must have its vanguard to carry it forward, while forcing its way into society, endure enormous heavy tasks and costly sacrifices . . . Al-Qaeda al-Sulbah [the solid base] constitutes the vanguard for the expected society.

—ABDULLAH AZZAM[1]

WHEN I arrived at the White House in February 2017, the reluctance to discuss Afghanistan reminded me of the reluctance to discuss Vietnam when I was a cadet at West Point about three decades earlier. Vacancies across the government in positions with responsibility for Afghanistan seemed to reflect America's lost interest in its longest war. Few Americans understood what our soldiers were still doing in that remote, mountainous landlocked country. Other issues took precedence, such as determining what to do about North Korea's nuclear and missile programs and the need to frame a China strategy in advance of Chairman Xi's upcoming visit to Mar-a-Lago. And why prioritize Afghanistan given President Trump's predisposition to withdraw? Those in favor of getting out did not want to provide the president with alternative options. Those who wanted to correct deficiencies in our strategy and continue supporting Afghan forces in their fight against the Taliban and other terrorist organizations were afraid to bring up the topic because it might trigger the president to order a precipitous disengagement. Would it not be preferable to let the war muddle along as it had for the previous sixteen years rather than ask for a presidential decision that might cut against U.S. interests?

To help the president fulfill his wartime leadership responsibility to the American people, we needed to develop options on Afghanistan that fit into a comprehensive strategy for South Asia. The region encompassed two nuclear powers, India and Pakistan, that viewed each other as enemies. India, the world's largest democracy, presented tremendous opportunities. But mainly because of the hostility between India and Pakistan, South Asia was the least integrated region in the world economically. Twenty foreign terrorist organizations were active in Afghanistan and Pakistan alone.[2] South Asia contained vast potential, grave dangers, and daunting challenges that were important to American security and prosperity.

Decades of war had traumatized Afghan society. American policy makers and strategists failed to appreciate how protracted conflict had divided and weakened the country. After the military successes of 2001, a complex competition ensued with an unseated, but not defeated, Taliban; an elusive Al-Qaeda; new terrorist groups; and supporters of those terrorist organizations, including elements of the Pakistan Army, a supposed ally. Plans did not anticipate political drivers of the violence in Afghanistan, especially how enemy organizations would capitalize on tribal, ethnic, and religious competitions. Paradoxically, a short-war mentality lengthened the conflict. The war had lasted nearly two decades, but the United States and its coalition partners had not fought a two-decade-long war. Afghanistan was a one-year war fought twenty times over.

By 2017, the Afghan war effort seemed like a plane crashing on autopilot. No one was paying attention. Years of incoherent policy and ineffective tactics had left our troops vulnerable while the Taliban, Al-Qaeda, and other terrorist organizations, often with the assistance of Pakistan, had become resurgent. I viewed the absence of a viable strategy in Afghanistan as more than a practical problem; it was an ethical failure. As in the Vietnam War, our nation's soldiers were taking risks and making sacrifices without understanding how those risks and sacrifices contributed to a worthy outcome. If the objective was withdrawal, why were soldiers still in harm's way? In the thirteenth century, Saint Thomas Aquinas wrote that to be just, war had to meet the criterion of right intention—that is, it must aim to reestablish a just peace.[3] Based on our fundamentally flawed and constantly shifting strategies, the Afghan War, I believed, no longer met that test. I thought it important to get the president options before the United States lost the war by default.

I planned a trip to South Asia for April to hear assessments of the situation and use what we learned to inform those options. Representatives from the relevant departments and agencies, including the Directorate of National Intelligence and the Departments of State, Defense, and Treasury, traveled with me and members of the NSC staff to foster collaboration and common understanding of problems and opportunities in the region. I had just hired a new senior director for South Asia, Lisa Curtis, a regional expert with an impeccable reputation, who would join us on the Pakistan leg of the trip. I would come to rely heavily on Lisa, who had two decades of experience on South Asia as a diplomat, intelligence officer, and think tank analyst. She had served in India and Pakistan where, in her mid-twenties, she met Jalaluddin Haqqani, whom the United States called a freedom fighter during the Soviet-Afghan War in the 1980s and who would later turn his organization against the United States and the Afghan government. The acting senior director for South Asia and director for Afghanistan was Fernando Lujan, an Army Special Forces officer. Lujan, a graduate of West Point with a master's in public policy from Harvard, had the combination of education and experience needed in that job during a period of transition between administrations. As an acting senior director on the national security staff, Fernando sometimes ran afoul of hierarchy-minded officials in the Department of Defense who resented a lieutenant colonel convening meetings with military and civilian officials more senior in rank. Fernando had been due to leave the White House and depart for Afghanistan, but I asked him to stay until Lisa took over the effort.

Since 2009, the Obama administration and Department of State had been trying to negotiate an acceptable peace agreement with the Taliban while executing an announced withdrawal. As I read my background papers on the plane, I concluded that some officials in Washington had convinced themselves that the Taliban was a relatively benign organization that, with the promise of power sharing in Afghanistan, could be persuaded to renounce support for jihadist terrorist organizations. It was an extreme case of strategic narcissism based in wishful thinking and a false premise that the Taliban was disconnected from terrorist organizations and open to a power-sharing agreement consistent with the Afghan Constitution. As usual with strategic narcissism, policy makers had created the enemy they wanted in Afghanistan and Pakistan.

We took off from Andrews Air Force Base on April 14, 2017. Our team met in the large office that doubled as a bedroom at the back of the plane. During our discussion, I described the Taliban as a reactionary, ruthless, inhumane, and misogynistic organization that was intertwined with Al-Qaeda and other jihadist terrorist organizations. The stakes were high because the terrorist ecosystem in Afghanistan and Pakistan could produce powerful organizations with access to a lucrative drug trade in relatively inaccessible territory in which they are able to plan, resource, organize, and train for attacks. The drug trade is a source of strength for terrorists, putting hundreds of millions of dollars per year into their coffers. It is also a source of weakness for the Afghan government and institutions, as many officials are unable to resist the lure of easy money. I suggested that we heed the advice of the Chinese philosopher of war Sun Tzu, who observed 2,500 years ago that "If you know your enemy and know yourself, you need not fear the result of a hundred battles. If you know yourself but not the enemy, for every victory gained, you will also suffer a defeat."[4]

I quickly realized that some members of our team had become so emotionally invested in the withdraw-and-talk strategy that fantasy had overwhelmed reality. They imagined a reformed Taliban that would forswear its goal of establishing an Islamic Emirate under brutal Sharia law. Additionally, talking with the Taliban had become a cottage industry in which academics and think tank analysts got a crack at wartime diplomacy. But they were talking with the enemy of their imagination rather than the barbarous terrorists who had aided and abetted Al-Qaeda in the murder of nearly three thousand Americans on September 11, 2001, and who were perpetuating violence against the long-suffering Afghan people. I hoped that our trip might foster a higher degree of strategic empathy for the situation across South Asia and a better understanding of our enemies and adversaries in the region. For too many years, we had suffered a defeat for every victory gained.

Foundational to strategy formation is the willingness to unearth and challenge assumptions. I asked the team to use discussions in Kabul to challenge four assumptions foundational to America's fantasy in South Asia:

- First, that a narrow, counterterrorism-only approach that used intelligence collection to cue strikes and raids was adequate to prevent terrorist organizations from threatening the United States.
- Second, that the Taliban was separate and distinct from Al-Qaeda and other transnational terrorists who threatened the security of the United States, its allies, and its citizens abroad.
- Third, that the Taliban, even as it gained strength and the United States withdrew, would negotiate in good faith and agree to end its violent campaign.
- Finally, that Pakistan would, based on U.S. assistance and diplomatic requests, end or dramatically reduce its support for the Taliban and other terrorist organizations.

* * *

AS OUR plane touched down in Kabul, I was happy to be back. I had departed the country five years earlier after a twenty-month-long tour of duty. When he was named commander of the International Security Assistance Force (ISAF) in Afghanistan, General David Petraeus had asked me to join him and help take on one of the greatest obstacles for success in Afghanistan, corruption and organized crime. In the job I left as I deployed, chief of concept development and learning for the army, I studied closely the wars in Iraq and Afghanistan. The two years in Afghanistan confirmed the principal lesson I had identified: the United States military and its civilian counterparts had failed to direct efforts toward achieving sustainable political outcomes. I was grateful for the opportunity to organize an effort that might, finally, develop a realistic strategy for Afghanistan. It was past time to act on the hard-won lessons of America's longest war.

Although the distance from the Kabul airport to the embassy was only 2.5 miles, we loaded onto helicopters and flew to a landing zone at the sprawling secure area that contains the U.S. Embassy and the headquarters for the U.S.-led military coalition. The fact that the city had become too unsafe for routine ground movement of U.S. officials was used by advocates for withdrawal as a sign of the war's futility rather than as an indication of a deficient strategy. The U.S. chargé d'affaires Hugo Llorens, our ranking diplomat in Kabul until the president appointed a new ambassador, greeted us as the Black Hawks touched down. We walked together to the embassy. Hugo was our tenth chargé d'affaires in Afghanistan since 2001.

Llorens was an experienced diplomat who, after thirty-five years of working mainly on U.S. relations in the Western Hemisphere, had decided to retire. Hugo was energetic; he looked much younger than his sixty-two years. He had told me in Washington two weeks earlier that his concerns were growing. It was neither the Taliban's mass murder attacks in Afghan cities, nor Taliban gains in the countryside, nor even large-scale offensives such as the September 2015 attack on Kunduz or the September 2016 attacks on Tarinkot that most concerned the ambassador. Rather, it was the lack of clearly stated U.S. goals and the ambiguity of U.S. strategy. Ambiguity emboldened the Taliban and shook the confidence of the Afghan government and people. Moreover, doubts about the future impeded reforms necessary to harden Afghan institutions against the regenerative capacity of the Taliban—regenerative capacity hidden across the mountainous border with Pakistan. The ambassador also worried about the depletion of U.S. will due to unrealistic expectations. Although the Afghan government needed to reduce the corruption that perpetuated state weakness, Afghanistan never would become, nor did it need to become, Switzerland.

Since my first visit to Afghanistan in 2003, I had felt the emotional impetus behind Afghan policy shift from over-optimism to resignation and even defeatism. Hugo agreed that we needed a realistic strategy based on an honest appraisal of the situation and of the degree of influence that the United States and its partners could exert to ensure that South Asia never again became a base for terrorist organizations that aimed to attack the U.S. homeland, U.S. allies, or our citizens abroad. And we needed a sustainable strategy that could be pursued over time at a cost acceptable to the American people. As with understanding the challenges posed by Putin's Russia and Xi's China, the appraisal of the present and what was needed in the future had to begin with an understanding of the recent past.

* * *

AS WE walked out of the embassy compound on our way to the military headquarters, I reflected on the history of the place I had first visited seventeen months after Afghan militias and U.S. forces drove the Taliban out of Kabul. In December of 2001, U.S. Marines reoccupied the embassy building that their predecessors had abandoned in January 1989 as the last Soviet troops pulled out of the country. In a locked vault in the basement, the marines found a folded American flag and a handwritten note addressed to "Marines" from Sgt. James M. Blake who had led the last marine detachment. Afghan caretakers had kept the old embassy like a time capsule, taking care of the grounds but never entering the building. In 2001, Hamid Mamnoon, one of those caretakers, remembered that cold winter of 1989. He was "very unhappy that the international community forgot Afghanistan." He told a reporter that he was "happy now that the international community is with us and they do not forget us anymore."[5] By 2017, the modest 1960s-era embassy building was dwarfed by large, gleaming structures. The scene reminded me of the classic children's book *The Little House*, by Virginia Lee Burton. At a construction cost of nearly eight *hundred* million dollars, the embassy compound should have been a tangible symbol that America would see the war through; the United States was not going to forget Afghanistan again. But instead, the extravagant embassy seemed to belie America's grave doubts about the mission in Afghanistan and the near-constant announcements of imminent military disengagement. It had come to symbolize the many inconsistencies and contradictions in U.S. policy that perplexed Americans and Afghans alike.

The secure area that contains the U.S. embassy, various headquarters, and the Arg, home to Afghan leaders since 1880, is a museum of Afghanistan's troubled recent history. In 1989, as the last Soviet troops departed and the Soviet Union began to collapse, support for their client government in Kabul dried up. In 1992, after anti-communist mujahideen militias unseated the last pro-Soviet leader, Mohammad Najibullah, a brutal civil war broke out. Najibullah and his brother fled from the Arg and were granted sanctuary in the United Nations compound. During the civil war, the Afghan people suffered as warlords and thugs preyed on them with impunity. Many Afghan tribes were led by criminals who not only extorted the population, but also engaged in murder, torture, rape, and egregious child abuse. The Taliban's appeal was based on its pledge to end the chaos and criminality. In 1996, the Taliban, with Pakistani support, took over Kabul. Fighters secured the Arg palace for their leader, the one-eyed cleric, Mullah Omar. After declaring Afghanistan a "completely Islamic state" in which a "complete Islamic system will be enforced," Omar gave his first order: seize Najibullah. A Taliban death squad dragged him and his brother from the UN compound, tortured them, murdered them, and hung their bloated bodies from a lamppost in the traffic circle outside the walls of the Arg.[6] The Afghan people wanted order, but the Taliban would inflict on them a new form of brutality based on a ruthless purity agenda.

As the Taliban took control of Kabul and most of the country, Ahmad Shah Massoud and thousands of militiamen, comprised mainly of ethnic Tajiks, Uzbeks, and Hazaras, hung on to territory in the north of the country and continued to resist the Islamist regime's attempts to subjugate them. The Taliban and Al-Qaeda tried for years to assassinate Massoud. But Massoud inspired fierce loyalty. His image is still ubiquitous in Kabul on billboards, taxis, and in government offices. Osama bin Laden knew that once the United States traced the 9/11 murderers back to Al-Qaeda's safe haven in Afghanistan, the combination of U.S. forces and Massoud's Northern Alliance would be potent. So, on September 9, 2001, two Al-Qaeda terrorists disguised as Arab television journalists entered a concrete bungalow used as a Northern Alliance office on the pretext of interviewing Massoud. The terrorists detonated explosives hidden in their camera equipment, inflicting fatal injuries on him and serious injuries on his longtime political aide Massoud Khalili.[7]

Osama bin Laden's plan backfired. Instead of neutering the Northern Alliance, Massoud's martyrdom rallied fighters determined to inflict retribution on the Taliban and Al-Qaeda. Forty-nine days after CIA operatives and U.S. Special Forces soldiers arrived in the Northern Alliance's camp in the Panjshir Valley, they liberated Kabul. U.S. casualties were low. It seemed that the "light footprint" approach, which combined special forces and CIA operators with U.S. airpower and anti-Taliban militias, had worked. The war, however, soon entered a new phase. To paraphrase Sun Tzu once again, that which depends on me I can do; that which depends on the enemy cannot be certain.[8]

The initial campaign removed the Taliban, but in December 2001, Osama bin Laden and approximately five thousand terrorists and Taliban fighters escaped to Pakistan. The commander of U.S. Central Command, Gen. Tommy Franks, and Secretary of Defense Donald Rumsfeld, had decided to conduct a swift, economical campaign to unseat the Taliban and destroy Al-Qaeda. Their unwillingness to deploy the troops required, which was based on the belief that larger numbers of troops could immerse the United States in a protracted insurgency, set conditions for what they hoped to avoid as bin Laden, Al-Qaeda, and the Taliban began to reconstitute with the assistance of Pakistan's Inter-Services Intelligence (ISI).[9] From 2001 to 2017, inconsistent, inadequate U.S. strategies gave Al-Qaeda, the Taliban, and other jihadist organizations the time and space needed to regain strength. When strategies did attempt to isolate the enemy from sources of support and harden Afghanistan against the Taliban's regenerative capacity, those strategies were resourced inadequately or abandoned prematurely.

In truth, the United States had been behind in the effort to stabilize post-Taliban Afghanistan from the very beginning. Paying little attention to the history of previous military interventions, U.S. planners did not prioritize the establishment of a replacement government as essential to preventing Afghanistan from becoming, once again, a terrorist sanctuary. Under the auspices of the United Nations, participants in a December 2001 conference in Bonn, Germany, planned a new constitutional order and elections for Afghanistan. A Loya Jirga, a grand assembly of representatives from all Afghanistan, chose Hamid Karzai as interim president. As the new government formed, U.S. officials kept the mission in Afghanistan focused narrowly on terrorism—military forces were there to hunt down Taliban and Al-Qaeda leaders.[10] As we drove to the headquarters, I reflected on my first visit to Afghanistan, in 2003, just after Secretary Rumsfeld, the principal architect of the "light footprint" approach that left commanders with insufficient troops to block Al-Qaeda's escape routes, had ordered further reductions that left fewer than one army combat brigade split between bases at Bagram and Kandahar. Rumsfeld stood with Commander of Coalition Forces Gen. Dan McNeil and Afghan president Hamid Karzai to announce that "we're at a point where we clearly have moved from major combat activity to a period of stability and stabilization and reconstruction activities."[11] The war in Iraq, launched two months earlier, was receiving the preponderance of the George W. Bush administration's attention at that point. The Iraq War would preoccupy the Bush administration and preclude the development of an effective strategy in Afghanistan and South Asia. Despite an international coalition that eventually grew to fifty-one nations, the war in Afghanistan remained under-resourced in those early years, even as the Taliban and Al-Qaeda were regenerating.

But after an anemic and fragmented effort on reconstruction in the war's early years, the Bush administration did an about-face and initiated large-scale programs and investments to help establish a functioning state. Money poured in from seemingly innumerable national, international, and nongovernmental organizations, but at unsustainable rates beyond the absorptive capacity of Afghanistan's economy; much of the assistance was stolen or wasted. The effort to build state institutions was erratic, with different NATO nations sponsoring individual ministries. Development programs were ill-conceived; many sought to create centralized national-level systems incompatible with the traditionally decentralized form of governance in Afghanistan.[12] The lack of transparency strengthened criminalized patronage networks that looted the assistance efforts, profited from the wartime economy, and preyed on the Afghan people. U.S. officials often averted their eyes from criminal activity, such as the theft of salaries for "ghost" soldiers and police, even though the diversion of aid and the extortion enriched not only corrupt leaders, but also the Taliban and terrorist organizations. In the absence of strong security forces and rule of law, many Afghans had no choice but to seek protection from powerful warlords, criminal networks, and militias, further fragmenting society and frustrating efforts to develop a common postwar Afghan identity and vision for the future. All were hedging against a return of civil war or the Taliban. So were some U.S. officials.

From summer 2010 to spring 2012, I encountered many U.S. officials who believed that efforts to strengthen Afghan state institutions were unnecessary, impossibly hard, and even counterproductive. As an army brigadier general, I commanded a multinational intelligence, law enforcement, and military task force with the mission of reducing the threat of corruption and organized crime to a level that was no longer fatal to the Afghan state. But in Kabul and Washington, U.S. officials tended to view corruption as immutable and endemic to Afghanistan rather than as a product of political competition among factions and weak institutions. That view sometimes seemed like bigotry masquerading as cultural sensitivity—Afghans were not culturally predestined to corruption and criminality.[13] Differences of opinion often reflected a false choice between unrealistic goals and doing nothing to encourage incremental reform. Afghanistan was not going to become corruption-free, but it was still possible to constrain corruption and the criminal actors that posed the biggest threats to a fragile state. The lack of U.S. action to do so often left Afghans perplexed. Some concluded that American officials were incompetent, complicit, or both.

In conversations with intelligence officials, I sensed nostalgia for the excitement of earlier campaigns. In the 1980s, CIA officers had supported mujahideen groups' resistance to Soviet occupation, delivering that support mainly through a local ally, Pakistan's ISI. In 2001, CIA officers were again on the front lines alongside U.S. special operations forces advising the Northern Alliance and other militias, including the militia of Hamid Karzai. They forged close relationships with mujahideen groups whom they empowered with money, weapons, intelligence, and air power.[14] Years later, some of those same officers prioritized close relationships with militia leaders over incentivizing the reforms necessary to counter corruption and strengthen the Afghan state.

It was a complicated dynamic—making assistance conditional on anticorruption measures might jeopardize relationships with those groups that could provide the intelligence we might need to fight the terrorists if the Afghan state did collapse. As we drove up to the headquarters, I wondered how our conversation would compare to conversations of five years earlier.

* * *

THE MILITARY headquarters, known by then as Resolute Support, was in a building with large Georgian columns that evoked its history under Mohammed Zahir Shah, the last king of Afghanistan, who reigned from 1933 until he was deposed in 1973. The building served as the Army Club. Officers gathered in its well-appointed salons, where the military band often entertained them and their families. Officers' children used to swim in the pool, which had been filled in to support the many temporary aluminum-sided buildings needed for the coalition staff under U.S. general John "Mick" Nicholson.

We met Mick at the front steps. He was an old friend and one of the finest officers with whom I had served. When I entered West Point as a "plebe," or freshman, in the class of 1984, Mick, a "firstie," or senior cadet, was the "First Captain of the Corps." He had an easy-going demeanor and a good sense of humor. And he was unflappable, having commanded multiple times in hotbeds of the Taliban insurgency from the agriculturally rich lowlands of Kandahar and Helmand, inhabited mainly by Durrani or lowland Pashtuns, to the steep mountains and lush valleys of the eastern highland region historically inhabited by Ghilzai or highland Pashtuns.

Mick understood the complex tribal dimension of the struggle for power in Afghanistan. Durranis consider themselves the rightful rulers of Afghanistan. They were the tribe of the monarchs since Emir Ahmad Shah Durrani founded the modern Afghan state in 1747. The Ghilzai Pashtuns, known mainly for their prowess as fighters, jealously guard tribal autonomy and challenged the Durrani claim to leadership. Divisiveness between those confederations and among the tribes within them caused shifts in loyalties between the government and the Taliban. Competitions within the Pashtun tribes and with other ethnic groups such as the Tajiks, Uzbeks, and Hazara were not a new feature of the Afghan social and cultural landscape.[15] But this factionalism had intensified in recent decades, especially as foreign fighters flooded into Afghanistan in the 1980s to fight the Soviet Army, bringing with them an extreme Islamist ideology from Saudi Arabia and the Deobandi school from Pakistan. Among them was Osama bin Laden, the seventeenth son of a Yemeni who had immigrated to Saudi Arabia and built a billion-dollar construction company in the kingdom.

Osama bin Laden preferred holy war to the family construction business. He got his break in the war against the Soviets by using construction equipment from his family's company to fortify guerrilla positions. He took part in selective battles to build his reputation as a *mujahed*. He was particularly skilled at raising money and providing logistical support, skills that would prove helpful after he and Palestinian cleric Abdullah Azzam founded the group he named Al-Qaeda (meaning "the base") in 1988.[16]

Perhaps most important, bin Laden and his fellow jihadists brought with them an extreme and perverted interpretation of Islam. In Pakistan, proselytizing Islamist extremists found a sympathetic audience among those of the Deobandi school, an orthodoxy with roots in nineteenth-century northern India. Deobandis joined Arab jihadists to promote religious intolerance and brutal enforcement of Sharia law. A growing number of Afghan refugees in Pakistan and those who lived in the mountains along the Afghanistan-Pakistan border were susceptible to their demagoguery.

Bin Laden built Al-Qaeda on hatred of those who did not adhere to its extremist interpretation of Islam. The hatred was directed at Sunni Muslims, or "apostates," who did not support Al-Qaeda's sanctioned cruelty and misogyny. Later, under Al-Qaeda in Iraq (AQI) and other terrorist organizations such as ISIS, hatred expanded to include Christians and Jews, or "unbelievers"; Shia Muslims, or "rejectionists," who regarded Ali, the fourth caliph, as Muhammad's first true successor; and Sufis, who reject violence in favor of introspection and spiritual closeness with God. Al-Qaeda's "near enemies" were governments in Muslim-majority countries that did not adhere to a severe form of Sharia law. Its "far enemies" were Israel, Europe, and the United States. Al-Qaeda believed that members of these groups had only two choices: either surrender and convert, or else be killed. Al-Qaeda was to serve as the vanguard for an Islamic revolution that would establish the caliphate.

Al-Qaeda and the Taliban were a match made in hell. Mullah Omar and Osama bin Laden shared complementary missions and ideologies as well as an affinity for extreme brutality. Bin Laden urged the *ummah* (Muslim community) to unite behind the Islamic Emirate of Afghanistan (the Taliban) as "the seed" of global jihad. After U.S. Navy SEALs killed bin Laden in 2011, his successor, Ayman al-Zawahiri, repeatedly swore allegiance to the Taliban's emir, Mullah Omar. Al-Zawahiri stated that many Al-Qaeda terrorists had heeded bin Laden's call and joined "together around this Islamic Emirate" to create "an international jihadist" alliance stretching from Central Asia to the Atlantic Ocean. In August 2015, after it was discovered that Mullah Omar had been dead for over two years, al-Zawahiri pledged his fealty to Omar's successor, Mullah Akhtar Mohammad Mansour.[17]

Mick walked me to his office for a one-on-one meeting and a video telephone conference with the commander of Central Command, Gen. Joe Votel. Mick was on his sixth tour of duty in Afghanistan. In 2006, as a brigade commander in the mountainous border region in the east, he and his fellow soldiers saw up close the Taliban's brutality.[18] And after serving on the border of Pakistan, they knew that South Asia was a geographic epicenter in our effort to defeat terrorist organizations that threatened the United States and the world. Mick highlighted the connection between counterterrorism and the need for Afghan institutions to be strong enough to withstand the regenerative capacity of the

Taliban. He and Votel agreed that counterterrorism from afar was problematic. Access to intelligence would evaporate as terrorists hid among the population, and a lack of pressure on their organizations would allow them to plan, organize, train for, and execute attacks as they had prior to September 11, 2001. Terrorist control of the narcotics trade had given them the financial strength to grow their organizations and improve the sophistication and lethality of their methods. Without a viable Afghan government and security forces, another large-scale terrorist attack might force the United States and coalition troops to return to Afghanistan in large numbers. Mick pointed out that Afghan forces were doing most of the fighting and making the preponderance of the sacrifices. If Al-Qaeda and the Taliban were to collapse the Afghan government, they would control territory that, like the Fertile Crescent of Ancient Mesopotamia, had great spiritual significance. Our conversation made clear that a narrow counterterrorism strategy using intelligence collection to cue long-distance raids or strikes was flawed.

We walked from Mick's office into the command center that, in the time of the king, had served as the large dining room. Now it was full of flatscreen televisions broadcasting video feeds from drones and digital maps superimposed with military symbols. The wide range of camouflage patterns on the uniforms in the room indicated that officers from many of the thirty-nine coalition member nations were present. The uniforms symbolized broad support for stabilizing Afghanistan and fighting jihadist terrorists, but they also represented wide variance in coalition members' willingness to take risks and engage in combat. Some nations were happy to train Afghan forces but unwilling to accompany them in battle. Many others fought courageously in difficult conditions only to leave the battlefield reluctantly after their nations were no longer willing to pay the price in blood. Inconsistencies in U.S. policy made the Coalition's incoherence worse. Nations were reluctant to make long-term commitments and share burdens if they doubted America's staying power. The experience in Afghanistan had validated Winston Churchill's observation that "there is only one thing worse than fighting with allies and that is fighting without them."[19]

Our team was welcome. The Americans and Coalition members present were hoping that the new administration might provide a sound and sustainable strategy to breathe new life into the effort. Some in the command feared that the war had been forgotten. Reporting on the war in U.S. and European media was scarce and shallow. Media business models, including newspapers, no longer supported professional, sustained coverage of wars. In the United States, television shows rarely cut to foreign correspondents in Afghanistan or Pakistan. Cable networks' news programs found they could save money and maximize profits by replacing foreign correspondents with pundits paid to sit around tables and talk with (or at) each other, mainly about White House intrigue, partisan politics, or popular culture.

But the Afghan war was *never* well understood by Americans. Most strained even to name the Taliban and other terrorist groups we were fighting there, let alone describe their goals and their strategy. The war garnered attention only sporadically, and then usually only after a spectacular enemy attack. Details about the numbers of troops deployed or casualties suffered appeared without context—what the United States and its allies should try to achieve in Afghanistan, why the outcome was important, and what strategies were that might deliver that outcome at an acceptable cost. Coverage portrayed American and Coalition warriors as passive recipients of enemy action. Casualties were mourned, but combat prowess and battlefield achievements went uncelebrated. The Afghan people and Afghan soldiers fighting to preserve their freedom from Taliban oppression were unknown to Americans. Some saw the problem as media bias, but the Afghan war seemed to be the most underreported and, therefore, least understood war in recent history.

I wished that the American people could have heard the briefing from Nicholson's staff that day. His intelligence officer explained how the terrorist ecosystem in Pakistan has global reach. As of 2019, there were twenty terrorist groups based in the

Afghanistan-Pakistan region. Operations against a particularly brutal and well-trained terrorist organization, ISIS-Khorasan (ISIS-K), demonstrated how those jihadist terrorist organizations were intertwined and thrived in the terrorist ecosystem in Pakistan and Afghanistan. ISIS-K was a branch of the terrorist group that had, in the summer of 2014, conducted a murderous offensive in Syria and Iraq that left it in control of territory larger than the state of Maryland. As ISIS lost control of its *wilayets*, or provinces, there, ISIS-Khorasan became more important to the global jihad.[20] Although it competed with Al-Qaeda and the Taliban, those organizations shared common goals and many of the same people.

One powerful ISIS ally is Tehrik-i-Taliban Pakistan (TTP), a group waging jihad against the Pakistani government. Around 2004, TTP began absorbing portions of thirteen terrorist organizations (many of which had been created by the Pakistan ISI's Directorate S) to fight against the U.S.-led coalition and Afghan security forces.[21] The TTP was one of ISI's Frankenstein's monsters; the organization turned on its creator, killing tens of thousands of Pakistanis from 2007 to 2014. On December 16, 2014, in what may have been its most heinous act, six TTP terrorists attacked the Army Public School in Peshawar, Pakistan, murdering 149 innocents and injuring 114 more, the vast majority of them schoolchildren between the ages of eight and eighteen. The attack demonstrated not only TTP's brutality, but also how it and other groups had evolved from an international network that breeds terrorists in the Pakistan-Afghanistan border region. The murderers included a Chechen, Egyptian, Saudi, Moroccan, and two Afghans. TTP and Al-Qaeda trained together in Pakistan. They and other groups shared resources and expertise in the terrorist ecosystem astride the Afghanistan-Pakistan border.[22]

If given the opportunity, many of the groups based in South Asia would commit mass murder in the United States. For example, TTP-trained Faisal Shahzad, a Pakistan-born U.S. citizen, aimed to set off a car bomb in Manhattan's Times Square. On May 1, 2010, the bomb malfunctioned, in part because Shahzad had had only superficial training in Pakistan. U.S. intelligence collection and strike capabilities based in Afghanistan had forced Shahzad and his trainers to stay on the move.[23] It was the effectiveness of the U.S. intelligence facility in Khost that led TTP on December 30, 2009, to use a Jordanian doctor to infiltrate the base and detonate a suicide vest, killing three security contractors, four CIA officers, and a Jordanian intelligence official.

Ensuring that terrorists remain preoccupied with surviving rather than plotting attacks on innocents requires a sustained effort against determined, adaptive, and ruthless enemies. Nicholson's staff summarized Afghan operations in mountainous Nangarhar and Kunar Provinces against ISIS-K with U.S. air power, Special Forces, and Army Rangers in support. A U.S. drone strike killed ISIS-K's first emir, Pakistani national Hafiz Saeed Khan, on July 26, 2016.[24] His successor, Abdul Hasib, had masterminded a heinous attack on Afghanistan's military hospital in Kabul a few weeks before my visit, killing thirty defenseless hospital personnel and patients and wounding over fifty more.[25] Two weeks after my visit, Afghan special security services and U.S. Army Rangers would hunt Hasib down and kill him.

The briefers' explanation of the ecosystem that sustained TTP, ISIS-K, Al-Qaeda, the Taliban, and other terrorist organizations exposed flawed assumptions that our policymakers had made about the enemy. U.S. leaders often imagined bold lines between terrorist groups—lines that simply did not exist. Although the Taliban and numerous terrorist organizations in Afghanistan and Pakistan sometimes clashed with one another, they more often formed alliances or shared resources to pursue their common cause. But the desire to simplify and shorten the war perpetuated self-delusion. Self-delusion about the enemy was the basis for America's South Asian fantasy, in particular regarding the Taliban and Al-Qaeda as completely separate organizations. An extreme version of this delusion, created during the Obama administration and initially debunked but resurrected later in the Trump administration, sustained the forlorn hope that conciliation with the Taliban could provide an easy way out of Afghanistan. The false hope was also based in a failure to realize that the terrorist ecosystem along the Afghan-Pakistan border had not developed organically; it was a creation of the Pakistani military to keep ethnic Baluch and Pashtun populations suppressed and to prevent them from seeking either independence or finding common cause with their kinfolk across the 2,430-kilometer-long "Durand Line" drawn by a British diplomat and Afghan Abdur Rahman Khan to delineate British and Afghan spheres of influence. Paradoxically, the effort to simplify the enemy to shorten the war not only obscured the stakes in Afghanistan and diminished the will necessary to sustain the mission, but also complicated and prolonged what had become America's longest war.

The first day's meeting reinforced my belief that narrow counterterrorism efforts focusing exclusively on Al-Qaeda would prove insufficient to protect the United States from jihadist terrorists based in Afghanistan and Pakistan. The Taliban, Al-Qaeda, and other terrorist organizations were interconnected both ideologically and physically, in the mountainous region along the Afghanistan-Pakistan border.

The briefings also revealed that, despite the Obama administration's 2014 declaration that the war was over, American soldiers, alongside Afghan armed forces, were still fighting against ruthless enemies that were determined to establish control of territory, people, and resources. Their twisted aim was to establish an Islamic Emirate based on a distorted interpretation of Islam. It was clear to me that time was running out. The Department of Defense was executing the Obama administration's policy of withdrawal. U.S. troop strength fell from a high of 100,000 in the spring of 2011 to 9,800 at the end of 2014 to roughly 8,400 in March 2017.[26] The "troop cap" had no connection to what military forces were meant to achieve. Restrictions on how the military fought limited its effectiveness and created opportunities for the Taliban and its terrorist allies. Moreover, military and diplomatic efforts were completely disconnected. Since 2009, the Department of State had tried to negotiate an acceptable peace agreement with the Taliban while executing an announced withdrawal.[27]

The Obama administration even declared that the Taliban was no longer an enemy force, which meant that the U.S. military stopped offensive operations against the Taliban and was unable to bring its considerable intelligence and airpower capabilities to bear offensively unless hostile actions were taken against them. Without U.S. advisors, airpower was less responsive; bombs directed from remote headquarters often fell far from intended targets and sometimes inflicted unintended losses on civilians caught between the Afghan Army and the Taliban. Emboldened, the Taliban stepped up attacks on Afghan National Security Forces and U.S. forces. "Insider attacks," in which Taliban infiltrators opened fire on Afghan or U.S. forces from within what were supposed to be secure locations, were meant to erode trust among Afghan soldiers and between Afghan and U.S. forces. The Taliban and associated terrorist networks also expanded mass murder attacks against Afghan civilians, often with the assistance of the Pakistan Army's intelligence arm, the ISI.[28] In 2015 and 2016, the Afghan National Security Forces lost at least 13,422 killed and 24,248 wounded. In those two years, the Taliban and other terrorist organizations murdered 4,446 innocent civilians.[29] Battlefield gains, insider attacks, and the murder of civilians were meant to exhaust the will of the Afghan people and the American public. It was working. Meanwhile, the Taliban gained control of more territory, intensified attacks, and inflicted more losses on U.S. forces and especially on Afghan forces and the civilian population.

As we left General Nicholson's headquarters, I was even more determined to provide the president with options, including a strategy based on strategic empathy rather than narcissism. Effective strategies require a clear-eyed understanding of the enemy and a harmonizing of the ends (or what is to be achieved) with the ways (the methods and tactics) and the means (the resources applied). If the aim was to ensure that Afghanistan could never again harbor terrorists who could attack the United States and our interests abroad, there was not only a persistent mismatch in South Asia strategy between ends, ways, and means, but also inconsistencies across all three over time.

* * *

THE SECOND day in Kabul started early, with the familiar drive past the headquarters and then down the tree-lined road to the Arg. The palace, constructed in the late nineteenth century, sat behind high walls on eighty-three acres in the center of Kabul city. The grounds contained gardens, a mosque, offices, and the official and private residences of the president.[30] I visited the palace and the National Security Council building many times from June 2010 to March 2012. I often accompanied U.S. ambassadors Karl Eikenberry and, later, Ryan Crocker, as well as commanders of the International Security Assistance Force, first Gen. David Petraeus and later Gen. John Allen, for meetings with then-President Hamid Karzai. The meetings were touchy. They dealt with the problem of how to reduce the threat of corruption and organized crime to Afghan institutions and key sectors of the Afghan economy.

Karzai was usually in a bad mood. By 2010, the days of amicable cooperation between him and U.S. leaders were gone; the relationship had succumbed to a combination of inept diplomacy, incoherent strategy, Karzai's exhaustion, and a successful ISI psychological operation to make Pakistan the irreplaceable power broker in the war-torn country.[31] A lack of trust and the tendency of Afghan and American leaders to suspect the worst of one another had contributed to the ineffective and inconsistent strategies. The lack of trust reinforced flawed assumptions, including that "counterterrorism only" was a viable strategy and that the Taliban could be easily separated from Al-Qaeda and would negotiate in good faith, allowing the United States an easy exit from the war.

Karzai had doubts about the dependability of the United States from the beginning. When President George W. Bush met with him in the Oval Office just three months after 9/11, Karzai told Bush that "the most common question I hear from my ministers and others in Afghanistan is whether the United States will continue to work with us."[32] Despite Bush's reassurance, over the next seven years, the United States consistently gave the impression that its forces were on the verge of departure. When, on May 1, 2003, Rumsfeld announced an end to major combat operations in Afghanistan, Gen. Dan McNeill stated that the small U.S. force of seven thousand soldiers would soon shrink to an even smaller force, one focused exclusively on training the Afghan Army. It was the same day that President Bush announced the end of major combat operations in Iraq on board the USS *Abraham Lincoln* with a "Mission Accomplished" banner as the backdrop. Of course, neither the Taliban nor insurgents in Iraq complied. When Rumsfeld gave his speech, the Taliban had already initiated offensive operations along the border.[33] The vast majority of casualties suffered in both Afghanistan and Iraq were yet to come. And American partners in both countries, confronted with the U.S. aversion to so-called "nation building" and an associated denial of the need to consolidate initial military successes into sustainable political outcomes, hedged against an American withdrawal.

By 2006, enemy gains compounded Karzai's concerns over the staying power of his American ally. Between 2006 and 2009, the Taliban gained control of territory in the southern part of Afghanistan, including Karzai's native province of Kandahar, which was also the birthplace and spiritual home of Mullah Omar and the Taliban. NATO forces, mainly from Canada, the United Kingdom, and Denmark, expected peacekeeping duty similar to operations in the Balkans in the 1990s. Instead, they found themselves in fierce combat. By 2008, the security situation in Kandahar and neighboring Helmand Province was dire. President Bush, recognizing that overconfidence based on initial military success against the Taliban had "left us short of the resources we needed" to stabilize Afghanistan, authorized an increase in the troop level to 45,000 as he left office.[34] But the earlier announcements of withdrawal had already undermined Karzai's trust, and as the security situation in his tribal homeland deteriorated, so did his relationship with the United States.

Between 2010 and 2012, it was clear to me that Karzai's stress made him susceptible to conspiracy theories. Pakistan's ISI took full advantage of the opportunity to drive a wedge between Karzai and the new Obama administration. During my visits to the palace, Karzai's chief of staff, Abdul Karim Khoram, a short, rotund, unfriendly man, lurked in the background in the president's office. Khoram had been imprisoned under Najibullah and later fled to Paris, where he received a master's degree in international law and diplomacy. He seemed to be under the influence of the ISI.

The Taliban and the ISI used Khoram and other agents in the palace to convince Karzai that the United States had ulterior motives in Afghanistan beyond defeating jihadist terrorists.[35] Khoram and others would bring Karzai initial, often inaccurate, reports of civilian casualties caused by U.S. and Afghan forces while softpedaling reports of Taliban atrocities. Karzai gradually began to oppose Coalition and Afghan military operations, especially night raids, which were designed to minimize risk to U.S. and Afghan troops as well as innocent civilians. By 2012, the relationship was irreparable. The commanders I accompanied to the meeting in the palace, Generals Petraeus and, later, Allen, did their best to control the damage.

The United States began to see the Afghan government as part of the problem rather than as an essential element of the solution. U.S. leaders made an already bad situation worse. President Obama and Secretary of State Hillary Clinton assigned the Afghanistan and Pakistan portfolio to Ambassador Richard Holbrooke, a man whom many saw as imperious and condescending. The relationship with Karzai spiraled downward.[36] Holbrooke actively opposed Karzai in the run-up to the 2009 Afghan presidential election. After the election, he suggested a second round, which, along with reports of fraud, led to two months of postelection turmoil. In April 2010, Karzai exploded: "If you and the international community pressure me more, I swear that I am going to join the Taliban."[37] By the time I arrived in Afghanistan to form a countercorruption task force, Karzai viewed U.S. complaints about corruption as another effort to undermine his presidency.

Oddly, despite the difficult subject matter, Karzai and I had a good relationship. I had listened to Afghans and learned about the internal tribal, ethnic, and political competitions that drove corruption and perpetuated state weakness. Karzai seemed to enjoy our discussions, which were based on mutual understanding that the political settlement in Afghanistan had become dependent on unchecked corruption and organized crime. In exchange for their loyalty, Karzai had given mujahideen-era elites, including some of the most rapacious warlords, what amounted to a license to steal. They enriched themselves, grew their power bases, and controlled state institutions and functions, often extorting citizens at checkpoints, borders, and airports. We discussed how unchecked criminality perpetuated state weakness and dependence on international assistance while frustrating donors, who were less willing to underwrite corrupt enterprises that wasted their money. Despite our rapport, Karzai and I made only halting progress. Stabilizing Afghanistan, I had concluded, would take not only a better relationship with Afghan leaders, but also a sustained effort to convince those leaders to undertake reforms essential to the state's survival. But the poor relationship with Karzai convinced some that reform of Afghan institutions and reduction of the threat from corruption were impossibilities. The lack of confidence in reform reinforced the alternatives: a narrow counterterrorism strategy and a deal with the Taliban.

During those meetings with Afghan leadership, I remember thinking that Khoram was playing the treacherous Iago to Karzai's Othello. Just as Iago convinces Othello that his most loyal captain is having an affair with Othello's wife, so Khoram (and Pakistani leaders) convinced Karzai that the United States was not a trusted partner. Like Karzai, Othello wins over audiences early in the play, but then, based in part on Iago's web of lies, makes poor decisions that lead to his eventual undoing. Othello kills his wife, Desdemona, in the final act. Karzai left the presidency before he killed the partnership with the United States, NATO, and the international community. However, afterward, he continued to undermine the U.S.-Afghan relationship from his new home, adjacent to the palace grounds.[38] The damage he did was significant, especially the impetus he gave to Americans' flawed assumptions concerning the nature of the war, assumptions he would help revive later in the Trump administration.

It was in 2009 that our strategic narcissism in Afghanistan seemed to morph into Stockholm syndrome, or at least what psychologists call reaction formation, the superficial adoption and exaggeration of ideas and impulses dramatically opposed to one's own. Disenchantment with Karzai and desperation for an easy way out of the war led to a strange phenomenon in which the Taliban was viewed as a partner in ending the war. Once U.S. leaders assumed that the Taliban (even as it continued to kill innocent civilians and Afghan, U.S., and Coalition soldiers) was not the problem, the Afghan government filled a newly available role in what was becoming a Shakespearian tragedy: the enemy. And once the Afghan government was portrayed as the problem, some Americans began to perceive the war as a popular resistance to U.S. occupation rather than a broad multinational effort to support a representative government against a terrorist organization that had brutalized the Afghan people. As the rift between the U.S. and Afghan governments grew, it was time for Pakistan to take advantage of the situation.

Frustrations with Karzai also contributed to the Obama administration's doubling down on the Bush administration assumption that Pakistan, if offered a long-term relationship, would change its behavior. In a 2009 phone conversation, Karzai warned Obama that "the military and political dimensions of achieving peace in Afghanistan can't be addressed unless the issue of sanctuary in Pakistan is made explicit and is a priority in the new strategy."[39] But shortly after that call, Obama's NSC deliberately leaked to the press a "shift in thinking" designed to rationalize fewer troops than Gen. Stan McCrystal, the new American commander in Kabul, had recommended. Because the administration's "reframe" of its war strategy included the self-delusions that the Taliban was separate from Al-Qaeda, and Al-Qaeda in Afghanistan was weak, it chose to focus efforts on Al-Qaeda in

Pakistan rather than on the Taliban in Afghanistan.[40] But the administration took self-deception to a new level with the assumption that Pakistan, whose army supported not only the Taliban but also a range of terrorist organizations, would be a willing partner in a counterterrorism campaign against Al-Qaeda.

* * *

MANY OF the analysts and bureaucrats who regarded the Taliban and the Pakistan Army as partners and the Afghan government as a foe clung to that bizarre formulation even after presidential elections that resulted in a contested but peaceful transition of power from Hamid Karzai to President Ashraf Ghani in 2014. The election, beset by charges of corruption between the two leading opponents, brought the new secretary of state, John Kerry, to Afghanistan to broker an arrangement between Ghani, a Ghilzai Pashtun from the mountainous southeast, and Dr. Abdullah Abdullah, the primary Northern Alliance candidate and former foreign policy advisor to Ahmad Shah Massoud. Ghani reluctantly agreed to place Abdullah in an unprecedented and extra-constitutional position of chief executive officer. The hope was to create at least the veneer of unity between north and south, as well as between Pashtuns, Tajiks, and other ethnic groups.

In pursuit of a negotiated settlement, the Obama administration had encouraged Qatar to sponsor the opening of a Taliban political office in Doha comprised of high-ranking Taliban members.[41] Through the negotiations, the Obama administration actually helped the Taliban retain a cohesive identity rather than fragment and weaken the organization. Administration and Department of Defense lawyers placed restrictions on how U.S. forces could target Taliban leadership and fighters. U.S. leaders deceived the Afghan government as well as themselves, speaking of any potential negotiation as Afghan-led while deliberately concealing negotiations from Afghan leadership. The effect was to give the enemy freedom of action while undermining the Afghan government's legitimacy. Holbrooke bypassed the Afghan government to establish contact with Tayeb Agha, who served as head of the Taliban's financial commission. Barnett Rubin, a New York University political scientist, was enlisted by Holbrooke to establish contact with Taliban leadership.

After attempts at negotiation foundered, the United States, in May 2014, released five Taliban detainees from the U.S. detention facility at Guantánamo Bay, Cuba, in exchange for the release of a U.S. Army deserter. The subsequent effort to use the so-called Taliban Five to start effective negotiations appeared as an act of desperation.[42]

* * *

AS WE drove through the arched sally port and entered the well-manicured grounds, I thought of the history of the Arg and the fact that until Karzai transferred power to Ghani in September 2014, Afghanistan had not seen a democratic, peaceful transition. Former prime minister Mohammed Daoud Khan ended the monarchy in 1973 by seizing power while his cousin King Mohammed Zahir Shah was visiting Italy. Five years later, Afghanistan's decades-long wars began with an act of horrific violence in the same home in which Ghani now lived. After Daoud initiated a strategic opening to the United States and its allies, including Pakistan, the Soviets and their Afghan Communist clients decided that he had to go.

After seizing control of Bagram Airfield, the coup leaders executed thirty air force officers and seized Soviet MiG fighters to support a tank offensive against the Arg. The palace guard surrendered. Daoud had gathered twenty-four of his family members in the living room in the hope that they might be spared. It was not to be. Soldiers gunned down Daoud and all but seven of his family, then transported the bodies in a palace truck to the grounds surrounding the notorious Pul-e-Charki Prison, outside Kabul city. The officer in charge ordered his men to dump them into a ditch that an army bulldozer had readied.[43] The bodies were discovered and given a proper burial only in 2009. Afghan leaders had a reason to feel insecure. I thought of how the compounded stress of historical memory and doubts about America's commitment must have weighed on Karzai and were weighing on Ghani.

I met Hanif Atmar, Afghanistan's national security advisor, outside the Arg. An honor guard in dress uniforms lined the walkway to the palace entrance. Atmar's calm demeanor masked the trials of a turbulent life in which service in the Karzai and Ghani governments was only the latest chapter. Born to an ethnically Pashtun family in Laghman, Afghanistan, in 1968, Atmar served in the KHAD (Afghan secret police) in the 1980s. He used a cane and had to kick his prosthetic right leg ahead of him. He'd lost his leg in the lead-up to the seven-month-long Battle of Jalalabad in 1989, in which Afghan forces defeated a combined assault of mujahideen and Pakistan Army forces on the eastern city. Following the Soviet withdrawal, he went to Britain, where he earned two degrees from the University of York. He had served for years as a staff member in a Norwegian nongovernmental relief organization. In 2008, Atmar was appointed minister of the interior and tasked with reform of a ministry regarded by some as "the most corrupt of all government organizations."[44] I first met him after he resigned from the Karzai government in 2010. We spent hours together discussing the cauldron of Afghan politics, involving as it does avarice, historical animosity, deep distrust, and ever-shifting alliances.

Over the years, discussions with Atmar, Ghani, Abdullah, and other Afghan leaders convinced me that U.S. policy and strategy in Afghanistan was the opposite of what was needed. Afghanistan required a long-term commitment; the U.S. constantly announced its withdrawal timetable. Diplomatic and military efforts should be aligned; instead, they cut against each other, such as President Obama's announcement of U.S. withdrawal while negotiating with the enemy. Pakistan and other regional actors must play a less destructive role; the U.S. sent mixed signals and increased aid to Pakistan while the ISI increased support to the Taliban and other terrorist organizations. Reforms to strengthen

Afghan institutions were critical; anticipated U.S. withdrawal incentivized corruption that weakened those institutions.

Atmar and I walked through the cordon of soldiers, entered the building, and walked slowly up the staircase. He would leave me with President Ghani, and I would see him later in the day. I had looked forward to the meeting with Ghani. We had a shared understanding of how the war had evolved and the flawed assumptions that had underpinned failed policies. I hoped that our discussion might crystalize the outline of the first sound, long-term, and sustainable strategy for the war.

CHAPTER 6

Fighting for Peace

We do not seek peace in order to be at war, but we go to war that we may have peace. Be peaceful, therefore, in warring, so that you may vanquish those whom you war against, and bring them to the prosperity of peace.

—SAINT AUGUSTINE

PRESIDENT GHANI greeted me in his office. We talked alone for an hour. I had known him since the early days of America's war in Afghanistan. We had often lamented together opportunities lost in a conflict that had lasted almost two decades.

The contrast between Ghani and Karzai could not have been starker. Ghani loved the United States because the country had given him the opportunity to escape the horrors of war and achieve success as an academic and development expert. Born into a wealthy Pashtun family in 1949, he received a liberal education at Habibia High School in Kabul and went on to attend the American University of Beirut, where he met his future wife, Rula, a Lebanese Christian.[1] After teaching at Kabul University, he was awarded a scholarship to pursue a master's degree at Columbia University in New York City. He left Afghanistan in 1977, just months before soldiers gunned down President Daoud and his family and just two years before Soviet forces invaded Afghanistan to keep a friendly Communist government in power after the assassination of Daoud's successor, President Nur Muhammad Taraki.[2] Taraki, like Daoud, was assassinated in the palace in which Ghani and his wife would live. His family, however, was spared, and the act was less bloody: Taraki was instructed to lie down on a bed, was suffocated by three men with pillows, and then was secretly buried at night. It was during this turbulent period that most of Ghani's family was imprisoned. Ashraf stayed in the United States and completed his PhD in anthropology.

On September 11, 2001, Ghani was at his desk in Washington when a passenger jet carrying six crew members and fifty-eight passengers, including five Al-Qaeda terrorists, slammed into the Pentagon across the river from his World Bank office. He knew that there would soon be dramatic change in his native country. An exuberant man who, while discussing almost any development-related topic, would exclaim, "I have a paper on that!" Ghani immediately drafted a five-step plan for political, social, and economic transformation in Afghanistan. Two months after 9/11, his work with a smart, soft-spoken British human rights lawyer, Clare Lockhart, influenced the formation of the Afghan government following the Bonn Conference. Ghani joined Karzai's government to try to put his ideas into practice as finance minister. He warned against empowering warlords who would perpetuate state weakness and the lawlessness that gave rise to the Taliban. But he was unable to convince Karzai, whose writ depended on accommodating those same warlords. Ghani departed Afghanistan in 2006 and cofounded with Lockhart the Institute for State Effectiveness. The pair authored a book entitled *Fixing Failed States: A Framework for Rebuilding a Fractured World*, in which they describe the essential functions necessary to strengthen states in distress.[3] In 2011, the man who literally wrote the book on what needed to be done returned to Afghanistan to help President Karzai and the Coalition transition secure provinces to Afghan government control. Ghani ran for president of Afghanistan in 2014 as someone who could bridge the worlds of a Western development expert and a traditional Ghilzai Pashtun from eastern Afghanistan. He came from a people, the Ahmadzais, known as warrior-poets who relished autonomy and practiced Pashtunwali, the traditional Pashtun code of honor and hospitality.

No one doubted Ghani's determination but some were concerned that his background in academia and at the World Bank had developed in him a pedantic style as well as impatience with those who could not grasp or embrace his often rapid-fire reform initiatives. Many of these reforms were sound and practical, while others were aspirational and lacked a clear bridge to implementation. Ghani could be mercurial, and his temper occasionally alienated those he needed to implement the reforms, but he brought in a strong team and was making progress. Our conversation focused on what would become the outline of President Trump's South Asia strategy.

We sat alone in armchairs that turned slightly toward each other in a large room normally filled by international delegations and their Afghan counterparts. Like Ambassador Hugo Llorens, Ghani was concerned about the psychological dimension of the strategy. A joint strategy would have to generate confidence among the American and Afghan people, as well as Coalition members, while communicating determination to our enemies and their supporters. This "inside-out" effort to strengthen Afghanistan against the Taliban should be matched with an "outside-in" sustained diplomatic effort to convince key regional actors to play a positive, or at least a less destructive, role in Afghanistan and across the region. Our conversation was reminiscent of the many others we had had in his Kabul home five years earlier, over delicious dinners of lamb and Afghan *pulau* (basmati rice infused with a mix of carrots, raisins, and onions).

I was candid with Ghani about the principal constraint to sustaining a long-term approach in Afghanistan and South Asia: the will of the Afghan and American people. I asked for his and his government's assistance in reaching American audiences to explain what was at stake not only for Afghans, but also for Americans and all humanity. Ghani and I believed that Afghanistan was a modern-day frontier between civilization and barbarism, but few Americans understood South and Central Asia as an ecosystem in which more than twenty U.S.- designated terrorist organizations thrived.[4]

That ecosystem emerged from ideal conditions such as state weakness, access to young male recruits, the support of Pakistan's ISI, the ability to hide in loosely governed spaces, rivalries that allowed terrorists to gain sponsorship from particular tribes, access to a lucrative drug trade and other criminal enterprises, and the ability to flow money, weapons, people, and narcotics across porous borders. Geographically, the Afghanistan-Pakistan border region's centrality and relative inaccessibility made it the ideal spot for basing a jihad and from which to project murderous campaigns outward to India, Central Asia, Russia, China, Europe, and the Middle East. Moreover, the region had a strong ideological draw due to a prophesy in the Hadith (a record of the sayings and actions of the Prophet Muhammad) that an Islamic army will emerge from Khorasan under black banners and ultimately conquer Jerusalem. The text encourages Muslims to "join that Army, even if you have to crawl over ice; no power will be able to stop them."[5]

Ghani knew that President Trump had been elected largely by people who did not understand what was at stake in that faraway place and who were skeptical about what more Americans were calling "forever" or endless wars. I asked Ghani to help the world understand, without obscuring the daunting challenges that Afghanistan faced, the good that their efforts, alongside those of courageous Afghans who were sacrificing every day, had achieved. Despite the difficulties encountered in that long war, Afghanistan, by 2017, was a transformed society. After the fall of the Taliban, hundreds of thousands of refugees returned to the country. The city of Kabul grew from a population of one million to close to five million.[6] Social services expanded with the returning population. In October 2018, over 45 percent of Afghan voters voted in parliamentary elections. In September 2019, voter turnout dropped due to fears of Taliban attacks, but despite threats to murder anyone who went to the polls, two million people, approximately 27 percent of registered voters, voted in the presidential election.[7] Although there was far too much fraud in the development effort in Afghanistan, the entire effort was not wasted, as some have claimed. Americans do not know that Afghanistan is a transformed society because they do not know Afghans who have benefited from the extraordinary changes that followed the defeat of the Taliban in 2001. The contrast between Afghanistan in 2001 and Afghanistan today in the areas of education, technology, and women's rights is stark.

The education of young people expanded rapidly, including women who had been denied education under the Taliban. Prior to 2001, it was estimated that fewer than 1 million children in Afghanistan were enrolled in primary and secondary education. In 2017, UNESCO estimated children's enrollment at a total of 9.3 million. Higher education has also expanded, with 300,000 students enrolled in private and public universities as of 2019, one third of them women.[8]

Because of access to technology and information, Afghanistan is no longer isolated from the world. Over 80 percent of Afghans have access to mobile phones and 400 percent more Afghans reported using the internet to access news and information in 2018 than in 2013. Access to technology is helping Afghans counter corruption through social media exposure and the use of mobile payments and banking. Afghanistan has the most open press in the region, a stark contrast to the complete blackout of the Taliban period and the state-controlled and influenced media in other countries, such as Pakistan, Iran, and Central Asian states. By 2019, there were 96 TV channels, 65 radio stations, and 911 print media in Kabul as well as 107 TV channels, 284 radio stations, and 416 print media outlets in other provinces.[9]

Under Taliban rule, women were denied education and brutally punished for actions such as venturing outside the home unaccompanied by a male relative or talking with men to whom they were not related, even by telephone. The Taliban enforce those oppressive measures in the areas they control; in 2018, women were lashed as punishment under Sharia law and a Taliban court ruling.[10] By contrast, the progress achieved for women under the Afghan Constitution is unprecedented in the region. Afghan law requires that 25 percent of parliamentary seats be held by women, and a record high of 417 female candidates ran for parliament in October 2018.

Soon after the visit, Ghani conducted an interview with *Time* magazine in which he told Americans that "their security depends on us."[11] He went on to assure them that Afghans would continue to shoulder the vast share of the burden, reminding Americans that the number of troops committed in Afghanistan and the cost of the war had been reduced by 90 percent. He also expressed gratitude for Americans' sacrifices. He stated that the fewer than 50 U.S. soldiers killed in Afghanistan in the past eighteen months was "still too many" and asked Americans to compare those losses to the 2,300 Americans lost between 2001 and 2014. But the American people needed to hear more from their own leaders about what was at stake and what had been achieved. They also deserved a strategy designed to protect them at an acceptable cost.

Ghani was also well aware that sustaining American will required confidence in the reliability and virtue of the United States' partners in Afghanistan. Accordingly, Ghani was working hard with Ambassador Llorens and General Nicholson on a compact between our countries to establish clear objectives for Afghan institutional reform and metrics to evaluate effectiveness. In contrast to Karzai, Ghani wanted the United States and other donors to impose conditions on assistance. Those conditions were meant to incentivize reforms and counter opposition from mujahideen-era elites who wanted to maintain their patronage networks.[12] Ghani told me that his priority was to strengthen the security ministries of defense and interior as well as the intelligence ministry known as the National Directorate of Security (NDS). Also vital were institutions and functions critical to establishing rule of law.

But Ghani needed the United States to use its influence. American diplomats and military commanders were sometimes reluctant to impose conditionality of assistance because either they underappreciated the severity of the problem or were oversensitive to being seen as neocolonialists in a country that had fought four wars against occupiers: three against the British and one against the Soviets. But influence exerted in coordination with elected Afghan leaders was supportive of Afghan sovereignty. Ghani described how the Afghan minister of defense was working with General Nicholson's command to lessen the influence of criminalized patronage networks and improved dramatically the quality of leaders in the Afghan National Army.

We spoke about how the United States and others might reinforce Afghan efforts to lessen divisions and foster unity. Afghanistan is a decentralized nation in which local and tribal leaders jealously guard autonomy. Ghani's dilemma was how to pursue reform without widening divisions that could weaken the state. While Afghans often profess their identity as Afghans rather than members of a particular ethnic group, decades of war and the Taliban's brand of Pashtun nationalism mixed with religious extremism had bred resentment and fear of Pashtun domination among the Tajiks, Hazaras, and Turkic minorities in the country. The collapse of the Najib government in 1992 ushered in an era of competition among rival Islamist sects. During the resistance to Soviet occupation in the 1980s, Afghanistan's mild indigenous forms of Islam, Hanafi Sunnism and Sufism, were overwhelmed by imported extremist Islamist sects. Divisions weakened popular will to resist the Taliban and created opportunities for foreign actors, such as Iran and Pakistan, to support proxies in return for political influence. There were opportunities for the United States and like-minded countries such as the United Kingdom, India, and Scandinavian countries that have long-standing relationships there to reduce those divisions and foster accommodations among Afghanistan's communities.

Afghanistan is an increasingly urban and connected society. Communities come together physically in its burgeoning cities and electronically on the internet and through social media. The country's population is young: 63 percent are under twenty-four years old. Afghanistan's young people are better connected than ever before not only to the world, but also to one another. Ethnic communities converge on the plains north of Kabul and in the city itself, including Tajiks, Pashtuns, and Hazaras.[13] Their knowledge is no longer confined to their ancestral lands.

One key destabilizing external force in these efforts was Pakistan. Ghani reminded me that, early in his presidency, he took significant political risk in engaging Pakistan and tried to convince the army leadership to pursue their interests in Afghanistan diplomatically rather than support terrorist organizations such as the Taliban and the Haqqani network. It did not work, and it was not the first time the Afghan president was burned.

Back when the Obama administration prioritized cooperation with Pakistan in the fight against Al-Qaeda in South Asia, Pakistan approached a despondent President Karzai. In May 2010, soon after he professed his commitment to his U.S. partners, Pakistan's army commander, Ashfaq Parvez Kayani, sent the head of ISI, Ahmad Shuja Pasha, to Kabul to suggest to Karzai a pact that excluded the United States.[14] Karzai, after exhorting the United States to confront Pakistan for years, seemed ready to try something different. Kayani's real motive was to weaken the war effort against the Taliban by driving a wedge between the United States and Afghanistan. The Pakistanis were smooth, and Karzai was vulnerable. In an October 2011 interview with Pakistan's Geo Television, Karzai stated, "God forbid, if there is ever a war between Pakistan and America, then we will side with Pakistan. If Pakistan is attacked, and if the people of Pakistan need help, Afghanistan will be there with you. Afghanistan is a brother."[15] In the year of Pasha's conciliatory visit to Kabul, the Taliban committed at least six major attacks with no military objective on Afghan civilians and four hundred other IED and suicide attacks throughout the country that resulted in civilian deaths.[16] By 2017, I had come to believe that continuing to expect Pakistan to change its behavior after a perfect record of duplicity across almost two decades was the height of folly.

As I arrived in South Asia, the Haqqani network was preparing one of the deadliest mass murders in Afghanistan's recent history, an attack that would kill more than 150 people in Kabul on June 6, 2017. At the time, the United States had over six billion dollars in military and economic assistance in the pipeline. It was past time for a different approach to Pakistan.

I talked with other Afghan officials that day, including Hanif Atmar and Abdullah Abdullah, both of whom would oppose Ghani in the October 2019 presidential election. That evening, I met with Amrullah Saleh, who would later join Ghani as his running mate and candidate for vice president in the presidential election and narrowly escape a Taliban assassination attempt. Saleh, who had served for six years as the head of Afghanistan's National Directorate of Security until he and Atmar could no longer abide Karzai's conspiracy theories, offended many of his American interlocutors with his direct and sometimes strident criticism of the war effort. But I appreciated his candor and his passion for his country. His unwavering opposition to the Taliban and his willingness to highlight flawed assumptions were refreshing. Saleh always made clear what was at stake.

My conversations with Afghan leaders left me with a sense of hope and regret. I hoped that we might finally align efforts not only to fight the enemy, but also to strengthen Afghanistan such that it could withstand the regenerative capacity of the Taliban. I regretted the missed opportunities due to short-term approaches we had taken to long-term problems over many years. Had we done irreparable damage to our will? Were we out of time? If only the United States and other nations had taken a long-term view in Afghanistan from the outset and not tried to turn the war into something alien to war's very nature, we might not be in this situation. I was determined to get the U.S. president options to achieve sustainable security in Afghanistan and South Asia, but I knew it would not be easy to gain approval for those options and generate the will necessary to implement them.

* * *

MY LAST meeting of the day was with students who had been on the campus of American University of Afghanistan on August 24, 2016, when Taliban gunmen blasted through the university's fortified wall with a truck bomb. The terrorists laid siege to the campus for nearly ten hours. They killed fifteen people, including students, faculty, guards, and crisis responders, in addition to wounding dozens more.[17] When the school reopened seven months later, all but one of those students returned immediately. That student, paralyzed from the waist down, returned after medical treatment in Germany. Our conversation reinforced to me the importance of education in the long-term effort to defeat jihadist terrorists. Those intrepid young people were part of a cohort of young Afghans who transcended ethnic identity, rejected Islamist extremism, and were determined to build better lives for themselves and future generations.

Jihadist terrorists depend on ignorance. Decades of war and the brutality of the Taliban denied education to a population that became susceptible to the demagoguery of the Taliban and terrorist groups. These groups brainwash young people and foment hatred to inspire and justify violence against innocents. They prey on those most vulnerable: adolescents and teenage men (and increasingly women) who are disenfranchised or seek affirmation. Many young people are indoctrinated into jihadist terrorist organizations through sexual and other forms of abuse. The trauma they suffer prepares them to be systematically dehumanized, often through participation in beheadings or other egregious acts of violence. Those who claim piety not only commit the most heinous acts of violence, but also run an immense and profitable criminal enterprise that enriches its leaders, who live in comfortable compounds in Pakistan. Their children go to private schools there while they bomb girls' schools in Afghanistan.

As I met with these courageous students, I wished that more Americans could get to know them and see the halting but real progress their country was making. I also wondered what those who advocated power sharing with the Taliban envisioned the result would look like. Would the Taliban be permitted to bulldoze only every other girls' school? Would music and art be banned in only parts of the country? Would mass executions occur in the soccer stadium only every other Saturday? The lack of understanding of the enemy and of the Afghan people led to the paradox of excessive sympathy for the Taliban and disregard for (or at least disinterest in) courageous Afghans: soldiers, police, students, journalists, and government officials who were willing to die to defeat our common enemy and secure a better future for their children and ours. I would keep these brave Afghans in mind as I proceeded on to the next leg of my trip: Pakistan.

* * *

THE SHORT flight took us into Islamabad airport. It was familiar to me; during 2003 and 2004 I was a backbencher for several visits on the personal staff of Gen. John Abizaid, the commander of U.S. Central Command. Central Command is the overall military headquarters with responsibility for the wars in Afghanistan and, by then, Iraq and for military operations across the greater Middle East. Our ambassador to Pakistan, David Hale, met us. Hale joined the Foreign Service around the same time I entered the army, in the mid-1980s. We both spent considerable time in the Middle East, he at the U.S. missions to Tunisia, Bahrain, and Saudi Arabia and as ambassador to Jordan and Lebanon. Hale was no stranger to tough, frustrating jobs; he was special envoy for Middle East Peace from May 2011 to June 2013. But his work in Islamabad might have been the toughest of all. That is because the policy he was asked to implement was fundamentally flawed, and because many of his interlocutors were the most duplicitous people on earth.

Over the years, Pakistani officials had taken good advantage of incoherent and inconsistent U.S. strategies. On September 12, 2001, Deputy Secretary of State Richard Armitage called the head of Pakistan's ISI, Gen. Ahmed Mahmud to deliver the message that Pakistan "faces a stark choice: either it is with us or it is not." Mahmud stated that he wanted to "dispel the misconception" of Pakistan's "being in bed" with terrorist organizations and promised his and President Pervez Musharraf's full, "unqualified support."[18] It was the beginning of a sustained campaign of deception and subterfuge. Three U.S. administrations fell for it. Pakistani leaders proved particularly adept at using Americans' vanity and naïveté against them. Hale and I agreed that previous approaches to Pakistan suffered from a bad combination of superficial understanding and overintellectualization of the regional dimension of the jihadist terrorist problem.

Understanding was superficial because Americans often skipped over the base motivations, goals, and strategies of the Taliban and other terrorist organizations as well as the Pakistani ISI. Overintellectualization occurred when key officials in the Obama administration developed a flawed logic that underpinned a self-defeating strategy. That logic was: because Pakistan is more important than Afghanistan (as Pakistan is a nuclear-armed nation with more than 212 million people), U.S. policy should prioritize the relationship with Pakistan over the outcome of the war in Afghanistan.[19] The consequences of failure in Pakistan would be far greater than a failure in Afghanistan, the logic went. If security in Pakistan collapsed or if Pakistan became completely estranged from the West, the jihadist terrorist problem would increase by orders of magnitude. An isolated and desperate Pakistan might initiate another war with India, a war that could lead to nuclear devastation in one of the world's most populous regions. The best strategy, therefore, would be to prioritize good relations with Pakistan to prevent the worst possible outcome.

But that approach to Pakistan rested on wishful thinking and the dubious assumption that the Pakistan Army and ISI were willing to reduce their support for the Taliban and the Haqqani network. Pakistani leaders used these groups to coerce the Afghan government and prevent a Pashtun nationalist call to adjust the border. The ISI also wanted control of at least portions of Afghanistan to provide the "strategic depth" necessary, in their minds, to prevent India from encircling Pakistan with an Afghan government friendly to New Delhi.

U.S. prioritization of the relationship with Pakistan even at the expense of stability in Afghanistan overlooked the interconnected nature of security in both places and encouraged the

Pakistani military to continue its self-destructive support of terrorist organizations. It was, of course, a terrorist safe haven in Afghanistan, enabled by the Pakistan-supported Taliban government, that led to the mass murder of 2,977 innocents on September 11, 2001.[20] Insecurity and violence in Afghanistan blows back into Pakistan, creating the very problem that downgrading the priority of security in Afghanistan was meant to avoid. When the Obama administration removed the Taliban as a designated enemy, Pakistani leaders concluded that their American counterparts were easily duped. How seriously should the Pakistanis take American appeals to do more to fight the Taliban and the Haqqani network in their own country when the United States had already announced its timetable to withdraw and seemed desperate to accommodate at least some of the Taliban's demands?

★ ★ ★

PAKISTANI POLICE cleared the route for our motorcade as we made the short drive to the government section of the city. To those unfamiliar with South Asian cities, Islamabad might seem chaotic as cars, motorbikes, and the ornately decorated jingle trucks vie for advantage in its jam-packed boulevards. But Islamabad is orderly compared to other cities in the region. Built in the 1960s, Islamabad enjoyed a name and an ambitious construction plan that reflected Pakistan's aspiration to be the second Medina, an Islamic state analogous to the city to which the Prophet Muhammad migrated in AD 622. The capital was also meant to express unity in a country that encompassed diverse territories and peoples, including Punjabis, Pashtuns, and Sindhis.

When we arrived at the prime minister's office, I sat opposite Prime Minister Nawaz Sharif. The sun shone brightly through the large windows overlooking the garden. He was serving his third nonconsecutive term at the time. Sharif had previously been prime minister from 1990 to 1993 and again from 1997 until the former army chief Pervez Musharraf unseated him in a bloodless coup in 1999. Sharif was a political survivor, but he was under constant pressure from the army and his political opponents. He seemed sympathetic to our main point that the United States could no longer bear the fundamental contradiction in our relationship in which a supposed ally supported our enemies, perpetuated violence, and was therefore at least partially responsible for the deaths of Coalition soldiers and innocent civilians. I told the prime minister that patience was running out and soon we might no longer be able to provide economic and military aid. The United States had recognized that we were essentially funding our enemies through a middleman—U.S. assistance allowed the Pakistan Army to allocate more money to recruiting, training, equipping, and sustaining the Taliban and other terrorist organizations.

As we walked past the manicured lawn and then drove away down the tree-lined drive, the contrast between the trappings of power in the prime minister's office and the actual powerlessness of that place and those who occupied it was striking. Despite U.S., British, and other nations' efforts over many years to reinforce the civilian government of Pakistan, army headquarters remained the place of power and authority. The sympathetic hearing from Sharif was inconsequential. He and his finance minister would soon be indicted on corruption charges related to the Panama Papers leaks and removed from office, but members of Sharif's party attributed the charges to pressure from the military.[21] In 2018, the army's choice for prime minister, Imran Khan, a former world-renowned cricket player and playboy with a visceral dislike for the United States, took office. It was clear that the U.S. approach to Pakistan had to be based on the reality that while most countries have an army, Pakistan's army has a country. To move from strategic narcissism to strategic empathy, we would have to pay attention to the emotions, ideology, and worldview of the Pakistan Army leaders.

* * *

SINCE THE Pakistani state's inception in 1947, the army defined itself by the threat from its vast neighbor India. Pakistan fought and lost four wars, including its Civil War (also known as the Bangladesh Liberation War), in 1971, during which Pakistan ceded 55 percent of its population and 15 percent of its territory to newly independent Bangladesh. Territorial disputes remained. In the northwestern Indian subcontinent region of Kashmir, Indian and Pakistani troops engage in high-altitude and high-stakes skirmishes. Conflicts have often escalated to even the threat of nuclear weapons use, as happened in the summer of 2002. Kashmir was the Pakistan Army's early proving ground for using jihadist terrorism as an instrument of state policy, and it remains a flashpoint for conflict between Pakistan and India. India's contentious move in 2019 to remove the semiautonomous status the Indian state of Jammu and Kashmir had enjoyed for the last sixty-five years could stoke indigenous militancy in the region, much like flawed 1987 state elections did there in the late 1980s and early 1990s.[22]

American flag officers (generals and admirals) have been at times susceptible to the charms of Pakistani officers, who've shared the manners and comportment of a Western army. Their landscaped posts would make any U.S. Army command sergeant major envious. Many of their officers were educated at the British Army's training ground at Sandhurst. Others attended U.S. Army schools at Fort Benning, Georgia, Fort Leavenworth, Kansas, and Carlisle Barracks, Pennsylvania. They spoke the Queen's English, played polo, studied the American Civil War, and drank good whiskey. But they grew up in an organization that saw itself as the arbiter of national interest and the protector of Pakistan's Islamic identity. The army had veto power over foreign and economic policy.[23] And Pakistan's generals would not let go of what they believed was one of their most effective tools: terrorist organizations and militias that allowed them to use violence while denying responsibility.

The Pakistan Army got its start in jihad in 1947 and never

stopped. In the 1980s and early 1990s, the U.S. support for Islamist groups fighting against the Soviets in Afghanistan flowed through Pakistan's powerful intelligence service, the ISI. The ISI-run resistance to Soviet occupation set conditions for the Taliban's rise and foreclosed on any chance of a moderate political evolution in Afghanistan. ISI involvement in the Afghan drug trade provided covert funding for Pakistan's use of terrorist proxies and the development of nuclear weapons.

But the ISI's Frankenstein's monsters, and the sons of Frankenstein they spawned, groups like Tehrik-i-Taliban Pakistan and ISIS-Khorasan, turned on their masters. Attacks like the massive bombing of the Islamabad Marriott hotel in September 2008 and the Pearl-Continental Hotel in Peshawar in June 2009 are examples of Pakistan's project to promote terrorism against its neighbors badly backfiring on the Pakistani state.[24] These groups also killed Pakistani civilians and, increasingly, attacked religious minorities such as Shia or Sufi Muslims, risking the kind of destructive sectarian conflict that had engulfed the Fertile Crescent of the Tigris and Euphrates River Valleys. By the 2010s, attacks were no longer limited to Pakistan's far frontier as terrorists began to operate in Swat and Buner Districts. They even extended their murderous campaign beyond the Pakistan Army and began to target their families.

In a particularly egregious attack in December 2014, seven terrorists armed with suicide vests and guns burst into the Peshawar Army Public School in the northwest of Pakistan. The toll was heart-rending: 141 dead, 132 of them children ranging in age from eight to eighteen. The attack demonstrated the interconnected nature of these groups. Tehrik-i-Taliban Pakistan, the primary driver of antistate violence in Pakistan since 2007, took responsibility. The fact that the perpetrators included a Chechen, three Arabs, and two Afghans demonstrated the international dimension of the problem and the folly of believing that these groups could easily be contained geographically.[25]

In response to the attack, Pakistan's defense minister, Khawaja Muhammad Asif, stated, "All Taliban are bad Taliban. Extremism of any kind—of thought, action, religious, or political extremism—is bad. We have to eliminate them wherever we find them."[26] He also vowed to regulate religious education. If there were to be an incident that convinced the Pakistan Army to stop using terrorist organizations as an arm of its foreign policy and to pursue all terrorist organizations on its soil, that should have been it. It was not. That is why Ambassador Hale and I believed that any strategy in South Asia should begin with the assumption that the Pakistan Army would not change its behavior.

* * *

DESPITE OUR deep frustrations with each other, meetings between U.S. and Pakistani officers evinced a sense of mutual respect based on experience in the profession of arms. Hale and I met with Gen. Qamar Javed Bajwa, chief of army, and Gen. Naveed Mukhtar, director general of ISI. I had spoken with Naveed at his request two weeks earlier; we shared a background in tanks and mechanized warfare. Having learned of his interest in the history of the American Civil War, I had sent him a copy of Williamson Murray's book, *A Savage War*. Naveed, having learned of my daughter's upcoming wedding, sent a beautiful hand-knotted carpet, which, because its expense far exceeded what would be appropriate to accept, went directly to a government warehouse. Naveed developed his interest in the American Civil War during his attendance at the U.S. Army War College at Carlisle Barracks, Pennsylvania. In his 2011 War College thesis, "Afghanistan—Alternative Futures and their Implications," he wrote:

> Establishing a viable context for Afghan stability and security involves key regional and global stakeholders. Towards that end, the United States needs to employ major diplomatic measures designed to ease regional tensions and prevent external players from derailing the strategy.[27]

He was right, but he might also have written that the United States should try to solve the problem that Pakistan was creating. Naveed and other Pakistan Army officers often sounded like they were diagnosing the situation in South Asia as dispassionate outside observers even as they drove much of the instability and violence that was the subject of their analysis.

I opened the meeting with condolences for the more than eight thousand soldiers who had lost their lives to terrorist groups since 9/11 and the more than twenty thousand Pakistani civilians whom terrorists had killed across those years.[28] Having studied their army's recent adaptation to the demands of counterinsurgency, I expressed admiration for their soldiers' courage, especially given that their families were also at risk. I told them that Americans were deeply saddened by the attack on the Peshawar Army Public School. I intended these opening remarks, in part, to expose the confounding contradiction of the Pakistan Army's asking its soldiers to sacrifice in battle against terrorist groups while a wing of the army nurtured terrorist organizations with connections to those same groups. I told the generals that Ambassador Hale and I wanted first to listen so we might learn from their perspective and apply what we learned to the new administration's policy toward Pakistan and South Asia.

General Bajwa took full advantage of the opportunity to orient a member of a new U.S. administration to his worldview. Qamar Javed Bajwa, born in Karachi to a Punjabi military family, joined the Pakistan Army in 1978. He received training at the Pakistan Military Academy, National Defense University, Canadian Army Command and Staff College, and the Naval Postgraduate School in Monterey, California. His comments seemed crafted to preempt what he expected to hear from me: exhortations to "do more" against the Taliban and the Haqqani network and hopeful expressions that we could, finally, work together to end the violence in Afghanistan. Like his predecessor, Gen. Ashfaq Parvez Kayani, Bajwa wanted me to accept the fundamental contradiction between denials that the Taliban and other terrorist organizations enjoyed safe havens in Pakistan and assurances that the

Pakistani Army could do more against these safe havens with sustained U.S. support.

The conversation with Bajwa and Naveed followed a recurring pattern I had witnessed over the years since September 11, 2001. Pakistani military leaders, army commanders and heads of ISI, were brilliant at manipulating earnest American counterparts. As always, the conversation began with a litany of grievances from the Pakistan Army perspective. The first was the U.S. "abandonment" of Pakistan after the end of Soviet occupation of Afghanistan and the subsequent collapse of the Soviet Union itself. Then there was the temporary suspension of all U.S. military and developmental assistance in 1979 and again in 1990, after the evidence of Pakistan's nuclear weapons program could no longer be denied.[29] From there, the conversation shifted to the portrayal of Pakistan as a victim: a victim of the massive influx of refugees as Afghans fled the Soviet occupation, the Afghan Civil War, and the brutality of the Taliban; a victim of sanctions imposed by the United States due to its nuclear program; a victim of what the Pakistani generals portrayed as a "U.S. war on terrorism" in which Pakistan had been a dutiful ally whose sacrifices exceeded those of the United States and all Coalition armies in Afghanistan combined; and finally, always, a victim of Indian aggression in pursuit of recidivist objectives in Kashmir and the encirclement of Pakistan through the establishment of an India-friendly Afghanistan.[30]

The Pakistan-as-victim gambit then shifted to a narrative of weakness and beleaguerment. Pakistan was strained economically and overcommitted militarily. If only the United States were more patient and provided more assistance, Pakistan would gradually extend security to the regions in which the Taliban and various terrorist organizations, including Al-Qaeda, were based. Ignoring the implicit admission of the existence of safe havens on its territory, offers to do more in the future were interspersed with flat denials of support for the Taliban, Al-Qaeda, and key facilitators of their operations such as the Haqqani network. For those Americans who remained skeptical, Pakistani officers would whisper the possibility that rogue or retired ISI officers might, out of habit and long-standing relationships, continue to advise some terrorist groups without the command's knowledge, even though the ISI is vertically integrated into the army's chain of command. The clincher was a play to the ego of the visiting U.S. official. That official, Pakistani leaders intimated, would be the one to convince them to do more to pursue Taliban, Haqqani network, and Al-Qaeda leaders. The United States would just need to be patient and share more information and provide more assistance.

The typical reaction from U.S. leaders was to believe that their Pakistani counterparts were sympathetic to U.S. positions but were simply powerless to effect change, especially on the time line Americans expected. Those Americans often flew home and parroted the Pakistan Army's main talking point: Pakistan needed more time and more money. And so, the United States wrote more checks and provided more weapons and supplies to those who were sustaining and helping orchestrate a war against them and their allies. It was serial gullibility as new U.S. officials rotated into military, intelligence, and diplomatic positions, and Pakistani officials repeated their lines. Hale and I joked that the relationship had become almost farcical, but the joke was on us. A fundamental shift in policy was overdue, one that incentivized Pakistani leaders to change their behavior and, over time, tried to convince them that it was in their interest to end their support for terrorist organizations that had inflicted so much suffering not only in South and Central Asia and across the Middle East, but also around the world from Libya to France to the Philippines. And while the United States and other nations should continue to encourage a strong Pakistani civilian leadership, a realistic policy must acknowledge that the Pakistan Army makes the important decisions and holds the reins of power.

I tried to break out of the typical pattern by not using the talking points Bajwa and Naveed expected. I told the generals that the new administration would pay far more attention to deeds than words and explained that President Trump, as a businessman, would see funding our enemies indirectly through aid to Pakistan as a bad return on investment. To address General Bajwa's complaint that the United States was always asking Pakistan to do more, underappreciating Pakistani sacrifices and the limits of Pakistan's control over its far frontier, I assured him that we would stop asking him to do more. We would instead ask him and especially General Naveed to do less—provide *less* support to organizations that were killing Afghan and U.S. soldiers and murdering Afghan civilians. I hoped to at least convince him of our determination not to repeat mistakes of the past. I observed that U.S. administrations' relations with Pakistan tended to go through a cycle that began with conversations like the one we were having. U.S. officials' expectations rose after hearing Pakistani promises of real cooperation. The cycle always ended in abject disappointment. I said that I intended to stay at abject disappointment until there was a demonstrable change in behavior.

Most of that abject disappointment was based on the Pakistan Army's failure to confront a particularly brutal mujahideen militia that sought control of territory along the Afghanistan-Pakistan border, the Haqqani network. The Haqqanis provided another compelling example of how the terrorist ecosystem in Pakistan is a danger to that country and the world. The Haqqanis are also an example of the Pakistan Army's unwillingness to stop using terrorists as an arm of Pakistan's foreign policy. The Haqqanis, with ISI support, provide a safe space for terrorists. The Haqqani network, led by Sirajuddin Haqqani, who is also the military commander of the Taliban, connects the ISI, Al-Qaeda, the Taliban, and other local and global terrorists, many of whom are also hostile to the Pakistan Army and state. The network is valuable to all because of its ability to mobilize tribes, raise funds internationally and through organized crime, communicate through multiple media, and develop and maintain a high degree of military expertise. The Haqqanis provide the Taliban with a seemingly inexhaustible supply of brainwashed adolescents and young men, many of them from the more than eighty Haqqani-run madrassas (religious schools) in the tribal areas of Afghanistan and Pakistan.

The Haqqani network's role as a military incubator for both Al-Qaeda and the Taliban demonstrates how various terrorist and insurgent groups located mainly along the Afghanistan-Pakistan frontier are interconnected and enjoy protection from each other as well as from the ISI. And it is the interconnectedness of these groups, as well as the dream of Khorasan's becoming the site of their future caliphate, that makes this frontier such a geographic center of gravity for defeating these organizations.[31] The network has orchestrated attacks on U.S. facilities, including the U.S. embassy in Kabul in 2011 and the U.S. consulate in Herat in 2013. The Haqqanis are particularly adept at the mass murder of Afghan civilians. I told Bajwa and Naveed that I was not going to reiterate the same pleas of countless commanders and diplomats to go after the Haqqanis. It was time to operate under the assumption that Pakistan would not do so, and we had no option remaining but to impose costs on Pakistan.[32] On the way out, I passed General Bajwa a handwritten note with a list of names of U.S. and allied hostages held by the Haqqani network. He asked me if I would like him to do something about them. I replied that I was confident that he could if he wanted to, and emphasized that, from this point on, we would be sensitive to actions rather than words. We said good-bye and pulled away from the pristine grounds of army headquarters.

Hale and I discussed the sad pattern of U.S. policy toward Pakistan since 9/11, a pattern consistent with the definition of insanity attributed to Albert Einstein: doing the same thing over and over again but expecting a different result. Many diplomats and senior military officers, despite decades of experience to the contrary, continued to assume that the Pakistan Army would be honest with them and become a true partner in the fight against jihadist terrorists.

As I met with our team to prepare for the final leg in New Delhi, it was increasingly clear that we needed a strategy that kept in mind long-term implications. I thought of U.S. support for Muhammad Zia-ul-Haq, who in 1977 deposed and later killed the first elected prime minister of Pakistan, Zulfikar Ali Bhutto. Zia's eleven years in power left a damning legacy on the army and the country, one from which Pakistan has not recovered. During the war against the Soviet occupation of Afghanistan, the United States had unintentionally helped Zia, who believed that he was on "a mission, given by God, to bring Islamic order to Pakistan," transform Pakistan into a global nexus of jihadist terrorism.[33] I also remembered how, after 9/11, U.S. support for Prime Minister Pervez Musharraf, a former commander of the Pakistan Army who had internalized the army's hatred for India as well as Zia's affinity for Islamist militants, bolstered Pakistan's ability to use terrorism not only in Afghanistan, but also against India.

As we boarded the plane bound for New Delhi, it was clear that from the Pakistan Army's perspective, everything was ultimately about India. A friendly government in Afghanistan would give Pakistan "strategic depth" in case of another war with India. And the ISI's support for the Taliban and Al-Qaeda was dependent on the Pakistan Army's vast terrorist infrastructure, built to sustain groups that it uses as an arm of its foreign policy toward India. One such group, Lashkar-e-Taiba (LET), was founded in 1987 as a joint venture between the ISI and terrorists later involved in the founding of Al-Qaeda. LET's leader, Hafiz Saeed, made clear that LET's mission was to "fight against the evil trio, America, Israel, and India."[34]

* * *

INDIA WAS a critical partner, not only in combating South Asia terrorism, but also in the competition with the Chinese Communist Party across the Indo-Pacific. India is a country with tremendous challenges and opportunities. Although it was, in the 1980s, a leader in the Non-Aligned Movement (a group of developing states not formally aligned with the Soviet or U.S. power bloc), and thus resists formal alliances, it was obvious that U.S. and Indian interests were converging. India is projected to surpass China as the most populous nation in the world by 2024. Its population is young, savings and investment rates are healthy, and its economy is growing. But the government struggles to provide social services as rapid economic growth combined with the burgeoning population is generating severe interrelated problems in energy, environment, and food and water security. From 1990 to 2020, India halved its poverty rate, but more than 365 million people remain in multidimensional poverty. India's diversity also presents challenges to governance and to maintaining a cohesive national identity, as it is home to the largest population of Hindus (79.8 percent) and the second-largest population of Muslims (14.2 percent) in the world, as well as Christians and Sikhs.[35] India recognizes twenty-two official languages, but approximately one hundred more are spoken, highlighting the country's tremendous geographic and cultural diversity. But India works. And its leaders generally share U.S. and Western democratic principles, and concerns about the Chinese Communist Party's campaign to promote its authoritarian state model. India is a country that the United States and the world needs to succeed. I hoped to learn more about how to expand our cooperation across diplomacy, economic development, security, commerce, and emerging technologies.

India is vulnerable both to Pakistan's sustained support for terrorists and the existential threat of Pakistan's nuclear weapons. The greatest latent danger may be the potential for ethnic or religious conflict. Violent interactions between Hindu nationalists and Islamic terrorists would be disastrous. When Narendra Modi became prime minister, some feared that elements within his Bharatiya Janata Party would stoke the embers of sectarian violence with Muslims. When Modi was chief minister of Gujarat state in 2002, Hindus and Muslims clashed in a series of riots after a train carrying Hindu pilgrims caught fire in a predominately Muslim area. In the ensuing orgy of violence, Hindus targeted Muslims for weeks in a mass atrocity that killed up to two thousand Indians, most of them Muslims. Some accused Modi of failing to quell the violence. Although the George W. Bush administration initially banned Modi from coming to the United States, both the Bush and Obama administrations eventually advanced the relationship with Modi and India.[36] In February 2020, as President Trump visited India, at least thirty-eight people were killed in the worst sectarian violence in decades, evidence that sectarian tensions persist.

Before my visit, Prime Minister Modi had allayed fears that he would pursue a Hindu nationalist (Hindutva) agenda. But during his second term, those fears seemed justified. In 2019, he scrapped the semiautonomous status of the Muslim-majority territories of Jammu and Kashmir. Then a Supreme Court verdict allowed Hindus to rebuild a temple at the site of the Ayodhya mosque, Babri Masjid, which had been destroyed by Hindu zealots in 1992. At the end of the year, massive protests followed the passage of the Citizenship Amendment Act, which allows people from neighboring countries of all faiths except Muslims to gain Indian citizenship.

Still, these events had not yet occurred, and as we started our descent into Delhi, Lisa Curtis reviewed with me the recent history of strengthening U.S.-Indian relations. George W. Bush prioritized improved relations with India, removing all remaining sanctions initially imposed on India after a 1998 nuclear test. In 2005, the two countries had signed an agreement to expand defense relations, and in 2008 the United States led an initiative within the Nuclear Suppliers Group to allow India access to civil nuclear cooperation, despite the fact that it remained a nonsignatory to the 1968 Treaty on the Non-Proliferation of Nuclear Weapons. A formal defense relationship and a nuclear cooperation initiative followed. The Obama administration expanded cooperation in defense, cybersecurity, and energy security. I looked forward to working with India's national security advisor, Ajit Doval, and then-Foreign Secretary Subrahmanyam Jaishankar, to maximize the potential in our relationship. I would be meeting Doval and his team for dinner and Jaishankar for breakfast, after which I would drive to the prime minister's residence to meet with Modi.

It was easy to tell that Doval's background was in intelligence. With his head tilted slightly to the right, he would speak in a hushed voice even about the most innocuous subjects. Jaishankar, by contrast, was a polished diplomat who delivered succinct analyses and tactful, indirect recommendations for how the United States might be more effective in South Asia and across the Indo-Pacific region. Jaishankar and Doval saw up close how China was developing exclusionary areas of primacy. The China threat was shifting India's attitude toward greater multinational cooperation. For example, the Japan-India relationship grew stronger as China attempted to intimidate Japan with maritime militias in the Senkaku Islands and to control shipping routes through the Indian Ocean with facilities in Sri Lanka and the Maldives. India's leaders saw China's One Belt, One Road initiative as a one-way street that would disadvantage them. Still, India, due to the legacy of leading the Non-Aligned Movement during the Cold War, was reluctant to give the impression of entering a formal alliance.

It was inevitable that our discussions would turn to the clear and present danger from ISI-supported jihadist terrorists. Events from 2008 are seared in the memory of Indians, not only those of the global financial crisis, but even more the devastating terrorist attacks that exposed the regional and global dimensions of the threat from Pakistan. Ten attackers from LET carried out a series of coordinated shooting and bombing attacks from Wednesday, November 26, until Saturday, November 29, 2008, most prominently at the Taj Mahal Palace Hotel in Mumbai. At least 164 civilians were killed, including 6 American citizens; more than 300 people were wounded. The sole surviving attacker, a Pakistani citizen, revealed that he and his accomplices were members of LET, came from Pakistan, and were controlled from Pakistan. These painful wounds were reopened when, in 2015, Pakistan released from jail (on bail) the ringleader of the Mumbai attacks, Zakiur Rehman Lakhvi. He promptly disappeared, further highlighting Pakistan's support for terrorists.[37]

I could tell that Doval and Jaishankar were worried about the United States' ability to implement a consistent foreign policy in South Asia. At every turn in our conversations, they made the case for the United States to remain engaged in the region. The 2008 financial crisis and President Barack Obama's desire to disengage from overseas commitments to focus on "nation building here at home," a goal that resonated with the majority of the American public, left Indian leaders doubtful about whether the United States would still pursue an active foreign policy.[38] President Trump's campaign rhetoric did not allay their concerns. India, a country shaped by its independence from colonialism and historically critical of U.S. interventions abroad, had become most fearful of U.S. disengagement at the moment when both Chinese aggression and jihadist terrorism were growing. Jaishankar and Doval believed that U.S. disengagement would embolden both threats.

India would be critical to the outside-in aspect of the South Asia strategy, and I told them that the United States would support an enhanced leadership role for India in the region and beyond. India was influential in multinational fora such as the Brazil–Russia–India–China–South Africa association BRICS and would soon join the Shanghai Cooperation Organisation, which includes China, Kazakhstan, Kyrgyzstan, Russia, Tajikistan, Uzbekistan, and Pakistan. Indian leaders could use long-standing relationships with Russian leaders to persuade President Putin to act in his country's interest and stop supporting Taliban groups, and instead support the Afghan government. And perhaps, together with Russia, India might help convince China to pressure Pakistan to crack down on terrorist organizations. Terrorists already threatened Russia and China directly and could impede China's ambitious infrastructure projects in Pakistan and across Central Asia. Moreover, India might join the United States to prevail upon Gulf States such as Qatar, the United Arab Emirates, and Saudi Arabia to cut off terrorist financing and make their considerable assistance to Pakistan contingent on its no longer serving as a nexus for jihadist terrorists.[39]

On the final day of our trip, Lisa Curtis; MaryKay Carlson, the U.S. chargé d'affaires and deputy chief of mission in New Delhi; and I traveled to the Rashtrapati Bhavan, the prime minister's official residence, where we met with Modi. He gave us a warm welcome. It was clear that deepening and expanding that relationship should remain a priority for both nations. He expressed concern over China's increasingly aggressive efforts to extend its influence at India's expense and over its growing military presence in the region. He was very supportive of the Trump administration's embrace of the Free and Open Indo-Pacific policy and suggested that the United States, India, Japan, and like-minded partners emphasize the concept's inclusiveness and make clear that it is not meant to exclude any nation. At the end of the meeting, the prime minister put his hands on my shoulders and gave me a blessing. I was happy for any help I could get as our team headed back to Washington.

The trip convinced me that we needed to present the president with the withdrawal option and others that were based on the realities of the region and not on the wishful thinking on which previous policies had been based. The long-term objectives were to ensure that jihadist terrorists were unable to attack the United States and its allies; to prevent a potentially cataclysmic conflict between India and Pakistan; and to convince Pakistan to end its self-destructive support for terrorist organizations and undertake the reforms necessary to prevent its internal security from collapsing. The near-term focus should remain on denying terrorist organizations that have a demonstrated global reach access to the resources, freedom of movement, safe havens, and ideological space they need to plan, organize, and conduct attacks.[40] We would base those options on a new set of assumptions:

- First, a counterterrorism-only strategy would be untenable if security collapsed in Afghanistan; Afghans were doing the toughest fighting. The Afghan state needed to be hardened against the regenerative capacity of the Taliban and be strong enough to control critical territory and operate effectively against the nexus of insurgent groups, narcotics-trafficking organizations, and transnational criminal networks.[41]

- Second, the Taliban, Al-Qaeda, and other terrorist organizations were intertwined, as were many of the other dangerous groups located in the Afghanistan-Pakistan border area; we had tried too hard to "disconnect the dots."[42]
- Third, the Taliban could not be trusted to negotiate in good faith, especially if they believed they were winning and the United States was withdrawing.
- Finally, Pakistan would not end or dramatically reduce its support for the Taliban, the Haqqani network, or jihadist terrorists like LET.[43]

* * *

WE WERE at war with no strategy. As I returned to Washington, DC, and the office occupied by Lyndon Johnson's national security advisor, McGeorge Bundy, during the escalation of the Vietnam War, I felt that it was past time to clarify what we wanted to achieve in Afghanistan and South Asia. It seemed the worst form of cynicism to express sympathy for the soldiers and families of those killed or wounded in action while perpetuating a war without direction. There were obstacles to overcome in South Asia and in Washington. It took longer than I had hoped. From the end of the trip on April 20 to the day President Trump made a decision on a new South Asia strategy at Camp David on August 18, the Taliban and the Haqqani network conducted more than one hundred devastating attacks on Afghan security forces and civilians.[44] The day after we returned to Washington, Taliban gunmen and suicide bombers killed more than 140 and wounded more than 160 people at an Afghan army base in the northern Balkh Province. The Haqqani network conducted another devasting attack on May 31, in Kabul, at an intersection near the German embassy. The massive truck bomb killed more than 150 and wounded more than 400. Refusing to be intimidated, President Ghani hosted a planned peace conference in the capital six days later.

A few days after the Camp David meetings, on August 21, the president explained his decision and the strategy to the American people and the world in a speech at Fort Myer, Virginia. On Afghanistan, he acknowledged that the "American people were weary of war without victory" and stated that his "original instinct was to pull out." He summarized the dangers associated with a premature withdrawal, including the growing strength of terrorist organizations including ISIS and Al-Qaeda as well as the other twenty U.S.-designated foreign terrorist organizations active in Afghanistan and Pakistan. He noted the presence of nuclear weapons in such a volatile area. He addressed the contradictions in Pakistan policy, stated clearly that Pakistan offers safe haven to terrorist organizations, and outlined an approach that would deny terrorists control of territory, cut off funding, and isolate them from sources of ideological support. The new policy recognized the war as a contest of wills and aimed at the defeat of those organizations that threatened the United States. There would be no more artificial time lines, and the United States would no longer fall over itself trying to talk with the Taliban. President Trump pledged to support the Afghan government and military as they fought the Taliban while making clear Afghan responsibility to "take ownership of their future, to govern their society, and to achieve an everlasting peace."[45] Emphasis would be on aligning the elements of national power to include diplomatic, economic, and military efforts. After sixteen years of war, the United States had articulated a realistic, sustainable strategy. On August 21, 2017, I met with the president and First Lady in the residence and then jumped into the presidential motorcade as we made the short trip across the Potomac. As we passed Arlington National Cemetery and entered the gates of Fort Myer, I thought it was a fitting venue for the president to give a speech that for the first time clearly articulated how the sacrifices of America's young warriors in Afghanistan would contribute to an outcome in that country and across South Asia worthy of those sacrifices.

* * *

AS I had feared, however, the strategy did not last. Those who were deeply skeptical about America's long war in Afghanistan convinced President Trump to abandon it. Soon after I departed the White House in 2018, those who misunderstood the nature of the war, underappreciated the threat, and were ideologically predisposed toward disengagement from "forever wars" convinced him that a sustained and sustainable military effort in Afghanistan was futile and wasteful.

In July 2019, Pakistan's prime minister, Imran Khan, visited Washington. Khan had strong anti-U.S. credentials and had been the clear choice of the Pakistan Army to be the country's nominal leader. When President Trump asked publicly for Khan's help in bringing the war to a conclusion, it seemed as if another American leader had fallen for a Pakistani counterpart's effort to pose as a partner in counterterrorism. Khan even got a bonus as the president offered to mediate between Pakistan and India on Kashmir. President Trump went so far as to claim that Prime Minister Modi had asked him to mediate. But any observer of South Asia knows that Modi would never have done any such thing, as India had long-opposed outside mediation of the Kashmir dispute, primarily because past initiatives by the United Nations on the issue had proven disadvantageous to India. India is the status quo power when it comes to Kashmir, while Pakistan seeks to weaken India's grip over the territory. The visit must have gone better than expected for Khan, as Pakistan regained its role of both arsonist and fireman in Afghanistan and South Asia.

Shortly after the Khan visit, President Trump, indulging in a common misconception of the war, said, "I could win that war in a week. I just don't want to kill 10 million people."[46] His statement betrayed a misunderstanding of the nature of the conflict in Afghanistan and Pakistan. America and the broad international coalition were not fighting against Afghans in an effort to subjugate popular will; they were fighting *with* the vast majority of Afghans to prevent the forcible return to power of a small, unpopular minority that had governed Afghanistan through terror and brutality. The Afghan peoples, like peoples across South and

Central Asia and the Middle East, were the principal victims of the Taliban and jihadist terrorists. Despite this obvious distinction between the twenty-first-century war in Afghanistan and Pakistan and historical Afghan resistance to Soviet occupation in the twentieth century and in the First and Second Anglo-Afghan Wars of the nineteenth century, the narrative associated with Afghanistan as the "Graveyard of Empires" persisted.

Sadly, the president's statement cheapened not only the sacrifices made by the more than 58,000 Afghan soldiers and police who died defending their country and their families from the Taliban, but also the sacrifices of the more than 2,300 American servicemen and women who lost their lives in the war.[47] The loss of will to sustain the effort in Afghanistan led to rationalization of the decision to withdraw and the resurrection of the flaws and contradictions that undercut U.S. policy there almost from the start.

In September 2018, Secretary of State Mike Pompeo appointed Zalmay Khalilzad, former ambassador to Iraq and Afghanistan, as special representative for Afghanistan reconciliation to negotiate a peace deal with the Taliban. The self-delusion that the Taliban would forswear any connection to Al-Qaeda and other jihadist terrorist organizations had returned. In January 2019, the same month that Khalilzad told the *New York Times* that the Taliban would restrain Al-Qaeda, a UN report observed that Al-Qaeda "continues to see Afghanistan as a safe haven for its leadership," and the U.S. intelligence community's annual threat assessment noted that Al-Qaeda continued to provide support for the Taliban. Another UN report, in June, warned that the Taliban "continues to be the primary partner for all foreign terrorist groups operating in Afghanistan."[48] Meanwhile, as the United States pursued talks, the Taliban intensified its attacks. In July, as Khalilzad sat down with Taliban leaders in Doha, Qatar, a truck bomb in Ghazni city killed 12 and wounded 179 others. At the eleventh hour, the U.S. administration avoided the embarrassment of hosting Taliban terrorists at Camp David to finalize America's surrender and abandonment of its partners on, almost incomprehensibly, the eve of the anniversary of the 9/11 attacks. The president canceled the planned meeting with a tweet on September 7 after learning of a Taliban car bomb attack that killed one American soldier and eleven others. Trump would go on to demand a Taliban cease-fire and a sharp reduction in violence as a precondition for resuming negotiations. Although there would be no immediate U.S. withdrawal, Trump made his intentions clear, stating that "We've been policemen there for a long time. And the government is going to have to take responsibility or do whatever it is they do."[49]

Later that month, the first emir of Al-Qaeda in the Indian Subcontinent (AQIS), Asim Umar, was killed in a joint Afghan-U.S. raid on a Taliban compound in Helmand Province. Haji Mahmood, a Taliban military commander, the AQIS chief for Helmand, and Umar's courier to Al-Qaeda leader Ayman al-Zawahiri, fell next to him. The Department of Defense suppressed a press release that would have announced the death of Umar because of concerns that it "would complicate further negotiations" by exposing once again the interconnected nature of the Taliban and Al-Qaeda. Self-delusion was back in full force.[50]

It got worse. Khalilzad resumed negotiations with the Taliban, and after a "one-week reduction in violence" to demonstrate the Taliban's control over its fighters and terrorists, would sign an agreement that offered the conditional withdrawal of U.S. forces in exchange for a sustained reduction in violence and other promises from the Taliban. These dubious promises include the commitments from the Taliban not to host, cooperate with, train, recruit, or fundraise for Al-Qaeda and other terrorist organizations that threaten the U.S. and its allies. The Taliban also agreed to take immediate action against threats that the U.S. deemed as urgent. It is difficult to understand how any American familiar with the nature of the Taliban, its ideology, and its record would believe those promises.

In the run-up to the agreement, the *New York Times* lent its editorial page to the deputy emir of the Taliban's Islamic Emirate of Afghanistan and specifically designated (by the FBI) global terrorist Sirajuddin Haqqani. The man who leads the Taliban military arm, facilitates Al-Qaeda, and who has the blood of perhaps hundreds of thousands of innocents on his hands blamed America for the war that began, in part, because his father helped cement the relationship between the Taliban and Al-Qaeda. The author's assurances to permit women's education and employment were tepid. And the failure, in the agreement or in the op-ed, to recognize the Afghan government or the Afghan Constitution indicates that the Taliban has not given up its ambition to regain power through violent means.

What was particularly tragic was that the Trump administration strategy announced in August 2017 was working. The Afghans were taking the brunt of the fighting. Although every U.S. loss is a tragedy, casualties per twelve-month periods fell from a high of 499 in 2010 to fewer than 20 in the twelve months prior to the president's statement that he intended to withdraw all forces.[51] The annual cost of the war fell from a high of $120 billion in 2011 to an estimated annual cost of $45 billion per year in 2018.[52] The plan was to reduce that cost by half again as allies pledged to assume more of the financial burden. Afghanistan had not become Denmark, but government reforms, especially in the security ministries, were progressing. Amrullah Saleh, a no-nonsense leader, had taken over as interim minister of the interior. The January 2018 suspension of security assistance to Pakistan meant that Pakistan, at least, could no longer have it both ways, posing as a U.S. ally while supporting our enemies. The Taliban was under significant military pressure and no longer could simply wait out the U.S. withdrawal time line. Afghanistan was not a pretty situation. It was still a violent place, and Afghan, U.S., and coalition soldiers were still taking risks and making sacrifices; however, the United States had set conditions for its entering a future negotiating process from a position of strength, not desperation.

The September 2019 Afghan presidential election was not pretty. The Afghan government conducted the election under duress. The Taliban threatened anyone who voted. In addition to charges and countercharges of fraud, voter turnout was low. After nearly five months, Ghani was declared the winner by a slim margin. But the election happened. The result presented the United States and other nations with an opportunity to support an elected government and the 82 percent of Afghans opposed to the return of the Taliban and to bolster the will of those fighting on a modern-day frontier between barbarism and civilization.[53] It seems the Trump Administration chose not to take advantage of that opportunity.

Khalilzad continued to pursue negotiations in 2020, perhaps to get the best deal possible given the American president's desire to withdraw. But any deal meant primarily to fulfill that desire was bound to rest on the self-delusion that the Taliban would be an effective partner in countering terrorist organizations. Such an agreement would also empower the Taliban and weaken the Afghan government and security forces as the U.S. exited. That is why if the decision is to withdraw all U.S. forces, regardless of what happens afterward, it would have been preferable to do so with no deal.

* * *

THE PROBLEM set in South Asia is connected to other security challenges. It is a region in which cooperation with Russia and China is possible, as both countries would suffer the consequences of a growing terrorist threat emanating from the region. Russia, China, and other states such as Saudi Arabia, the UAE, and European nations have an opportunity to pose Pakistani leaders with a clear choice between isolation as a pariah or partnership in an effort to take advantage of the region's tremendous latent potential and address its grave problems in the areas of energy, climate change, environment, and food and water security. These are problems that affect Pakistan, India, Bangladesh, and the majority of the subcontinent. Multinational cooperation on South Asia's problems is also important to convince Iran to play a less destructive role not only in South Asia, but also across the greater Middle East, where a sectarian civil war created a humanitarian catastrophe and a cycle of violence that strengthened jihadist terrorist organizations globally.

Conflicts evolve based on interactions with enemies and on other factors, such as declining public support and the will of leaders. By 2019, it was clear that the person most critical to sustaining the first long-term strategy for Afghanistan and South Asia was not determined to give the strategy he approved an opportunity to succeed. Responding in large measure to a vocal group within his political base, President Donald Trump abandoned the psychological gains that had strengthened the will of America's partners and diminished the will of the Taliban enemy and their supporters. The mantra to "end the endless wars," however, expressed a sentiment that was gaining momentum. As one who not only fought in the wars in Afghanistan and Iraq and bore witness to the horrors and sacrifices made there, but who also saw my daughter and son-in-law deploy to those same conflicts, I, too, wanted them to end. But I also knew that there was no short-term solution to South Asia's long-term problems. What we owed our nation and the sons and daughters who fight in our name was a long-term strategy capable of delivering an outcome that would keep our nation safe at an acceptable cost.

Tragically, as American soldiers who were not yet born on September 11, 2001, deployed to Afghanistan, America was still fighting its longest war one year at a time. But the results of striking a deal with the Taliban for the purpose of withdrawing from America's longest war are likely to be far worse than a sustained commitment under a sound strategy.

PART IV

Middle East

This will not stand . . . **WAR IN THE GULF: U.S. BOMBS BAGHDAD** . . . *KURDS IN CLASHES WITH IRAQ TROOPS* . . . **NO FLY-ZONE IN IRAQ. WHY? . . .** There is no doubt that Saddam Hussein now has weapons of mass destruction . . . **WAR'S ON** . . . military operations to disarm Iraq, to free its people and to defend the world from grave danger . . . *There are known knowns . . . there are known unknowns . . . there are also unknown unknowns* . . . **THE CAPTURE OF HUSSEIN: INSURGENCY ATTACKS GO ON** . . . It is clear that we need to change our strategy . . . **AMERICA'S MILITARY EFFORTS IN IRAQ WILL END** . . . **MALIKI GIVEN 30 DAYS TO FORM GOVERNMENT IN IRAQ** . . . We're leaving behind a sovereign, stable and self-reliant Iraq . . . **IN TUNISIA, ACT OF ONE FRUIT VENDOR SPARKS WAVE OF REVOLUTION** . . . Egypt will never be the same . . . *After a lot of blood spilled by Iraqis and Americans, the mission of an Iraq that could govern and secure itself has become real* . . . **HOSNI MUBARAK RESIGNS** . . . **QADDAFI, SEIZED BY FOES, MEETS A VIOLENT END** . . . *Qaddafi's grip of fear appeared to give way to the promise of freedom* . . . **500,000 PEOPLE FLEE IN NORTHERN IRAQ** . . . *A red line for us is we start seeing a whole bunch of chemical weapons moving* . . . America will lead a broad coalition to roll back this terrorist threat . . . **U.S. CONCLUDES SYRIA USED CHEMICAL WEAPONS** . . . *Are you truly incapable of shame?* . . . **MOSUL WAKES UP TO A DAY WITHOUT ISIS** . . . *GREEK REFUGEE CAMPS NEAR CATASTROPHE* . . . **U.S. STRIKES ON SYRIA** . . . It is time for us to get out of these ridiculous Endless Wars . . . **TRUMP THREATENS TO "DEVASTATE TURKEY ECONOMICALLY" IF IT ATTACKS KURDS** . . . *Baghdadi's demise demonstrates America's relentless* . . . we're talking about sand and death . . . **PROTESTERS STORM US EMBASSY COMPOUND IN BAGHDAD** . . . The United States will protect and defend its people . . . **TRUMP TO OPEN MIDDLE EAST PEACE DRIVE** . . .

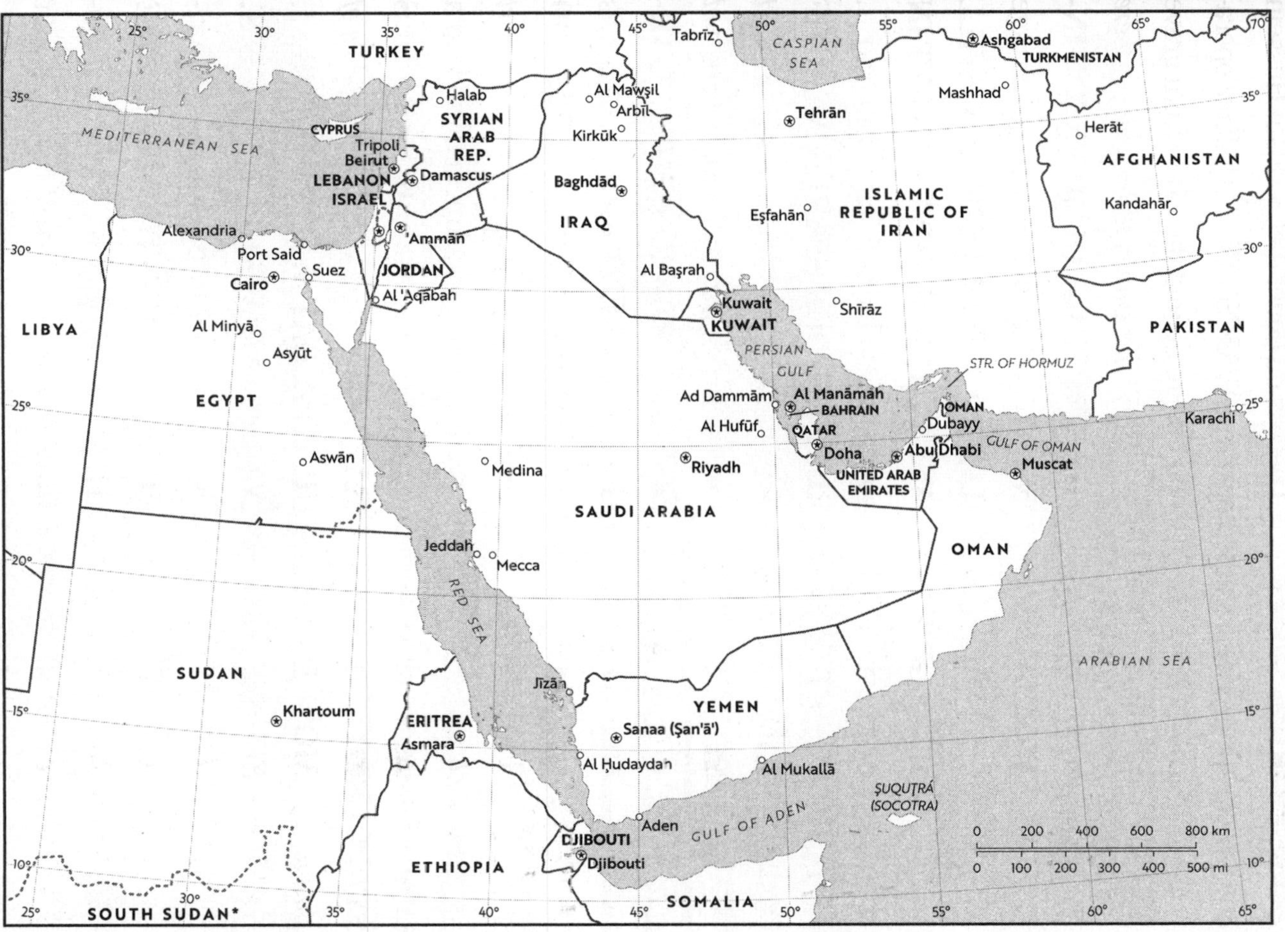

TURKEY
CYPRUS
SYRIAN ARAB REP.
LEBANON
ISRAEL
JORDAN
IRAQ
KUWAIT
ISLAMIC REPUBLIC OF IRAN
TURKMENISTAN
AFGHANISTAN
PAKISTAN
SAUDI ARABIA
BAHRAIN
QATAR
UNITED ARAB EMIRATES
OMAN
YEMEN
EGYPT
LIBYA
SUDAN
SOUTH SUDAN*
ERITREA
ETHIOPIA
DJIBOUTI
SOMALIA
MEDITERRANEAN SEA
CASPIAN SEA
PERSIAN GULF
STR. OF HORMUZ
GULF OF OMAN
ARABIAN SEA
RED SEA
GULF OF ADEN
ṢUQUṬRĀ (SOCOTRA)
Halab
Tripoli
Beirut
Damascus
'Ammān
Al 'Aqabah
Al Mawṣil
Arbīl
Kirkūk
Baghdād
Al Baṣrah
Kuwait
Tabrīz
Tehrān
Eṣfahān
Shīrāz
Mashhad
Ashgabad
Herāt
Kandahār
Karachi
Ad Dammām
Al Manāmah
Al Hufūf
Doha
Abu Dhabi
Dubayy
Muscat
Riyadh
Medina
Mecca
Jeddah
Jīzān
Sanaa (Ṣan'ā')
Al Ḥudaydah
Al Mukallā
Aden
Djibouti
Asmara
Khartoum
Aswān
Asyūt
Al Minyā
Cairo
Suez
Port Said
Alexandria
0 200 400 600 800 km
0 100 200 300 400 500 mi

CHAPTER 7

Who Thought It Would Be Easy? From Optimism to Resignation in the Middle East

This is a long twilight war, the struggle against radical Islamism. We can't wish it away. No strategy of winning "hearts and minds," no great outreach, will bring this struggle to an end. America can't conciliate these furies.

—FOUAD AJAMI

ON MARCH 20, 2017, exactly one month after I entered the White House, Prime Minister of Iraq Haider al-Abadi visited Washington. I looked forward to seeing him. He was a respected friend whose country was in turmoil. Iraq was at the center of the fight to defeat ISIS, the latest version of the Sunni jihadist terrorist organizations that had inflicted so much pain and suffering not only in the Middle East, but across the globe. The country was also on the front line of Iran's effort to extend its influence across the region. And along with Syria, Iraq was at the epicenter of a sectarian civil war between Shia and Sunni Muslims that was strengthening terrorist organizations, perpetuating state weakness, and inflicting human suffering on a colossal scale. The 2003 invasion of Iraq did not create the centripetal forces that were tearing the Middle East apart. But the invasion, lack of preparation for the war's aftermath, and precipitous withdrawal from Iraq in 2011 contributed to a contagious breakdown of security. Abadi's visit was important because a stable and secure Iraq that was not aligned with Iran was essential if the Middle East was to emerge from multiple crises and we needed to understand better how the United States could support Iraqi leaders who shared these goals.

The day of Abadi's visit was typical. There was a routine mid-morning Oval Office session with President Trump, Director of National Intelligence Dan Coates, and Director of the Central Intelligence Agency Mike Pompeo. I then sat down with Doug Silliman, U.S. ambassador to Iraq, in preparation for the Iraqi delegation's arrival at 3 p.m. From 2011 to 2013, Silliman had been political counselor and deputy chief of mission in Baghdad as Iraq unraveled after the U.S. withdrawal. He returned to Baghdad as ambassador toward the end of 2016 in the midst of the campaign to defeat ISIS, the terrorist group that had gained control of large portions of Iraq and Syria to establish its so-called caliphate.

I returned to the Oval Office for the usually rushed briefing in advance of the visit. I made sure the president, who always stressed the need for partners to pull their weight, knew that Iraqis were taking the brunt of the battle against ISIS. The Iraqis had lost approximately 26,000 soldiers since 2014, compared to seventeen U.S. soldiers and marines killed in action against ISIS in Iraq during that same period.[1] The president was concerned about Iranian influence in Iraq. I told him that Prime Minister Abadi was trying to strengthen Iraqi sovereignty and reduce malign Iranian influence. Abadi knew that if the Iraqi government aligned with Iran, another version of ISIS would portray itself to Sunni communities as a protector from Iranian-sponsored Shia militias. I stressed that a positive, long-term relationship with Iraq would not only assist with the defeat of ISIS, but also counterbalance Iranian influence there.

Iran was intent on keeping Iraq weak and divided. After the collapse of the Iraqi Army in 2014 and ISIS's rapid seizure of territory there, the Iraqi government deepened its reliance on Shia militias to maintain stability. These militias were mostly commanded by Iranian agents, giving Iran coercive power over the Baghdad government. Militias under the influence of Iran's Islamic Revolutionary Guard Corps (IRGC), the so-called Popular Mobilization Forces (PMF), included groups that had killed American soldiers.[2] For example, Asa'ib Ahl al-Haq (AAH), a group that operated in Iraq and Syria, had been equipped, funded, and trained by the IRGC. Shia militiamen were young; they and their commanders formed a new base of power. To further destabilize Iraq, Iran's IRGC and its Ministry of Intelligence and Security (MOIS) also took advantage of divisions among Kurds and between Kurds and Arabs by supporting Kurdish factions sympathetic to Tehran.

As Secretary of State Rex Tillerson and Ambassador Silliman joined us in the Oval Office, I suggested that two members of the delegation typified the promise and peril of Iraq. Abadi represented promise. He worked with all of Iraq's communities to reduce ethnic and sectarian divisions and shared the U.S. desire for a strong and independent Iraq. He was a unifying leader in a country whose complex quilt of ethnic, tribal, and religious communities had been torn apart many times since Saddam Hussein took power in 1979. Abadi's foreign minister, former prime minister Ibrahim al-Jaafari, personified peril. Jaafari advocated for Iraq's Shia community at the expense of others. He did Iran's bidding, sowed division, and perpetuated conflict.

To illustrate the point, I told them about how Abadi had supported the regiment I commanded more than a decade earlier in Tal Afar, a city that contained the complexities of Iraq and the Middle East in microcosm. In 2005, Tal Afar was a training ground and staging base for Al-Qaeda operations across Iraq. The terrorists used sectarian or religious conflict between Turkmen Shia and Sunni populations, as well as ethnic rivalries among Kurds, Turkmen, Yazidis, and Arabs, to embed themselves in communities that needed protection. As U.S. forces reduced their presence in northern Iraq, the city had become a sectarian battleground between Shia police and Al-Qaeda terrorists. Civilians were caught in the crossfire. Sunni and Shia families who had been friends and neighbors were forced to choose sides. Normal life stopped. Schools and markets closed. People barricaded themselves in their homes. The police morphed into a death squad that operated out of a sixteenth-century Ottoman castle in the center of the city and ventured out at night to murder, indiscriminately, Sunni men of military age. The police actually helped Al-Qaeda terrorists portray themselves as protectors and rationalize the brutal form of control they had established. Terrorists forced parents to give up their adolescent and young teenage boys to join their organization. So-called imams who rarely had more than an elementary school education put young recruits through initiations often involving sexual abuse and systematic dehumanization. For example, in one unexceptional case in Tal Afar, a young teenager was repeatedly raped and given the assignment to be the leg holder in beheadings of Shia or uncooperative Sunnis.[3] It was a dystopian scene.

Soon after the Third Armored Cavalry Regiment arrived in Tal Afar in May of 2005, I called Abadi to ask for his assistance in arresting the cycle of violence. Abadi, then a member of Iraq's parliament, was influential in the powerful Shia Dawa (Islamic Call) Party. After I told him that the chief of police was fueling the cycle of violence in Tal Afar, Abadi arranged for his transfer to Baghdad and cleared the way for a Sunni Arab general who transformed the police from predators to protectors of civilians regardless of their sectarian, ethnic, or tribal identity. That chief, Maj. Gen. Najim Abed Abdullah al-Jibouri, was an extraordinarily courageous leader and effective mediator. After a successful counterterrorism offensive in 2005, Jibouri fostered understanding among Tal Afar's ethnic and sectarian groups and a common commitment to preventing Al-Qaeda from returning. The city came back to life as schools and markets reopened once again, barricaded houses were no longer the norm, and the Iraqi Army and police barred Al-Qaeda from regaining a foothold. I wanted the president to know what often got lost in the reports of horrific violence and human suffering in Iraq: the perseverance of unifying leaders like Abadi and Jibouri, who wanted to forge a better future for all Iraqis regardless of religion, ethnicity, or tribal affiliation. But not all in Abadi's party were conciliators.

I advised the president that some members of the Iraqi delegation might look rough, but the thin, balding medical doctor with the close-cropped gray beard was the most coldblooded. Jaafari was driven mainly by a depraved desire for revenge against former members of Saddam Hussein's Baath Party. He was the perfect agent for Iran because he fueled the violence that debilitated Iraq. In the run-up to the Iraqi election that resulted in Jaafari's prime ministership, Iran provided millions of dollars to establish Shia sectarian-driven political parties. At the same time, the United States stood aside and did little to counter Iran's influence. As prime minister from May 2005 to May 2006, Jaafari used policies and actions to marginalize Iraq's Sunni Arabs and Turkmen and aid Iranian infiltration of Iraqi institutions.

Jaafari spent the 1980s in Iran as a member of the Supreme Council for the Islamic Revolution in Iraq, an anti-Saddam organization thoroughly penetrated by Iranian intelligence.[4] Later, he moved to London, where he became the spokesperson for the Dawa Party. Jaafari was pro-Iran and anti-United States. From 2007 to 2008, the former prime minister twice hosted me and then-Major Joel Rayburn at his home in Baghdad's Green Zone, the fortified neighborhood at the center of the Iraqi capital. As Rayburn and I drank sweet tea, the doctor gave long lectures about the West's many flaws. Jaafari shared the Iranian mullahs' strange ideological blend of Marxist-Leninism and Shia millenarianism with an added dose of material from leftist American academics. He would occasionally pull books off his shelves and quote from them, as if to enter evidence in his case against the United States, one favored source being linguist Noam Chomsky. Chomsky and Jaafari shared the belief that all the world's (and Iraq's) ills derived from colonialism and "capitalist imperialism." Jaafari seemed unaware that had it not been for the American invasion of 2003, he would have been quoting from Chomsky in exile.

As prime minister, Jaafari blamed the United States even as he helped the Badr Organization, a Shia militia, gain control of the Iraqi Ministry of the Interior.[5] The new minister, Bayan Jabr, who shared Jaafari's deceptively gentle appearance as well as his inner ruthlessness, used the National Police to abduct, systematically torture, and sometimes kill Sunni prisoners.[6] The atrocities were part score settling for Iran. Victims were often former members of Saddam's government, or Iraqi pilots who had bombed Iranian territory during the Iran-Iraq War, or even university professors who were critical of the Iranian Revolution. But, then, Iran was seeding nefarious people across the Iraqi government, with the help of people like Jaafari. One example was Abu Mahdi al-Muhandis. In 2005, U.S. officials discovered more than one hundred suffering people whom Iraqi national police had illegally arrested and crammed into his basement. Previously, Muhandis had been sentenced to death in Kuwait for terrorist bombings in 1983. He was later killed in a U.S. strike in January 2020 that targeted Iranian IRGC Quds Force commander Qasem Soleimani. And there was National Police general Mahdi al-Gharrawi, in whose headquarters U.S. Army soldiers uncovered a holding cell with 1,400 malnourished prisoners who clearly had been tortured repeatedly.[7] Jaafari protected Muhandis and Gharrawi from prosecutors. While Abadi saw people like Gharrawi and Muhandis as corrupt and a liability to Iraq, Jaafari enabled their efforts to foment the sectarian civil wars that helped give rise first to Al-Qaeda in Iraq (AQI) and, later, to its offspring, ISIS.[8] As we walked to receive the delegation, I predicted that Jaafari would reveal his antipathy toward the United States by demanding more aid. He and others who acted as agents of Iran thought of American leaders as dupes whom they could fleece before turning Iraq fully toward Iran and against their erstwhile patrons.

Much of the discussion across the Cabinet Room's long mahogany conference table was about military progress in the campaign against ISIS. Just four months earlier, Iraqi forces had liberated the ancient city of Sinjar from ISIS, where in August 2014, terrorists massacred an estimated five thousand Yazidis, tortured thousands of girls and women in rape camps, and sold women as slaves, forcing them to marry those who killed their fathers and brothers.[9] After the White House and Defense Department removed unnecessary restrictions on U.S. forces put in place during the Obama administration, such as how far forward advisors could operate in battle and how many helicopters could be in Syria, the pace of operations against ISIS in Syria and in Iraq increased.[10] Although tough fighting was still ahead, success in wresting control of territory and populations from ISIS seemed inevitable.

Toward the end of the meeting, the president asked if I had anything to add. I thought that the key questions were how to ensure the enduring defeat of ISIS and prevent Iran from extending its influence across Iraq, into Syria, and to the borders of Israel. I asked Prime Minister Abadi what more we and others might do to help him break the cycle of conflict. He spoke of the need to pull Iraq's traumatized society back together and convince all Iraqis that the government could secure them and provide them with a better future. When Jaafari, true to form, called for more U.S. aid, President Trump ignored him and ended the meeting. But before the table had cleared and the policy makers had dispersed, one of the prime minister's aides slipped me a note. Abadi was asking me to meet him at his hotel after hours.

At around nine o'clock that evening, I went to see Abadi. It was late, but late-night meetings are an Iraqi custom. Joel Rayburn came with me. We had served together many times in Iraq and Afghanistan. A colonel and a senior director on the National Security Council staff, Joel had recently written an excellent book that put Iraq's contemporary fragmentation into historical perspective. He had also edited a seminal critical study of the U.S. Army's experience in Iraq from 2003 to 2011. We had both taught history at West Point and shared a conviction that to understand the present, one must first understand the past. On the way to Abadi's hotel, we discussed how the prime minister had borne witness to the tragedy of Middle East politics and had somehow transcended the sectarianism that embroiled the region. Abadi could be candid away from his own delegation, many of whom, including Jaafari, he could not trust. He could help us understand what the United States might do to help him and other leaders overcome the decades of trauma in the Middle East.

* * *

BORN IN 1952, Abadi came of age in the postcolonial period of growing cultural awareness among Arab states. Socialist ideas were popular, especially state control of resources, such as oil, to achieve social justice and equitable income distribution. As new political movements emerged in Egypt, Syria, and Iraq, dictators gained power and fostered cults of personality with pervasive propaganda. In Egypt, Gamal Abdel Nasser nationalized the Suez Canal in 1956 and, with the support of the United States, prevailed over European powers in the crisis that ensued. He built the Aswan High Dam on the Nile, defended the Palestinian cause, and promised social reform. From Iraq, a young Abadi watched Pan Arabism (which envisioned a unified Arab world) and socialist experiments collapse as nationalist and Islamist ideologies gained strength.[11] In 1958, as Abadi entered grade school in Baghdad, waves of nationalist sentiment washed over Syria and Iraq. In Syria, a series of weak civilian governments, military coups, and counter-coups had thrust the left-leaning Baath Party into power. The Syria Baath, in pursuit of a single Arab state, entered into a union with Egypt that would last only two years.

In Iraq, a coup ended the Hashemite monarchy that the British had established in 1921. On July 14, 1958, Faisal II, a soon-to-be-married twenty-three-year-old monarch who had ascended the throne at age three, found himself and other members of his family lined up facing a wall in the palace courtyard. A machine gun opened fire. The king's bullet-ridden body was strung from a lamppost. Abadi remembered people celebrating the demise of the corrupt government but feeling remorse over the brutal murders of the king and his family. And he recollected how, after the king's death, the freedoms, such as multiple political parties and a free press, began to evaporate. Over the next decade, authoritarian dictatorships were ascendant in the Arab world; foremost among them was Nasser, whose brand of Arab nationalism included strong anti-Israel sentiment. In 1967, when Abadi was a student in Baghdad Central High School, he witnessed a war motivated by that sentiment, the outcome of which would sound the death knell for Arab nationalism.

Abadi recalled high expectations as Egypt, Jordan, and Syria initiated a war against Israel, and a huge letdown across the Arab world when, six days later, the Israel Defense Forces not only defeated the combined armies, but also seized Egypt's Sinai Peninsula and Gaza Strip, the Syrian Golan Heights, and the Jordanian West Bank. The rush to assign blame generated conflicts within and among Arab nations that would go on for decades. In Syria, Minister of Defense Hafez al-Assad, a member of the socialist Baath Party and the minority Alawi sect, a secretive Shia offshoot that blends religions, blamed Syrian president Salah Jadid for the defeat and loss of territory.[12] As the president's position weakened, Assad activated a network of friends and relatives. In 1970, he mounted a bloodless coup, but the three decades that followed were far from bloodless. The defeat in the Six-Day War empowered Islamists, who used the religious dimension of the Israeli-Arab conflict to grow support. Islamist challengers to the Iraqi Baath Party grew, spurred on by the brutality of the Baaths. In 1966, Abadi had met the founder of the Dawa Party, Muhammad Baqir al-Sadr. Al-Sadr had founded the party in 1957 to combat secularist, Arab nationalist, and socialist ideas; to promote Islamic values and ethics; to increase political awareness; and, ultimately, to create a Shia Islamic state in Iraq.

Abadi remembers being "overtaken" by the charismatic imam and being struck by his "humanness." Al-Sadr presented Islam as an alternative to Marxism and capitalism. Abadi joined Dawa in 1967, one year before the Baath Party seized power in Iraq, and he became part of a growing Islamist movement that encompassed the Sunni Muslim Brotherhood as well as Dawa and other Shia groups. Abadi's educated, pious Shia family, like many others, rejected a secular government driven by power and avarice.

Abadi was completing a doctorate in London when events in Iran and Iraq thrust the Middle East into a new era of conflict. In February 1979, Grand Ayatollah Ruhollah Khomeini, the son and grandson of mullahs, overthrew the Shah of Iran to become supreme leader of the Islamic Republic. Abadi, who had become the leader of the Dawa organization in Europe, was hopeful. He thought that if the all-powerful Shah could be deposed, the same might be possible in Iraq, Syria, and across the Middle East. Abadi and other Dawa members met in cafés to talk about prospects for overthrowing the Baath in Iraq. Just four months later, there was a change in government in Iraq, but not the kind that Abadi and Dawa sought. Saddam Hussein, a man whose father died before he was born and whose mother abandoned him, mounted a coup in Baghdad and declared himself president. In Iran, the revolutionaries executed the Shah's senior civilian and military officials at a steady pace while, in Iraq, Saddam executed hundreds in one week. His internal opposition either dead or quelled, Saddam then turned to the threat that the Islamic Revolution in Iran posed to his secular nationalist dictatorship. Just four months after seizing power, he ordered the invasion of Iran, anticipating a short war against a country still in the throes of revolution. The ensuing eight-year war took a heavy toll on both nations; it would end in stalemate after an estimated one million people perished.[13]

Those were hard years for the Abadi family. As Saddam invaded Iran, he ordered a brutal crackdown on Dawa and other Shia opposition groups. In April 1980, the founder of the Dawa Party and the man who inspired young Haider al-Abadi, Muhammad al-Sadr, was arrested by the Baath. The forty-five-year old cleric was tortured and forced to watch as his sister was raped. Saddam's agents then drove a nail into his skull.[14]

Ayatollah Khomeini framed the stakes of the Iran-Iraq War as more than the defense of Iran's territorial integrity—it was a war to uphold Shia Islam and spread the Islamic Revolution. The Iran-Iraq War rekindled ancient animosities rooted in historical memory of a battle in Karbala, Iraq in AD 680. After the prophet Muhammad died in 632, Islam split between Shia (which means followers of Ali, those who believed that Ali ibn Abi Talib, a cousin and son-in-law of the Prophet, was Muhammad's designated successor) and Sunnis (those who believed that Muhammad did not appoint a successor and who considered Abu Bakr to be the first rightful caliph after the Prophet). The Battle of Karbala pitted these factions against each other. The subsequent brutal defeat of the Shia forces is an emotional touchstone of their history, tradition, literature, and theology. Shia commemorate the battle over ten days every year, which culminate with the Day of Ashura, a day of mourning and processions that often entail self-flagellation. Khomeini made powerful use of Shia identity, especially historical grievances, to galvanize the masses.[15]

Religion served as justification for atrocities on both sides of the Iran-Iraq War. On the Iranian side, Ayatollah Khomeini sent unarmed teenagers into certain death, with instructions to pick up the rifle of the boy who fell in front of them. These young men wore red headbands printed with the words *Sar Allah* (Warriors of God). The Ayatollah gave them small metal keys that he promised would gain them admission to Paradise when they were martyred. Many were bound by ropes to prevent their desertion. On the Iraqi side, Saddam Hussein portrayed himself as the defender not only of Iraq, but also of the entire Arab world against Iran's Shia Islamist revival and designs for expansion. His forces used chemical weapons and waged war directly on Iranian civilians. In January 1987 alone, his forces launched more than two hundred missile strikes on Iranian cities, killing nearly two thousand and wounding more than six thousand innocents. More than sixty thousand Iranian civilians died in the war.[16]

Abadi was lucky. He missed the maelstrom. He ran a small design and technology firm in London and continued to host Iraqi exiles in a popular café. When he became more vocal in calling for the liberation of Iraq, the Baath Party rescinded his passport.[17] Meanwhile, in Iraq, Abadi's family was under duress. Two of his five brothers had been imprisoned under Saddam, and their fate was still unknown.

Outside Iran, the Islamists did not get very far. Nationalist dictators in Syria and Iraq brutally repressed opposition. In Syria, Assad's army put down a 1982 Muslim Brotherhood Islamist uprising in Hama, a city of around two hundred thousand, with a siege, bombings, and a tank assault. Between seven thousand and thirty thousand people died in the span of twenty days.[18] Those not killed were imprisoned under horrible conditions; many were never seen again. The blood spilled made Syria fertile ground for growing future generations of Islamist terrorists. Hama was a harbinger of the brutality to come to Syria under Assad's son, Bashar, an ophthalmologist turned dictator who would surpass his father's grim record of murder, imprisonment, and torture.

The legacy of the Iran-Iraq War and earlier Syrian uprisings are apparent in the sectarian civil wars that engulfed the Middle East in the first decade of the twenty-first century. Iran cultivated relationships with Shia opposition groups and militias that became part of Iraq's political and military landscape. During the Iran-Iraq War, Ibrahim al-Jaafari and many of Haider al-Abadi's Shia contemporaries fled to shelter in Iran. Iran exacted a price, recruiting young Iraqi men into anti-Saddam groups, such as the Badr Organization. Iran used Badr and other organizations for assassinations and guerrilla attacks in a low-level war that went on for fifteen years after the end of the conventional conflict in 1988. The terrible costs of the war and the historical animosities it resurrected divided Sunni and Shia Muslims. The sectarian divide fueled violence and complicated efforts to develop coherent national identities and legitimate governance across the Middle East. And it intensified the political and religious power struggle between Shia-majority Iran and Sunni-majority Saudi Arabia while setting conditions for a sectarian civil war in post-Saddam Iraq. As Islamism rose from the ashes of Arab and secular nationalism, it brought discord rather than unity, pursuit of particular rather than general interests, and a preference for violence rather than representative political processes to settle differences or compete for power.

Some, like Abadi and Jibouri, would emerge from that era as

conciliators who sought to bridge the sectarian divide. Others, like Jaafari, would deepen and widen that divide and, along with Sunni jihadist terrorist organizations like AQI, accelerate the destructive cycle of violence.

* * *

RAYBURN AND I arrived at the hotel and ascended the last two flights of stairs past several armed U.S. Secret Service agents to Prime Minister Abadi's floor. He welcomed us into his suite, and we began a relaxed conversation. I reminisced about meeting him on my first day in Baghdad in May 2003. I had left a fellowship at the Hoover Institution at Stanford University to serve on Gen. John Abizaid's staff at U.S. Central Command's forward headquarters in Doha, Qatar. As I flew with Abizaid from Doha to Baghdad, I met an Iraqi American who would become one of Abadi's assistants in the new Iraqi government. We talked about how, after the successful Coalition offensive that removed Saddam from power, the most difficult tasks lay ahead. Later, at the gaudy Republican Palace that Coalition forces occupied, I had helped Abadi receive his badge in time to attend a security committee meeting.

Now, in his Washington hotel suite, I began by telling him how much I appreciated his leadership under the most difficult circumstances. In 2014, it was under duress that the previous Iraqi prime minister, Nouri al Maliki, abandoned a postelection effort to remain prime minister and ceded leadership to Abadi. I asked him about our friend General Jibouri. When he assumed the prime ministership, Abadi had asked Jibouri who was then living and working in Northern Virginia and was on a path to U.S. citizenship, to return to fight the terrorists who had taken Fallujah and Mosul and were threatening the capital city. Abadi told me that Jibouri, as we had done in Tal Afar over a decade earlier, was helping to reconcile communities that had been torn apart. Abadi related the sad story that, when ISIS terrorists learned that Jibouri had returned to northern Iraq, they rounded up members of his extended family and his tribe, tortured them, and murdered them. Just a few months earlier, ISIS put six of Jibouri's family members in a metal cage, chained them to its bottom bars, and submerged them in a pool until they drowned.

Our conversation then turned to the regional dynamics that were perpetuating conflict. Having witnessed the failures of Arab nationalism and Islamism, Abadi believed that only representative governments that were regarded as legitimate by the population could allow the people to escape the suffering associated with the cycle of violence. The sectarian civil war would not end unless there was accommodation among religious, ethnic, and tribal communities, yet outsiders were continuing to pour fuel on the flames of communal conflict. Away from the rest of Abadi's party, Abadi and I discussed Iranian infiltration of the Iraqi government and especially of the formidable militias, the Popular Mobilization Forces (PMF), or Al-Hashd al-Shaabi, formed by and under the control of Iranian agents. The Iranians had pressured Abadi to appoint a titular commander of the PMF, and Muhandis as the more powerful deputy commander. Muhandis was also commander of the Iranian-supported anti-American terrorist organization Kata'ib Hezbollah. Members of the PMF have additionally attacked local diplomatic and energy targets, carried out drone attacks in Saudi Arabia, and provoked Israeli air strikes through efforts to assist Iranian missile proliferation.[19] Abadi knew that it would be very difficult for the United States to have a positive long-term relationship with Iraq under the conditions of Iranian state capture. It was clear to me that the United States and our partners in the region had to do more to strengthen Iraq's sovereignty and counter the Iranian efforts.

Abadi described how the Syrian Civil War was the epicenter of the sectarian violence afflicting the region. He highlighted the Assad regime's long history of sponsoring Sunni and Shia terrorist organizations, including Hamas, Hezbollah, Al-Qaeda in Iraq, Palestinian Islamic Jihad, and the Kurdish terrorist organization the Kurdistan Workers' Party (PKK). After the U.S. invasion of

Iraq in 2003, the Assad regime created a pipeline of foreign fighters to AQI to fight U.S. forces. I added that in Syria, Assad acted as an agent of Iran, stoking the flames of a broader civil war in the region designed to keep Iraq and the Arab world weak and divided.[20]

Abadi discussed how Assad, with the help of Iran and Russia, used radical Salafi jihadists to portray all opposition groups as terrorists aiming to create an Islamic caliphate in Syria. Assad released thousands of hard-line Islamists from Syria's jails in early 2011 who would later become major leaders in AQI and ISIS. I responded that, tragically, Assad's strategy seemed to have worked. As my friend the late Professor Fouad Ajami had observed as the war intensified, Assad was able to convince the Obama administration that his, Assad's, tyranny was preferable to the opposition.[21]

We discussed the unresolved legacy of the so-called Arab Spring and how the Syrian insurgency began in March 2011, after antigovernment protests spread from Tunisia, Egypt, and Libya into Syria. The Assad regime's arrest and brutalization of teenage demonstrators sparked riots across the country nine days later, the most violent of which occurred in Homs. And thus began an escalating cycle of growing opposition and brutal repression.[22]

We discussed how what we did together in Tal Afar, Iraq, was relevant to the problem set across the greater Middle East. While the United States was not going to deploy armored cavalry regiments across Iraq, Syria, and Yemen, Abadi stressed how the overall strategy must aim to help break the cycle of violence through diplomacy as well as military and intelligence support for local partners.

Jibouri and Abadi were threats both to Al-Qaeda and Iran because they were effective mediators, true humanists whose empathy allowed them to broker accommodation among parties in conflict. But the two men could go only so far without jeopardizing their lives and the lives of their families. The United States uses diplomacy and economic incentives to exert influence; Iran uses these tools as well as assassination.

* * *

UPON HIS arrival in Baghdad in April 2003, Abadi and his mother went immediately to the prison, where they hoped to find his long-lost brothers. After racing past the prisoners pouring out of the gates, he rifled through abandoned file cabinets in the office. There he found a document revealing that his brothers had been executed soon after their arrest three decades earlier. Across those decades, Abadi's mother had borne witness to three wars. The first, Iraq's war with Iran, killed an estimated six hundred thousand Iraqis.[23] Saddam killed over a quarter of a million more of his own people, including Abadi's brothers, in a country of twenty-two million. The second war began in 1990, instigated by Saddam's belief that Kuwait, Saudi Arabia, and other Arab states owed him a debt of gratitude—and cash to pay off his war debt. It was anger over their ungratefulness that led to the Iraqi invasion and annexation of Kuwait in August of that year. After the George H. W. Bush administration gathered a coalition of thirty-five nations and marshaled a force of 750,000 in Saudi Arabia, a thirty-seven-day-long bombing campaign and a one-hundred-hour ground war defeated Saddam's army and evicted it from Kuwait in February 1991. The third war, the U.S.-led invasion of Iraq in 2003, ended Saddam's brutal rule, but would not bring peace to Iraq's traumatized society.

Abadi and I disagreed about what should have happened after the 1991 Gulf War. Abadi welcomed the liberation of Kuwait, but he wanted the United States to unseat Saddam so that he and fellow Iraqis could replace the brutal dictatorship with the virtuous Islamist government they envisioned. The vast majority of the 1.5 million Iraqi exiles at the time of the Gulf War, 500,000 of whom were in Iran, viewed the Gulf War as an opportunity for revenge against a weakened Saddam and his Baath Party. Ideological divisions, however, weakened the exile opposition and the opposition inside Iraq. I witnessed those divisions firsthand in the wake of the Gulf War, and would confront them again during my service in Iraq across five years from 2003 to 2008.

Unlike Abadi, I did not believe that the United States should have unseated Saddam in 1991. After our cavalry troop's experience in An Nasiriyah at the end of the Gulf War, I wrote an editorial arguing that "although any alternative to Hussein would be an improvement, replacing him ourselves would have been problematic." I cited the divisions among Iraq's competing ethnic, sectarian, and tribal groups. In a post-Saddam Iraq, "justice and a responsible national leadership would remain elusive if not unobtainable," and "we would assume at least partial responsibility for establishing a new government under these conditions." In so doing, we would certainly encounter armed opposition, and "a new government would have to face the constant attack" from die-hard Baathists and Islamists who would attempt to cast our motives as "'Zionist' and 'neo-imperialist.'" I concluded that removing Saddam would result in "a post-Hussein commitment of tremendous proportion for a project of indeterminable time" with "no guarantee of success."[24]

In the wake of Operation Desert Storm, President George H. W. Bush encouraged a Shia uprising to topple Saddam, but then withdrew U.S. forces out of southern Iraq. Saddam's Mukhabarat and Special Republican Guard swiftly inflicted large numbers of casualties on the Shia and weakened the opposition. Abadi called the 1991 experience in southern Iraq a "Bay of Pigs Moment," referring to the failed CIA-supported invasion of Cuba thirty years earlier. Moreover, the subsequent sanctions placed on Iraq actually strengthened the Baath as they took control of smuggling networks to circumvent those sanctions. Sanctions made more Iraqis dependent on Saddam's patronage as the Baath reduced social services to communities deemed disloyal. Caught in the midst of a rebellion and the brutal repression of it, Abadi's father fled in 1994 to the United Kingdom, where he died of heart failure less than two years later. Abadi's mother, originally from Lebanon, remained in Baghdad. She would live to see what must have been unimaginable after watching Saddam's thugs drag three of her sons to prison in the 1980s: another one of her sons, Haider, sworn in as prime minister of Iraq in 2014. When I first met Abadi in Baghdad, neither one of us expected that we would work together in Iraq over the next several years in the midst of an insurgency and civil war, let alone meet again in the White House nearly fourteen years later.

Abadi shared my surprise over the lack of preparation for the ambitious endeavor to remake Iraq. In the years following the 2003 invasion, as the cost and difficulty of stabilizing Iraq became apparent, Americans would debate endlessly whether the United States and United Kingdom's invasion was misguided. Much of that debate centered on whether Saddam actually possessed weapons of mass destruction, the primary casus belli that the United States presented to the United Nations. But as the war morphed into an insurgency, a civil war, and a sustained counterterrorism campaign, I thought that a more useful question to debate might be, who thought "regime change" in Iraq would be easy, and why did they think so?

* * *

ON THE day I first met Abadi, I was serving as General Abizaid's executive officer. When Abizaid assumed command of Central Command (CENTCOM) two months later, I would become the director of the Commander's Advisory Group, a small team charged with helping the commander understand better the complex challenges in the Middle East and how U.S. military forces in the region could best contribute to U.S. policy goals.

I carried with me a dozen copies of a study that Col. Conrad Crane and Professor Andy Terrill had recently completed at the U.S. Army War College. The report warned that "the primary problem at the core of American deficiencies in post-conflict . . . is a national aversion to nation-building, which was strengthened by failure in Vietnam."[25] The study listed tasks that, after the invasion of Iraq, the military would have to perform initially and then hand over to civilian authorities or to a new Iraqi government and security forces. It was clear from the moment I arrived in Baghdad that the U. S. military and the hastily assembled civilian

Office of Reconstruction and Humanitarian Assistance (ORHA) were woefully unprepared for these tasks. Failure to consider what was required to turn military gains into ambitious political objectives led to many of the difficulties encountered after the invasion of both Afghanistan in 2001 and Iraq in 2003.[26] A post-Vietnam emotional aversion to long-term military commitments combined with faith in America's technological military prowess had overwhelmed historical experience, suggesting that the consolidation of military gains was an optional, not an essential, part of war. But those lessons of history that I carried with me lay inert in that unread study as those unprepared for the stabilization of Iraq reacted reluctantly, slowly, and often too late to circumstances much different from those they imagined in that complex, traumatized country.

Some seemed unaware of how ambitious their goals were: Operation Iraqi Freedom was to free the Iraqi people of a brutal regime, ensure that a hostile dictator did not possess weapons of mass destruction, and create a democratic government in the Middle East that would serve as an antidote to extremism. Leaders optimistically pursued those objectives despite the lack of broad international support for preventive military action. Warnings that a state-building effort in Iraq would be difficult, costly, and long-term were ignored. Optimists listened to the wrong people—many of whom were Iraqi expatriates with agendas of their own.

* * *

AS IN Afghanistan, the neglect of the very nature of war made an already challenging mission more difficult. Strategic narcissism is what conjured up the pipe dream of an easy win in Iraq. It did not imagine war as an extension of politics and a profoundly human endeavor in which the future course of events is uncertain. Although the Coalition and the Interim Governing Council adopted a United Nations–sponsored time line for the political transition of Iraq, the politics of the ballot were unfamiliar to those who had only experienced the politics of the gun. The Coalition vanquished Saddam's government, but a clandestine Baathist network and a well-funded external Baathist organization survived the invasion. What began as decentralized, hybrid, localized resistance against occupation coalesced over time into a highly organized insurgency. Disparate Sunni insurgent groups drew on the intelligence infrastructure that Saddam had designed to subjugate his own people. Agents of the former regime benefited from external support from Gulf states as well as relationships with Islamist groups. Foreign fighters and suicide bombers flowed across Iraq's unguarded borders.[27] The U.S. mission was far from accomplished.

U.S. leaders did not consider the degree to which Iraq's people were traumatized or Iraq's social fabric torn by the brutality of the Saddam regime and costly wars. The country's youth were vulnerable to recruitment into terrorist organizations due to poor education and Saddam's "Return to the Faith" initiative, designed to direct public anger toward Israel, the United States, and a grand "Zionist-Crusader Conspiracy" to keep Iraq, the Arab world, and Islam down. The middle class had eroded under the pressure of UN sanctions as the regime used corruption and patronage to stay in power. In the wake of the Coalition invasion, the institutions of Saddam's government fell apart. As looters tore down the remnants of the Iraqi state, criminals and a developing insurgency moved into the vacuum. Removing Saddam without a plan to secure the country unleashed the centripetal forces of sectarian violence that grew out of the failure of Arab nationalism, the Iranian Revolution, the Iran-Iraq War, the aftermath of the 1991 Gulf War, and the brutality of Baath Party dictatorships.

Sunni Arabs feared Shia Arab and Kurdish encroachment and extrajudicial retribution. Those fears grew due to the composition of the mainly Shia and expatriate Interim Governing Council; severe "de-Baathification" that disenfranchised former military officers, government officials, and even schoolteachers; and the potential for retribution from vengeful Kurdish and Shia militias. Some feared the new government would simply divide the country's oil reserves and richest agricultural lands between Kurds and Shia and leave a destitute Sunni population in an impoverished rump state. Many Sunni Arabs concluded that they could protect their interests only through violence; the insurgency strengthened.

Sometimes it seemed to me that those in Washington and the newly formed Coalition Provisional Authority in Baghdad were deliberately making the ambitious mission as hard as they could. In addition to severe de-Baathification, decisions not to recall at least a portion of the Iraqi Army, limits on the numbers of U.S. troops similar to those imposed in Afghanistan, delayed justice for the Baath party's worst criminals, and the composition of an Interim Governing Council that Iraqis regarded as inept and corrupt fueled discontent and prevented an effective response to a growing insurgency.

As in Afghanistan, the U.S.-led Coalition was too slow to adapt to the evolution of the enemy and its strategy. For the first year of the conflict, insurgents directed violence primarily at Coalition forces, nascent security institutions, and political leaders. Initial attacks were not very effective, but the insurgents learned to employ deadly roadside bombs and car bombs. In the summer of 2003, insurgents also destroyed infrastructure to frustrate progress and foster popular discontent. Terrorism was central to the strategy; civilian-targeted attacks aimed to undermine international support for the Coalition effort, such as the August 2003 bombing of the Jordanian embassy and UN headquarters.

As in previous wars, there was insufficient civilian capacity to stabilize the country. Establishing rule of law, providing basic services, and building local governance fell into the military's purview. Military units had not been trained for those tasks or for counterinsurgency operations; many were overcome by the unanticipated scope of responsibility. A few overreacted to the growing violence and generated more enmity through heavy-handed tactics and breaches in discipline. The Abu Ghraib prison scandal and other instances of Coalition abuses reinforced insurgent propaganda that Coalition forces were twenty-first-century

"Crusaders" or "Mongols" who intended to subjugate or destroy the country. Over time, the growing Baathist resistance subsumed nationalist and tribal recruits and forged an alliance with Islamists affiliated with Al-Qaeda in Iraq. By 2004, the conflict was morphing into a sectarian civil war.

I witnessed the evolution of the conflict firsthand as I traveled across Iraq multiple times during the first year of the war. In February 2004, the Commander's Advisory Group I led prepared a memo for General Abizaid that began with the observation that "the specter of civil war is haunting Iraq." A month later, the leader of AQI, Abu Musab al-Zarqawi, proposed "the Afghan model" for Iraq. Just as the Taliban's Islamic Emirate arose from the Afghan Civil War, Zarqawi's emirate would rise from the chaos in Iraq. Zarqawi would jump-start civil war with mass murder attacks on Iraq's Shia communities, inviting retribution. He would then use retribution attacks on Sunnis to portray AQI as the protector from Shia militias and Shia-dominated security forces. Ultimately, the Islamic Republic of Iraq would become a "jihadist state" for Al-Qaeda to use as a launching pad for attacks against the "near enemies" of Israel and the Arab monarchies as well as the "far enemies" of Europe and the United States.[28] To gain strength, nascent insurgencies require time and space when security forces either are not aware of them or are unable to quash them. Tragically, our strategic narcissism—which resulted in lack of preparation, an inability to consolidate military gains politically, and poor adaptation to an evolving conflict—gave the insurgency in Iraq such a respite.

* * *

A YEAR later, in February 2005, I returned to Iraq in command of the "Brave Rifles," the U.S. Army's Third Armored Cavalry Regiment. I visited Abadi in Baghdad as our regiment received a new mission to conduct counterinsurgency operations and develop Iraqi Security Forces in Nineveh Province across a 22,000-square-kilometer area that included a 220-kilometer-long border with Syria. In and around the city of Tal Afar, our troopers would experience in microcosm the problems that dominated the next and most destructive phase of the war. Abadi had friends among the Shia Turkmen tribes in Nineveh. He described the situation as consistent with Zarqawi's "Afghan Model," explaining how Al-Qaeda took advantage of sectarian, tribal, and ethnic divisions to foment violence. The terrorists used the ensuing chaos to gain control of territory, populations, and resources. That is how the city of 250,000 people astride an ancient route that Alexander the Great used on his conquest of Persia quickly became the main training ground and support base for Zarqawi and AQI.

When our regiment arrived in Nineveh, our troopers alongside Iraqi Army soldiers isolated AQI from sources of support in the region, interdicted the flow of foreign fighters and supplies from Syria, and eliminated support areas in the surrounding countryside. We strove to counter enemy propaganda and clarify our intentions with our deeds by pursuing the enemy relentlessly while protecting civilians. It was also important to expose the terrorists' inhumanity, criminality, and hypocrisy. Our regiment, alongside the Iraqi Army's Third Division and U.S. and Iraqi special operations forces, arrested the cycle of sectarian violence by moving into the city and protecting the population. Once the pall of fear over the city had been lifted, our soldiers gained access to intelligence. Locals were able to cooperate with U.S. and Iraqi forces. Terrorists could no longer hide in plain sight as our forces tracked them down in Tal Afar and the surrounding areas. In late 2005, after the enemy was defeated, mediation between Tal Afar's communities restored trust among them. Those returning to the city had renewed confidence in reformed local government and police forces. Schools and markets reopened. Abadi and Najim al-Jibouri, who had become mayor of Tal Afar, helped forge the accommodations between the Sunni and Shia communities essential to sustaining security. The sectarian civil war centered on Tal Afar in 2004–2005 was a harbinger for what was to come across the entire country from 2006 to 2008 and the cycle of sectarian violence that would again afflict the region after 2011.

* * *

PRE-INVASION WILLFUL ignorance about the complexity of stabilizing Iraq evolved into post-invasion denial about the growing insurgency and, later, into the refusal to acknowledge the evolution of the conflict into a sectarian civil war. Strategic narcissism produced the same depressing behaviors. Some leaders based strategies on what they preferred to do rather than what the situation demanded. The short-war mentality persisted; some generals and admirals seemed more interested in getting back to peacetime priorities than winning the war. By 2005, commanders had shifted the strategy in Iraq to one of "transition" to Iraqi forces as an end in and of itself. It was a rush to failure, as it vastly overestimated the Iraqi government's and security forces' ability to assume full responsibility. The drive to transition even as the security situation worsened was based in part on the rationalization that more Coalition troops were actually part of the problem rather than part of the solution; more troops would incite additional natural opposition to a Western occupying force, the thinking went. Yet, AQI would use the occupation as a recruiting tool as long as there was even one Coalition squad in Iraq; the numbers did not matter. Despite the country's clear drift toward civil war and the example in Tal Afar of what approach was necessary to forestall that drift, strategic narcissism persisted.

In February of 2006, AQI bombed the al-Askari, or "Golden," mosque in Samarra, one of Shiism's holiest sites. Until that bombing, the presence of Coalition forces and Shia restraint acted as brakes on the downward slide toward civil war. Afterward, Shia militia attacks on Sunnis mounted. Death squad killings became nightly occurrences. Sunni militias sprouted and affiliated with Al-Qaeda. Mixed neighborhoods underwent ethnic cleansing as one group or the other moved out or was forced to leave. More than 1,300 Iraqis died in sectarian killings in March alone. By midsummer, the death toll was over 100 per day. U.S. and government forces could not control the violence even after the killing of AQI leader Abu Musab al-Zarqawi in June. By the end of 2006, the chaos of civil war had strengthened the influence of Iranian-sponsored Shia militias and accelerated the cycle of violence. The Iraqi government not only lacked the capacity to help pull the country back together; it had become a sectarian battleground itself and lacked the will to do what was necessary to stabilize the situation.

It was during this period that Rayburn coined the term "Myraq," to describe the tendency of U.S. civilian and military officials to describe the situation in "I-raq" as they would like it to be. It was a funny play on words, but strategic narcissism is dangerous and costly in wartime. The transition strategy continued to brief well on PowerPoint; bulletization of the situation in Myraq masked the reality that the strategy was failing. Metrics measured the successful execution of a flawed strategy. Ultimately, the Iraq strategy was wrested from the Department of Defense as the disconnect between Pentagon briefings and the actual situation in Iraq was undeniable. By the fall of 2006, several reassessment efforts were under way in Washington.

I was part of one of those assessments, one that became known as the "council of colonels." Chairman of the Joint Chiefs of Staff (JCS) Peter Pace assembled a team of us to help the Joint Chiefs prepare recommendations to the broader White House assessment run by Deputy National Security Advisor J. D. Crouch. General Pace, a gentleman of great integrity, was unable to shake his and the JCS's tendency to base recommendations on their preferences rather than what the situation in Iraq demanded. Our deliberations revealed that the JCS and the secretary of defense suffered from cognitive dissonance in not grasping that the transition strategy was utterly inconsistent with the deteriorating situation in Iraq.

On Veterans Day weekend, my colleagues asked me to write the memo for the chairman's recommendation to the National

Security Council staff. I tried to expose for the chairman and others the disconnect between the nature of the problem in Iraq and the recommendation to continue with the same strategy. The memo made what had previously been implicit assumptions explicit.[29] They included:

- Iraqi security forces possessed not only the strength but also the legitimacy with the Iraqi people to defeat the insurgency and provide sustainable security.
- The Iraqi government was capable of providing basic services on a nonsectarian basis and convincing Sunni Arab and Turkmen populations that they could protect their interests through a political process rather than through violence.

I concluded the memo with the observation that the principal advantage of the recommended strategy was that it did not require any additional U.S. troops or resources. When General Pace's executive officer, Mike Rogers, who would later become commander of Cyber Command and director of the National Security Agency, came into the Pentagon to read the draft, the always calm navy captain was exasperated with me. He wanted me to redraft the memo, but I argued that there was not anything in the memo that was untrue or did not reflect the chiefs' discussions. I asked him to forward it to the chairman as drafted. I would take full responsibility and, of course, make any edits or changes that General Pace directed. General Pace, an officer committed to giving his best military advice, forwarded the memo to the White House as it was drafted.

Meanwhile, Vice President Dick Cheney was developing his own reassessment to provide an alternative perspective to the government process. The American Enterprise Institute, a think tank, conducted a war game and planning session under the direction of Drs. Frederick and Kimberly Kagan, good friends of mine, fellow historians, and particularly astute critics of Iraq War strategy. I would not participate with the Kagans' project, but what our regiment learned from the experience in Tal Afar informed their effort. Two recently retired officers, our cavalry regiment's deputy commander, Col. Joel Armstrong, and plans officer, Maj. Dan Dwyer, structured a war game and developed a "surge" option to adapt the Iraq strategy to the evolving nature of the war. Soon after I drafted the memo for General Pace, Vice President Cheney invited me to his home. In response to his questions, I described how our strategy in Iraq had been, from the very beginning, disconnected from the character of the conflict there. During a second meeting, I brought with me a handwritten list of the "Top Ten Reasons We Do Not Deserve to Win the War in Iraq," to highlight the effect of strategic narcissism on the war.

As I got home to London for Christmas, where I was working at the International Institute for Strategic Studies, President George W. Bush directed a dramatic shift in the strategy for the war in Iraq. I returned to Iraq less than two months later. Along with Ambassador David Pearce and a talented interdisciplinary and multinational team, we helped draft the strategy for the new commander in Iraq, Gen. David Petraeus, and the new U.S. ambassador, Ryan Crocker. Prior to the invasion in 2003, Pearce, an exceptionally talented and experienced diplomat, had produced a largely ignored study of Iraq that predicted many of the difficulties we would encounter after Saddam's removal. Pearce knew a civil war when he saw one. Before joining the State Department, he had been the chief Middle East correspondent for United Press International during the Lebanese Civil War. Pearce and I agreed that we needed a political strategy to address the causes of the violence and ensure that all activities, programs, and operations in Iraq, as well as diplomatic efforts in the region and beyond, were directed at addressing those causes.

★ ★ ★

I HAD last seen parliamentarian Haider al-Abadi in Tal Afar in early 2006. As we in the Joint Strategic Assessment Team, or J-SAT, started our work in March 2007, I borrowed Petraeus's plane and took Abadi back to Tal Afar. Rayburn, by now my trusty sidekick on these projects, joined us. The city was flourishing. Despite Al-Qaeda's best efforts to restart the cycle of sectarian violence, the peace held. Residents in Tal Afar did not want the horror to return. Mayor Jibouri was a superb leader and mediator who kept communications open between the communities and insisted on professional police forces. A small U.S. force, less than one battalion, remained to continue advising and assisting Iraqi security forces. Iraqi soldiers and policemen had good relations with each other and with the population. We walked through neighborhoods that had, two years earlier, been battlegrounds. But this time, we wore no helmets and no body armor. Marketplaces were thriving. Schools were full of happy children. Mothers watched their children play together in the city's parks and playgrounds. The people greeted us and thanked us. Tal Afar, once wracked by horrible violence, now seemed a sanctuary from the death and destruction across mixed sectarian areas of Baghdad and the Tigris and Euphrates River Valleys.

Tal Afar was a model for what was needed across Iraq: political accommodation between communities. To foster this accommodation, the United States needed more than reinforcements—it needed a fundamental change in strategy. That strategy would aim to extend a Sunni tribal "awakening" against AQI, defeat that terrorist organization, and nurture local cease-fires. Isolating not only AQI but also Shia militias from popular support required convincing communities that they could protect their families and their interests through a political process rather than through violence.

But it is difficult for political processes to take hold when factions are shooting at one another. The military headquarters in Iraq, III Corps, under Gen. Raymond Odierno, was already directing its forces to physically break the cycle of violence, establish security, and help broker local cease-fires as a first step toward "bottom-up" reconciliation efforts. Special operations forces under Lt. Gen. Stanley McCrystal were focusing on irreconcilable factions within AQI and Shia Islamist militias. The idea was that reconcilable factions, witnessing the fate of those targeted, would be incentivized to join the political process. The utter brutality of AQI helped. Rather than patrons and protectors, Zarqawi and those who flocked to Iraq to fight with him were beginning to be seen by the Sunni population as a foreign pathogen that needed to be excised.[30] All programs, from development assistance to development of security forces, were aimed at reducing malign sectarian and foreign (i.e., Iranian) influence. It worked. In the year following the complete deployment of surge forces, violence in Iraq fell to the lowest levels since 2004.[31]

Abadi, Rayburn, and I returned to Tal Afar again in early 2008. Abadi believed that the surge had pulled Iraq away from the precipice of utter mayhem. As we walked through a peaceful Tal Afar, we recognized that the success of the surge was fragile and reversible. What was needed to ensure that Iraq remained stable and was not aligned with Iran was long-term military, diplomatic, and economic engagement. Unfortunately, this was not to be.

* * *

AT A crucial moment in 2009–2010, the new administration would drive U.S. policy based on a new version of strategic narcissism, one based in pessimism and resignation to any outcome in Iraq as long as the United States disengaged. Candidate Barack Obama campaigned on a sixteen-month time line for withdrawal from the country. When the Iraqi prime minister, Nouri al-Maliki, endorsed that time line and Obama won the election, the momentum behind complete disengagement proved irresistible. Whereas the Bush administration's overconfidence led to underappreciation for the risks and costs of intervention, the Obama administration's pessimism led to underappreciation for the risks and costs of extrication. Whereas the Bush administration's strategy betrayed the conceit that U.S. decisions in war would produce the desired outcome, the Obama administration saw U.S. presence in Iraq as the principal cause of problems. Both approaches lacked strategic empathy because they failed to consider the agency that others, especially enemies and adversaries, had on the future course of events. In December 2011, President Obama declared, "We're leaving behind a sovereign, stable and self-reliant Iraq, with a representative government."[32] Like the announcement of the withdrawal of U.S. forces in Afghanistan even as additional troops deployed there in 2009, the Obama administration equated American withdrawal to the equivalent of the end of the war. Vice President Biden called President Obama from Baghdad to say "thank you for giving me the chance to end this goddamn war."[33] Key officials in the Obama administration saw American disengagement from not only Iraq, but also the Middle East, as an unmitigated good.

Sadly, U.S. diplomatic and military disengagement from Iraq removed checks on the worst inclinations of Iraqi prime minister Nouri al-Maliki's government and cleared the way for the return of large-scale sectarian violence, the extension of Iranian influence over the Iraqi government and Shia population, and the growth of a new version of Al-Qaeda in Iraq—ISIS, the most potent jihadist terrorist organization in history.

As it disengaged, America failed to contest the Iranian-led effort to hand the 2010 Iraqi national election to Prime Minister Nouri al-Maliki, even though a secular Shia with broad appeal among the Sunni community, Ayad Allawi, had won the popular vote. Maliki seemed to do all he could to restart the civil war. His government purged Sunnis from the army and key governmental positions and reneged on the promise to keep Sunni tribal militias on the payroll, instead dismissing them and providing Al-Qaeda with well-trained recruits who were convinced that the political process was hostile to their interests. After his reelection, Maliki arrested the Sunni vice president, Tariq al-Hashemi, on terrorism charges based on coerced testimonies, and in 2012, Hashemi was sentenced to death in absentia. Even as he alienated Sunni Arabs and Turkmen, he released large numbers of prisoners who, upon return to their home villages and cities, elicited a hostile response from those who feared the return of Al-Qaeda. The alienation of Iraq's Sunni population, Iran's subversion of the

Iraqi state, and the effects of the Syrian Civil War propelled the rise of ISIS and the collapse of Iraqi security forces during ISIS's June 2014 offensive.[34]

Anyone paying attention should have seen ISIS, or AQI 2.0, coming miles (and years) away. For those of us who knew Iraq, the ISIS train wreck had already happened, but the world had not yet heard the sound. In 2011, I was in Afghanistan when the situation in Iraq was unraveling. After receiving calls and emails from concerned Iraqi friends who saw the accommodation between Sunni and Shia populations crumbling, I read a report predicting exactly what would happen in the coming months and years. The continued sectarian policies of the Maliki government, combined with the release of Al-Qaeda prisoners, would result in the Sunni communities again concluding that they could protect their interests only through violence. A new version of Al-Qaeda would again portray itself as a protector of those communities and replicate the Al-Qaeda offensive of 2004 to 2006. The report predicted that this renewed offensive would collapse Iraqi security forces already hollowed out by sectarian purges.

From my office in Kabul, I forwarded the paper to the chairman of the Joint Chiefs of Staff, Gen. Martin Dempsey, at the Pentagon. Dempsey, who understood Iraq and regional dynamics in the Middle East well, found the paper compelling and circulated it across the JCS staff and Department of Defense. Senior intelligence officials gave it a tepid response. Its predictions did not conform to U.S. leaders' self-delusion that AQI had been defeated and that Iraqi forces were resilient. Some in the U.S. military were unwilling to believe that after over a decade of training, Iraqi security forces would collapse. Strategic narcissism turns fantasy into perceived reality even if a preponderance of evidence is presented to the contrary. Less than thirty months after the December 2011 "end of mission" ceremony for the U.S. command in Baghdad, Mosul fell and Abu Bakr al-Baghdadi, who had in December 2004 been released from Camp Bucca prison in Iraq, climbed the stairs of the Great Mosque of al-Nuri, in Mosul, to declare the formation of a new caliphate.[35]

* * *

TWO THOUSAND eleven was a pivotal year not only in Iraq, but also across the Middle East. Eight years after the invasion of Iraq, the United States withdrew its forces from that country and declared its intention to leave the troubled region behind. The so-called Arab Spring began in Tunisia in late 2010, when a Tunisian fruit vendor named Mohamed Bouazizi publicly self-immolated to protest socioeconomic injustice and autocracy in his country. Bouazizi's act inspired a series of antigovernment protests, uprisings, and armed rebellions across the Arab-majority regions of the Middle East. The protests highlighted the severe power imbalance between heads of state in populations such as Tunisia, Egypt, Libya, and Syria—eventually leading to the resignation of the Tunisian and Egyptian presidents, a full-scale insurgency that toppled Muammar Gaddafi's government in Libya, and a civil war that nearly felled the Assad regime in Syria. It was in response to the Arab Spring that the Obama administration, as it disengaged from the Iraq War in a final repudiation of George W. Bush's invasion in 2003, intervened in Libya in a way that replicated the fundamental folly of Iraq: the use of military force without a plan to influence what happened next. The protests in Tunisia toppled President Zayn al-Abidine Ben Ali, who had ruled Tunisia since 1987. The political awakening spread via Twitter and Facebook across Tunisia and then moved east to Egypt, home to eighty-six million, the largest Arabic-speaking population in the world.

Egypt had been run by Hosni Mubarak, a powerful president who assumed office in 1981, and who maintained power through rigged elections and imprisonment of political adversaries. Stirred by Bouazizi's self-immolation in Tunisia, on January 25, 2011, Egyptians took to the streets to protest the oppressive security regime that denied them freedom of speech and assembly. On February 1, Mubarak conceded to the protest movement. Appearing on national television, he announced that he would step down when his six-year term ended in late 2011. It was not enough to satisfy the protesters, who refused to disband until he resigned. Mubarak did so on February 11, 2011. In eighteen days, the wave of demonstrations rocking the Middle East had forced the resignation of the most powerful leader in the Arab world.

On February 15, the protest movement poured over Egypt's western border into Libya, an Alaska-size nation of six million inhabitants. Since taking power in 1969, Gaddafi had maintained power in Libya via an iron fist, control of revenue, and an extensive patronage network. Unlike Egypt and Tunisia, where presidents were toppled with a relatively minor death toll, Libya descended into a full-blown civil war.

After the fall of Ben Ali, the resignation of Mubarak, and Arab Spring-inspired uprisings throughout the Middle East, Arab strongmen in palaces, such as Gaddafi, suddenly appeared seated within houses of cards. Arab monarchies such as those in Saudi Arabia, Kuwait, Bahrain, and the United Arab Emirates grew nervous while many in the United States and the West saw the Arab Spring as an inevitable burst of freedom from long-suppressed peoples. The relatively new medium of social media was credited with catalyzing the protests and changes in government and extolled for enabling a wave of freedom across the Middle East.

* * *

ON MARCH 17, 2011, the Obama administration succeeded in getting the UN Security Council to pass Resolution 1973, authorizing military intervention in Libya with the goal of saving the lives of pro-democracy protesters from the forces of Libyan dictator Muammar al-Gaddafi. Besides, Gaddafi was endangering the momentum of the nascent Arab Spring. Like the initial invasions of Afghanistan and Iraq, the NATO-led air campaign was at first declared a tremendous success.

With the help of NATO airpower, the Libyan rebels over the course of several months were eventually able to depose Gaddafi, who was forced out of a hiding place on October 20, 2011. While his convoy raced across a potholed Libyan highway, American fighter jets zoomed in overhead. American aircraft fired air-to-surface missiles that disabled the convoy. As the shock of the explosion wore off, Gaddafi's entourage departed their vehicles. Gaddafi's son hurried him through the desert to the shelter of a nearby drainage pipe underneath a road. Rebels soon pulled Gaddafi from the ditch and tore off his shirt. Two rebels held him upright while another plunged a sharp metal rod through his pants and into his anus. A fourth man videotaped the sodomy while onlookers jeered the violence. Finally, the seventy-year-old dictator, bloodied, bruised, and disoriented, was led to a waiting ambulance. Accounts vary as to what happened next, but Gaddafi did not depart the vehicle alive.[36]

The day after Gaddafi's death, President Obama addressed the world regarding U.S. operations in the Middle East. The president was not committing additional ground forces to the region. Rather, turning to Iraq, he declared that "the rest of our troops in Iraq will come home by the end of the year." The president elaborated: "Over the next two months, our troops in Iraq, tens of thousands of them, will pack up their gear and board convoys for the journey home. The last American soldier[s] will cross the border out of Iraq with their heads held high, proud of their success, and knowing that the American people stand united in our support of our troops. That is how America's military efforts in Iraq will end."[37]

Due to the success of the Libyan aerial campaign, American denial syndrome returned. The purported lesson was not to repeat the mistake of Iraq and put "boots on the ground" in Libya. But without any effort to establish security in post-Gaddafi Libya, NATO actually repeated the mistakes of Afghanistan and Iraq in the extreme, again neglecting the very nature of war as an extension of politics and as driven by human emotion. Chaos and a smoldering tribal conflict ensued. Islamist groups and terrorist organizations thrived in the lawless environment. Eleven months after Gaddafi's death, a branch of Al-Qaeda commemorated the anniversary of that organization's 9/11 mass murder attacks on the United States eleven years earlier with an attack on the U.S. consulate in Benghazi, Libya. The well-respected Middle

East expert and newly appointed ambassador to Libya, Christopher Stevens, was visiting from the capital, Tripoli. Terrorists attacked the consulate and another compound. Security was inadequate, in part due to the desire to keep a light footprint in the country. Ambassador Stevens and U.S. Foreign Service information management officer Sean Smith were killed at the consulate. CIA contractors Tyrone Woods and Glen Doherty died at a government annex about one mile away. Stevens was the first U.S. ambassador killed in the line of duty since 1979. Libya became both a source and a waystation for refugees desperately seeking the safety of Europe's shores. Meanwhile, the Syrian Civil War was exploding into a cacophony of violence and, in Egypt, Mohamed Morsi was consolidating power to replace Hosni Mubarak's nationalist dictatorship with a Muslim Brotherhood Islamist dictatorship. Seasons change. In the Middle East, spring was over.

★ ★ ★

DURING THE Cold War, Iran and Saudi Arabia served as the "twin pillars" of the U.S. effort to contain Soviet influence in the Middle East. The Iranian pillar collapsed during the 1979 revolution, and the Saudi pillar proved to be structurally flawed based on Saudi Arabia's production and export of a virulent strain of Islam. The countries that had been seen as sources of stability created conditions for a sectarian civil war that would wrack the region for decades. Abadi and other leaders across the Middle East tried to bring communities together, but the legacy of the Iranian Revolution, the Iran-Iraq War, brutal dictatorships, and the emergence of a destructive perversion of Sunni Islam overwhelmed their efforts.

As frustrating as the American experience in the Middle East was after the invasion of Iraq in 2003, disengagement from the region made a bad situation worse. The United States and other nations are not going to solve the region's problems, but outsiders can support those, like Abadi and Jibouri, who are determined, despite the daunting tasks before them, to create a future for the region that allows it to emerge from the serial failures of the twentieth and early twenty-first centuries. If the free and open societies of the world turn their backs on the people of the region, they will not be immune to the problems that emanate from it.

CHAPTER 8

Breaking the Cycle

Terrorism is inseparable from its historical, political, and societal context, a context that has both a local and a global dimension.

—**AUDREY KURTH CRONIN, *HOW TERRORISM ENDS***

IN 2020, the situation across the greater Middle East, the region spanning Morocco in the west to Iran in the east and encompassing the northern countries of Syria and Iraq to the southern countries of Sudan and Yemen, remained as confounding as it was wretched. The inability of the United States to develop and implement a sound, consistent policy in cooperation with like-minded nations contributed to the scale and duration of the catastrophe in the region and diminished American influence there. The policies of the George W. Bush, Obama, and Trump administrations were consistent with America's tendency to engage the region episodically and pursue short-term solutions to long-term problems.[1] As the experience in Iraq demonstrated, treating symptoms rather than causes of violence perpetuates conflict and magnifies threats to national and international security. Our efforts in the Middle East should focus on ending the sectarian civil war that is at the root of the humanitarian crisis and the threats that emanate from the region. To succeed, those efforts must be executed at a cost acceptable to the American people.

If Americans are to view the Middle East other than as a mess to be avoided, our strategy must begin with an explanation of what is at stake and how sustained engagement in the region is important to citizens' security and prosperity. As Middle

East analyst Kenneth Pollack observed, the region is important to Americans because problems there do not conform to the city of Las Vegas's motto—that is, what happens in the Middle East does *not* stay in the Middle East.[2] The failure of the Arab Spring, the Syrian Civil War, and the rise of ISIS reached far beyond the region. From March 2011 to October 2018, the Syrian Civil War alone caused the deaths of more than 500,000 people and displaced more than a quarter of Syria's 21 million people.[3] Bashar al-Assad's regime forcibly disappeared more than 98,000 men, women, and children and arbitrarily imprisoned more than 144,000 in Syria's despicable prisons.[4] No one knows how many civilians the Syrian regime murdered, but estimates are in the hundreds of thousands.

As violence intensified, President Barack Obama and U.S. allies in Europe disengaged from the region and decided against options that might have limited the scale and reach of crises in Libya and Syria. A multinational security force in Libya after the fall of Muammar Gaddafi, for example, might have prevented the fragmentation of a potentially wealthy country that spans a wide geographic area but has a population of only less than seven million people.[5] In Syria, no-fly areas and safe zones such as those established for Iraq's Kurds in the aftermath of the 1991 Gulf War might have mitigated the humanitarian crisis, stemmed the flow of refugees, constrained Iranian and Russian influence, impeded the growth of ISIS and other jihadist terrorist organizations, and put pressure on the Assad regime and its sponsors to seek a political resolution of the civil war.[6] Viable options to address Syria's problems diminished over time as moderate opposition to the regime lost ground to Islamist extremists and Russia intervened to prevent Assad's collapse in 2014, the same year that ISIS conducted its massive offensive across Syria and Iraq. Indeed, the intensifying war in Syria, large-scale sectarian violence in Iraq, chaos in Libya, and, after 2014, a civil war in Yemen created conditions ideal for the growth of ISIS and Al-Qaeda-related groups (such as Hayat Tahrir al-Sham and Nusra Front). Just over a year after al-Baghdadi gave his sermon in Mosul, ISIS directed or inspired attacks, including the mass murder of Parisians in November, mosque bombings in Yemen, attacks on tourists in Tunisia, suicide bombings in Ankara and Beirut, the destruction of a Russian airliner over the Sinai Peninsula, and the San Bernardino shooting in the United States in December. On March 22, 2016, three ISIS-directed suicide bombings struck the airport and a Metro station in Brussels, killing two civilians. Meanwhile, ISIS's external plotting against Western targets, including the aviation sector, continued. The more than one hundred Americans who traveled to Syria to join terrorist groups, and the thousands of Europeans who did so, represented a new danger from returning fighters to the United States or to countries with visa waivers for travel to North America.[7] ISIS's reach was not limited to the physical world. The terrorist organization mastered the internet and social media to recruit and inspire attacks. It was apparent that what happened in the caliphate did not stay in the caliphate.

The argument to war-weary U.S. citizens as well as NATO and EU nations sharing the burden in the Middle East is that preventing the rise of terrorist organizations is less costly than responding after they become an inescapable threat. When the U.S. military was forced to return to the region in support of the Iraqi Army and mainly Kurdish militias in Iraq and eastern Syria, the subsequent five-year campaign to wrest control of territory the size of Britain from ISIS cost far more in lives and treasure than sustaining the effort in Iraq and mitigating the mass atrocity of the Syrian Civil War would have cost.

Another aspect of that argument is that it is wiser to address the causes of violence than continue expensive treatments of the symptoms. Syria became the worst humanitarian crisis since World War II and has generated a refugee crisis that has overburdened neighboring states of Lebanon, Jordan, and Turkey and extended into Europe. Between 2014 and 2018, approximately 18,000 refugees who, in desperation, trusted their lives to criminal traffickers drowned in the Mediterranean Sea, over 24.3 percent of them children. By 2020, more than 1 million refugees made it to

Europe's shores or across its borders from Turkey. Syrian refugee populations reached 900,000 in Lebanon, 600,000 in Jordan, and 3.6 million in Turkey.[8]

Statistics, however, were numbing, and reports rarely covered real victims' experiences. Only occasionally did people in the United States and Europe appear struck by the human cost of the crises, such as when, in September 2015, heartrending photos circulated of the body of the drowned three-year-old boy, Alan Kurdi (born Aylan Shenu), on a Turkish beach; or, in August 2016, a video of a child victim of Assad's bombing of Aleppo, Omran Daqneesh, showed him in shock, bleeding from the face in the back of an ambulance. The refugee crisis grew worse as the war intensified. In late December 2016, counting on continued U.S. and European indifference, Russia had expanded the indiscriminate bombing of Aleppo. In what before the war had been Syria's most populous city and cultural center, thirty thousand Syrians perished.[9] The bombing campaign inflicted mass casualties on civilians, targeting hospitals and funerals. An assault by Iranian proxy Shia militias followed. On December 13, U.S. ambassador to the United Nations Samantha Power asked the Russians rhetorically, "Have you no shame?" But President Obama and his administration were not sufficiently moved to reverse their policy of resignation toward what they regarded as an intractable mess. In an op-ed in the *New York Times,* National Security Council senior director for the Middle East and North Africa Steven Simon wrote, "the truth is that it is too late for the United States to wade deeper into the Syrian conflict without risking a major war."[10] Americans' continued view of the crisis in the Middle East was as a protracted episode of mass homicide limited to faraway places and therefore not an American problem.

Refugees generated financial and political as well as physical burdens, weakened already fragile political orders in the Middle East, and fueled political polarization and the rise of nativist sentiment across Europe. In countries hosting refugees, initial outpourings of sympathy and willingness to provide assistance gave way to vitriolic debates between those who favored continued support and others who resented the diversion of resources from their own populations. European fears rose that a large influx of Muslims could alter the social and religious character of their nations or that jihadist terrorists would infiltrate the throngs of refugees.

Despite the overwhelming evidence that crises in the Middle East do not stay there, it has proved difficult for the United States to sustain efforts, beyond billions of dollars in humanitarian assistance, that actually addressed the disease of sectarian conflict. In the summer of 2019, President Donald Trump resurrected the Obama administration assumption that the United States could disengage from the region and remain insulated from the conflicts there. As he announced the withdrawal of the small American special operations force that had been fighting alongside the People's Protection Unit (YPG)–dominated Syrian Democratic Forces (SDF) in northeastern Syria, Trump tweeted that it was time to "get out of these ridiculous Endless Wars." He described Syria as nothing but "sand and death."[11] While he later stated that two hundred to three hundred troops would remain to secure the oil in eastern Syria, the damage to U.S. credibility and influence compounded the damage done by the Obama administration's withdrawal from the region.[12] Potential consequences of disengagement included a return to low-level Kurdish-Turkish conflict, YPG accommodation with the Syrian regime and its Russian and Iranian sponsors, the return of ISIS or a new-and-improved version of it, and the extension of Iranian influence across Syria in a way that threatens Israel and perpetuates the sectarian civil war. As horrific as the wars in Iraq, Syria, and Yemen have been, disengagement made them worse.

In late 2019 and early 2020, the Assad regime and their Russian and Iranian sponsors intensified their murderous campaign against Syrian civilians in the northeastern province of Idlib. Over 900,000 people, including half a million children, battled the indiscriminate bombing and freezing-cold conditions. Like the Obama administration's response to the destruction of Aleppo in 2016, the mass murder and forced displacement of innocents in Idlib elicited statements of condemnation but no direct action, even as the Turkish armed forces began to take casualties. The muted response to the mass atrocity in Idlib revealed that the already catastrophic conditions in the Middle East can get worse. Convincing Americans to support sustained engagement in the region should begin with leaders explaining why the Middle East is important to the American people.

Many who advocated for withdrawal from Syria argued that the Middle East was no longer important to American security and prosperity because the United States had become the world's largest oil producer and a net energy exporter.[13] But the Middle East has always been and will remain an arena for competitions that have consequences far beyond its geographic expanse. Competitions with revisionist powers, rogue regimes, and jihadist terrorists converge and interact there. Enemies and adversaries often operate in parallel, but they also cooperate when their interests align. For example, Russia and Iran aid, abet, and sustain the murderous Assad regime in Syria. Russia used the crisis in the region as a way to weaken Europe and present itself as the indispensable power broker that can ameliorate the problems that it is helping to create. Iran, too, has taken advantage of regional chaos to its strategic advantage against its foes Israel, Saudi Arabia, and the United Arab Emirates. Even North Korea was a participant in destabilizing the region as it contributed to Assad's nuclear weapons program, an effort that went undetected until 2007.[14]

Beyond the physical threats associated with the confluence of revisionist powers, hostile states, and terrorist organizations, the Middle East, along with the historical "Khorasan" region (which comprised what is now northeastern Iran, southern Turkmenistan, and most of Afghanistan), is foundational to jihadist terrorists' psychological and ideological strength. Their plan is to reestablish the caliphate in the Fertile Crescent that runs from the Persian Gulf, through Iraq, Syria, Lebanon, Jordan, Israel, and northern Egypt. And their mission is to fight to liberate the *ummah*, or Muslim community, everywhere from what the jihadis see as foreign control. Moreover, world economic growth still remains dependent on the free flow of oil through the Strait of Hormuz just as it did during the oil embargo and crisis of 1973 and the "tanker wars" of the 1980s.[15]

* * *

THE REFUGEE crisis and the humanitarian catastrophes in the region are connected to the trafficking of drugs and other illicit goods as well as people. Perhaps most important, the crisis is generating a large-scale, multigenerational threat from jihadist terrorism. By the second decade of the twenty-first century, that threat was more severe than it was on September 10, 2001. It was alumni of the mujahideen resistance to Soviet occupation of Afghanistan in the 1980s who built Al-Qaeda, declared war on the United States, and executed a series of attacks prior to 9/11. The terrorist alumni from twenty-first-century wars are orders of magnitude larger than their Afghan War alumni predecessors. ISIS was an improvement on Al-Qaeda in Iraq, as it came with a sophisticated propaganda machine, recruiting agency, organized crime network, and proto-state. ISIS attracted more than thirty thousand fighters, not only from the Middle East and the greater Arab world, but also from developed countries such as EU nations, the United States, Canada, and Australia. Within two years following its formation in the heart of the Middle East, ISIS spawned franchise organizations in states ranging from Algeria to Nigeria to Yemen to Somalia and even the Philippines.[16]

Twenty-first-century terrorist organizations not only have global reach; they are also pursuing technologies and destructive weapons previously associated only with nation states, including chemical, biological, radiological, nuclear, and high explosives. There is broad agreement on the worst-case scenario: a terrorist organization in possession of a nuclear device either stolen, purchased, or provided to it by a hostile state. A dirty bomb, which combines commonplace explosives with highly radioactive materials, would inflict far less damage and fewer casualties than a nuclear device, but it would be easier to obtain. Its detonation in a densely populated area would incite fear in cities across the globe. Other emerging capabilities that are readily available, such as offensive cyber and weaponized drones, are threats to aviation and people on the ground.[17] The terrorist's most powerful weapon, however, may be the computer, camera, and communications device that every one of them carries in his or her pocket. Encrypted communications improve terrorists' ability to evade intelligence collection and coordinate their actions. Also, twenty-first-century terrorists produce and distribute slick propaganda to attract susceptible young people to their depraved cause.

* * *

AT BEST, it will take time and effort for the United States to regain influence lost due to its lack of a sustained and consistent policy in the Middle East. Adversaries have stepped in to advance their interests at the expense of the United States and its traditional partners, including Israel, Turkey, the Kingdom of Saudi Arabia, Jordan, the United Arab Emirates, Egypt, Kuwait, Qatar, and Bahrain. In 2017, I found that these partners were hedging, working with U.S. adversaries to limit their exposure to another sudden change in U.S. policy. They were falling for empty promises from Russia under what might be labeled Putin's Potemkin Peace Plan. In exchange for keeping Assad in power and guaranteeing Russian interests in a post–civil war Syria, Russia promised our partners that it would gradually attenuate Iranian influence in Syria. What Russia desperately needed in return was for Gulf States to pay for the reconstruction of the rubble-strewn Sunni Arab cities and towns that Russia, the Assad regime, and the Iranians had destroyed. Russia's promise was a lie because Assad had become far more reliant on Iranian than on Russian support, but Israel and the Gulf States suspended disbelief because if the United States disengaged from the region, influence with Russia might become vital to preventing even greater threats from Iran.

Turkey joined with Russia and Iran in sham peace processes for Syria that undercut the legitimate UN effort to end the war there. False Russian promises and the reality that Moscow had forces in position to influence the situation is why King Salman of Saudi Arabia flew to Moscow in October 2017 and pledged to buy Russian air defense systems.[18] Meanwhile, Jordan purchased more Russian military equipment, and Israel expanded high-tech partnerships with Russian companies. Between 2017 and 2019, Israel pretended to believe Putin in exchange for Russian forces turning a blind eye as the Israel Defense Forces conducted more than two hundred strikes against Iranian facilities and Islamic Revolutionary Guard Corps of Iran operatives in Syria.[19] By playing Syria both ways, as Iran's principal enabler and the best potential check on Iranian hegemony in the region, Russia saw its influence grow.

When I confronted our partners about the contradiction between their grave concern about growing Iranian influence in the region and behavior that aided Iran's principal enabler, Russia, they protested. They pointed to the Obama administration's withdrawal from the region and enablement of Iran with sanctions relief associated with the 2015 Iran Nuclear Deal and professed the need to retain some influence with Russia as a way to mitigate the damage. I tried to assure them that the Trump administration had developed a long-term strategy for breaking the cycle of sectarian violence in the region that would, over time, influence an outcome consistent with not only U.S. but also their vital interests.

Secretary of State Rex Tillerson summarized that strategy in a speech at the Hoover Institution at Stanford University in January 2018. First, U.S. military presence and engagement in the region was essential for deterring actions that magnified the humanitarian crisis. He recalled that, the previous April, President Trump had responded to Assad's use of sarin nerve agent against innocent civilians with strikes that destroyed 20 percent of Assad's air force. He then summarized the need for a long-term effort to ensure the enduring defeat of ISIS and other terrorist organizations that represented "continued strategic threats to the U.S." He highlighted the need to counter the IRGC's attempts to control routes from Iran to Lebanon and the Mediterranean and the need to reduce Iranian influence such that it could no longer perpetuate sectarian civil wars in the region. He reassured allies and partners that the president would not repeat the 2011 mistake of premature disengagement from Iraq and pledged to "continue to remain engaged as a means to protect our own national security interest." Finally, the United States would implement a comprehensive regional strategy for ensuring jihadist terrorists did not threaten the United States and its allies, ending the Syrian Civil War under the UN political process; checking Iranian ambitions; and ending the humanitarian crises across the region such that refugees could return to secure environments and start rebuilding their lives. He acknowledged that accomplishing those goals would take a long-term diplomatic and military effort, but assured all listening that our military mission in Syria and elsewhere in the region would remain conditions-based.[20]

Our partners' skepticism frustrated me, but it turned out to be well placed. They knew that, in 2016, even as candidate Trump vowed to accelerate the defeat of ISIS, he shared the Obama administration's sentiment that continued military engagement in the Middle East was futile and wasteful. On October 13, 2019, after a series of announcements of his intention to withdraw from Syria and reversals of that decision, Trump ordered all U.S. forces out immediately, in part to clear the way for a Turkish offensive to take control of a "safe zone" south of its border with Syria.[21] Subsequent to the withdrawal, Russian forces raised flags over former U.S. bases and tens of thousands of Iranian-backed militias occupied territory that was formerly held by ISIS in eastern Syria. Turkish-supported militias poured into Northern Syria. Many committed war crimes against Kurdish civilians, including the murder of Hevrin Khalaf, a female Kurdish politician.[22] The U.S. abandonment of its Kurdish YPG partners and its withdrawal from northeastern Syria validated the wary approach of our allies in the region, but the regional accommodation to the Iran-Syria-Russia axis was as unnatural as it was detestable.

The people of the region know the cause of their suffering. In September 2019, major protests broke out in the eastern Syrian province of Dayr al-Zawr as Sunni Arabs demanded the withdrawal of Iranian militias. In southern Syria, protests against the Assad regime continued, and insurgent attacks grew more frequent. In October, in Iraq, antigovernment protests directed at the Iraqi political leaders evolved into a revolt against Shia political parties and increasing Iranian influence. That same month, Lebanon had the biggest protest since its independence, where demonstrators demanding political reform and an end to corruption forced Prime Minister Saad Hariri to resign. Then, in December, more than two hundred thousand protesters in Iraq raged against the Iraqi government and a foreign occupier—not the United States, but Iran—chanting, "Free, free Iraq" and "Iran get out, get out."[23] They demanded the resignation of the Iraqi prime minister, Adel Abdul Mahdi. He resigned on November 30, 2019, but remained at the head of a caretaker government until February 2020, when Iraqi president Barham Salih appointed Mohammed Tawfiq Allawi, a former minister of communication under Prime Minister Maliki, as prime minister.

At the end of 2019, as Iraqi protests intensified, Iranian proxies stepped up attacks on U.S. forces in Iraq. On December 27, a rocket attack on a U.S. base in Iraq killed an American contractor, Nawres Waleed Hamid, and wounded several soldiers.[24] Iran's IRGC had clearly used Iraqi proxy militias under the command of Abu Mahdi al-Muhandis. After the United States retaliated with airstrikes on five militia outposts along the Syrian border, Iranian-backed Shia militias mobilized mass protests and an attack on the U.S. embassy in Baghdad. The images of the angry mob on December 31 were reminiscent of Iranian attacks on the U.S. embassy in Tehran forty years earlier. Simultaneously, commander of the IRGC's Quds Force, Qasem Soleimani, was coordinating broader and potentially deadly attacks on U.S. facilities in the region. On January 3, at around 1 a.m., soon after

Muhandis picked up Soleimani at Baghdad Airport to coordinate their next move, a U.S. missile struck their vehicle, destroying it and its passengers. President Trump decided that the action was necessary to restore deterrence against Iran and to prevent the attacks Soleimani was planning. Although Shia protesters in Baghdad and other Shia-majority cities, such as Basra, lamented the deaths of Muhandis and Soleimani and the Iraqi parliament, with Sunni and Kurd members abstaining, passed a nonbinding resolution for U.S. forces to withdraw, Iraqis continued to call for Iran to withdraw. In early 2020, Iran did its best to increase the pressure on the United States in Iraq by mobilizing a million-man march and a resistance front that would integrate rival militias. The Iranian regime grew increasingly concerned that protests in its own country and in Lebanon and Iraq, combined with severe economic problems, represented a grave threat to the Islamic Republic's legitimacy. Iran attempted to use the Soleimani and Muhandis killings to co-opt the Iraqi protest movement and quash opposition to its subversion of the Iraqi state. That effort failed because Iraqi alignment with Iran is unnatural, and the vast majority of the Iraqi people equate Iranian influence with corruption and failing governance. In early 2020, Iraqi protesters rejected the appointment of Tawfiq Allawi as prime minister and continued to demand an end to corruption, ineffective governance, and malign Iranian influence.

But despite the growing resistance to Iran and its proxy militias, the prospects for stabilizing Iraq remained dim due to the fragmentation of Iraqi society along sectarian lines, Iran's sustained campaign of subversion, and America's vacillation between intervention and disengagement. In 2020, U.S. influence, not only in Iraq but also across the Middle East, had diminished due to the United States' demonstrated inability to develop and implement a consistent, long-term policy toward that vexing region.

* * *

DEVELOPING A long-term strategy requires a big dose of strategic empathy and a more sophisticated understanding of the dynamics playing out in the region. The problem in the Middle East is, at its base, a breakdown in order associated with the serial failures of colonial rule, postcolonial monarchies, Arab nationalism, socialist dictatorships, and Islamist extremism to provide effective governance and forge a common identity across diverse communities. Decades of conflict fragmented societies along ethnic, sectarian, and tribal lines and perpetuated competitions for power, resources, and survival. The resulting violence strengthens jihadist terrorists and extends Iranian influence.

The United States and its partners should identify and then strengthen groups that will contribute to enduring political settlements, break the cycle of sectarian violence, and prevent terrorists from establishing support bases. The experience in Tal Afar in 2005 and the Iraq surge are examples of how support for political accommodation among communities and the reform of security forces and local governance can isolate extremists from popular support. By contrast, the Obama administration's efforts to fight ISIS in Syria and thc Trump administration's announcement of withdrawal from eastern Syria revealed a failure to see military efforts as a means to a political end. Contrary to the belief among some in Washington that there were no Sunni Arab partners to work with, effective support for Sunni anti-Assad opposition early in the Syrian Civil War would have been the best way to fight Al-Qaeda, ISIS, and other jihadist terrorist organizations. In January 2014, five months before the Obama administration intervened against ISIS in Syria, Sunni Arab rebels dealt ISIS a devastating blow. Opposition protesters and moderate rebels launched an offensive, expelling ISIS terrorists from Idlib, the Damascus suburbs, and major parts of Aleppo Province. These Arab Sunnis who opposed Assad had suffered greatly from ISIS. The largest ISIS mass atrocity in Syria was its brutal murder of approximately one thousand members of the anti-Assad Al-Sha'itat tribe in one day in August 2014.[25]

But the Obama administration's five-hundred-million-dollar "train-and-equip" program to support Arab opposition forces was disconnected from the political struggle. Assistance focused narrowly on anti-ISIS operations, even though it was Assad's army's and Iranian militias' attacks on Sunni Arab communities that allowed ISIS to portray itself as a protector. Fighters receiving training and assistance had to sign a contract pledging to fight only ISIS and not to attack Syrian or Iranian forces, the people who had driven them out of their homes and killed their families and friends.[26] It should have been no surprise when the program collapsed. The Obama administration's reluctance to support forces that were also in opposition to Assad and his Iranian sponsors was due in part to the fear of spoiling negotiations with Iran over its nuclear program. Defining the effort in Syria, Iraq, and across the region narrowly as a military campaign against ISIS jihadist terrorists does not address the sectarian conflict that strengthens those very terrorist organizations and helps Iran extend its influence. For example, when Iran placed Lebanese Hezbollah at the vanguard of an offensive against Syrian opposition forces at Al Qusayr in 2013, Sheikh Yusuf al-Qaradawi, a spiritual leader of the Muslim Brotherhood, declared that jihad in Syria against Alawites and Shiites was now obligatory, calling Alawites "worse infidels than Jews or Christians."[27]

Curbing Iranian influence requires a multiprong approach and long-term commitment. One key way to limit Iranian influence is to integrate Iraq and, eventually, a post-Assad Syria and a post–civil war Yemen into the region diplomatically and economically. When Prime Minister Abadi visited the White House in March 2017, he had recently hosted Saudi Arabia's foreign minister, Adel al-Jubeir, in Baghdad.[28] In June, Abadi visited Riyadh. Those were important visits because strengthening Iraq's majority-Shia government's diplomatic and economic relationships with the Arab world diminishes Iran's ability to sow division and fuel violent conflict. Peace in Iraq depends on reducing Iranian influence such that Sunni and other minorities no longer feel marginalized and beleaguered. The effort to erode Iranian influence will take far more than a few precision strikes or even the killing of Soleimani and Muhandis.

Good relations between Baghdad and Riyadh might persuade all Gulf states to build strong diplomatic relations with Iraq rather than support Sunni militias and terrorist organizations that attack the Shia-majority government. Diplomatic engagement with Arab states that have Shia majorities (such as Bahrain and Iraq) and those with significant Shia minorities (such as Saudi Arabia and Kuwait) should emphasize the importance of equal rights under the law, responsive governance, and the delivery of social services on a nonsectarian basis as essential to forestalling violence and building common identity across communities.

In the fall of 2019, protests in Lebanon and Iraq demonstrated that Iranian influence in pursuit of Iran's nefarious designs for hegemony in the region was unnatural. But it also revealed the hunger people have for competent, fair governance. In Lebanon and Iraq, people were asking for representative government to meet their basic needs and allow them to build a better future for their children. The models for postcolonial governance in the Middle East failed. Therefore, the United States and other nations should support the development of representative governance consistent with the culture and traditions of the peoples of the region. That includes supporting reforms necessary to establish rule of law and improving the state's responsiveness to the needs of all peoples, regardless of religion, ethnicity, or tribe.

Haider al-Abadi once told me that, after his time in the United Kingdom, he wanted to foster in Iraq democratic processes that permitted the anticipation and resolution of problems before they became crises. Because the Arab and North African states that experienced the Arab Spring of 2011 lacked legitimate opposition parties or civil society organizations, previously clandestine Islamist organizations (e.g., the Muslim Brotherhood), criminal organizations (e.g., human trafficking and transnational organized crime networks in North Africa and the Maghreb), militias (e.g., Iranian proxies in Syria and Iraq as well as Khalifa Haftar's and various Misratan militias in Libya), terrorist organizations (e.g.,

ISIS, Hayat Tahrir al-Sham, and al-Nusra Front in Syria), and foreign intelligence operators (e.g., IRGC in Syria and Iraq) were in the strongest positions after the collapse of authoritarian regimes. That is why the long-term strategy for ending the sectarian and tribal conflicts in the region begins with strengthening governance and democratic institutions and processes.

Tunisia was a success story that deserves more attention as the Al-Nahda Party there is Islamist and transitioned power peacefully to a rival party after elections. In countries such as Egypt, where an authoritarian Islamist leader replaced a nationalist dictator and then, in turn, was replaced by a general in a military coup, long-term security and stability will not be possible unless the government, which claims that it is transitioning to representative government, encourages rather than discourages the development of legitimate opposition parties and civil society organizations to participate in the political process.

Support for minority religious sects and ethnic groups across the Middle East (such as the Druze and the Baha'i) is also important to enduring peace in the region. Because minorities have often been a force for tolerance and moderation, protecting minorities and the participation of minorities in political processes, security forces, and government institutions can act as a buffer on extremism or help foster the accommodation between communities necessary to end religious and ethnic warfare. Since the 1980s, the growth of Islamist politics and parties; political instability; war; and the rise of jihadist terrorists have encouraged indigenous religious minorities of the Middle East to flee the region. ISIS's genocidal campaign against the Yazidis and mass murder of Christians in Syria, Iraq, and Egypt victimized communities that have promoted moderation through diversity for thousands of years. Minorities such as Yazidis in Iraq, Baha'i in Iran, and Druze in Lebanon faced violence as traditions of tolerance collapsed. The United States and its allies should encourage and incentivize policies and actions that protect minorities across the region. Special attention is due to the needs of the 25 million to 35 million Kurds spread across northern Syria, eastern

Anatolia, northern Iraq, and northwestern Iran. Persecution of the Kurds and a growing sense of ethnic nationalism among their diverse tribes has been a source of conflict for decades. And the Kurds have been long victimized by external repression. Despite tremendous gains in Iraq, prospects for a sovereign Kurdish nation are dim due not only to opposition from the countries in which Kurds live, but also to their own tribal and ideological differences.

* * *

ADDRESSING ALL these regional issues would be much easier with the cooperation of Turkey. Our halting disengagement from the Middle East has created space for Russia and Iran to accelerate Turkey's drift away from NATO, Europe, and the United States. President Recep Tayyip Erdogan and his Justice and Development Party (AKP) are increasingly authoritarian and anti-Western, so improved cooperation may not be possible. A strategy for the region should therefore consider how to mitigate the loss of Turkey as an ally while moving toward a transactional relationship. After a 2016 coup attempt failed to oust Erdogan, he consolidated power and drifted further away from his NATO allies. The AKP conducted massive purges of the military, judiciary, law enforcement, the media, and universities. Most of those purged were sympathetic to the pillars of Kemalism—the philosophy of modern Turkey's founder, Mustafa Kemal Ataturk: secularism, Westernism, and nationalism.[29]

As Erdogan accused the United States of complicity in the coup, it became clear that Erdogan and the AKP were ideologically opposed to Kemalism. The AKP, founded by Erdogan himself in 2001, has roots in conservative Islamist ideology connected to the Muslim Brotherhood and has fostered a form of anti-Western nationalism. The Turkish government even resorted to a form of hostage taking, imprisoning on baseless charges U.S. citizen Rev. Andrew Brunson and Turkish citizens who worked for the U.S. embassy.[30] Trump tried to develop a positive relationship with

Erdogan, but grew frustrated over the Turkish president's intransigence, especially with respect to Reverend Brunson. In August 2018, the Trump administration levied sanctions and doubled tariffs on Turkish steel and aluminum, causing the value of the Turkish lira to plummet. Erdogan released Brunson on October 12, 2018, but it was clear that the relationship had become transactional; the alliance with Turkey seemed to be in name only.

But a smarter U.S. strategy should recognize that Turkey's interests do not naturally align with those of Russia and Iran. Russia will not be a trustworthy partner in Erdogan's effort to make Turkey part of an alternative world order. Decisions to import Russian arms and deepen Turkey's dependence on Russian energy will give Moscow leverage to use against Ankara that is likely to cause resentment and revive memories of Turkey's unhappy historical experience with Russia and Iran's hegemonic aspirations in the region. The early 2020 Russian- and Iranian-backed offensive in Syria's Idlib Province killed approximately sixty Turkish soldiers and drove nearly one million refugees toward the Turkish border.[31]

One wonders if, as Turkish soldiers died in the Russian-backed onslaught, Erdogan, or even those around him, felt at least a twinge of buyer's remorse for those S-400s. In the near term, the United States and Europe should try to avoid a complete rending of the alliance with Turkey while developing relationships outside the AKP, including an effort to reach the Turkish people outside their state-controlled media. Diplomats might help Turkish leaders realize that their long-term interests run counter to those of Russia and Iran. Although the AKP dreams of a post-Western international order, Turkey needs the transatlantic community to overcome its formidable economic challenges. Despite Erdogan's purges of institutions and his assault on freedom of the press and freedom of expression, elections still matter in Turkey. By 2020, the AKP had been in power for eighteen years and was struggling to maintain its hold. It lost the mayoral race in Istanbul in 2019 by an even greater margin in a revote that Erdogan ordered because he was unhappy with the initial results.

Americans and Europeans should try to encourage Turkish leaders to reverse course and reestablish close relations in recognition that Turkey's European and transatlantic relationships are vital to its overcoming the humanitarian, geopolitical, and economic challenges it faces.

⋆ ⋆ ⋆

JIHADIST TERRORIST organizations remain the major destabilizing force in the region, and defeating them will require new thinking and renewed efforts. Our strategies should begin with broad questions, such as what is the identity of the group and how does it fit into the constellation of other terrorist organizations? What are the organization's goals and more specific objectives? What is its strategy? What are its strengths and its vulnerabilities? And finally, how can intelligence, law enforcement, military, financial, informational, cyber, and economic efforts be integrated and isolate the organization from sources of strength and attack vulnerabilities? The failure to ask and answer those questions sometimes leads to rushed actions that are inadequately coordinated and not clearly connected to objectives.

Terrorist organizations need to mobilize support. They become more dangerous when they acquire funds from states, from control of territory, from illicit activities, or from wealthy individuals such as Osama bin Laden. That is why integrated intelligence, law enforcement, and financial actions are important to restrict terrorist organizations' ability to move and access their money. Strategies should ensure that short-term efforts such as intelligence, military, or law enforcement pursuit of terrorist leaders or facilitators are aligned with long-term efforts such as building local intelligence and law enforcement capabilities, supporting educational and economic reform, and expanding communications efforts to discredit terrorist organizations, especially in communities from which they recruit.

Although U.S. government agencies and multinational efforts

have expanded counterterrorism cooperation dramatically since 2001, there are still opportunities for improvement. One person in the U.S. bureaucracy should be responsible for integrating intelligence with all the U.S. and international tools available for use against particular terrorist organizations. The mentality of that person and all persons engaged in the fight must be offensive, due to the unscrupulousness of the enemy and the inability to defend everywhere.

Preventing and preempting attacks at their origin is the most effective approach; terrorists under relentless pressure have to prioritize their own survival over planning and preparing for attacks or developing new, more destructive capabilities.

Despite calls to bring American troops and intelligence and law enforcement officials home from the "wars of 9/11," working with partners overseas reduces the cost and increases the effectiveness of sustained offensive efforts to disrupt jihadist terrorists. For example, African Union troops in Somalia complement the small U.S. force there that relentlessly attacks not only Al-Qaeda's Somali affiliate, Al-Shabaab, but also Al-Qaeda in the Arabian Peninsula (AQAP), based in Yemen. AQAP is determined to attack Americans and U.S. interests abroad. From 2017 to 2019, U.S. operations killed two leaders of AQAP as well as the expert bomb maker involved in at least three plots to destroy U.S. airlines.[32]

Military operations overseas will remain important not only for attacking terrorist leadership when forces have legal authority to do so, but also for denying terrorist organizations safe havens and support bases. Control of territory, populations, and resources was important to ISIS's psychological and physical strength. And ISIS consolidated this control through not only intimidation, but also the establishment of governance, the provision of basic services, and the preemption of internal threats by a sophisticated internal security organization, Amniyat. ISIS generated revenue through a wide range of criminal activities, including illicit trafficking (mainly of oil), theft, and extortion. Jihadist terrorism is a multigenerational threat. Recall that it was in Pakistani safe havens that the "alumni" of the resistance to Soviet occupation in Afghanistan learned the skills and were indoctrinated with the ideology that would later give rise to Al-Qaeda. The threat from ISIS alumni is orders of magnitude larger. Breaking up a would-be caliphate militarily is important for destroying the perception of its invincibility and reducing its ability to attract recruits.

There must also be an unwavering focus to prevent successor groups from emerging. In October 2019, when President Trump announced the withdrawal of U.S. troops from Syria and U.S. Army special operations forces killed ISIS's leader Abu Bakr al-Baghdadi and its spokesman Abu Hassan al-Muhajir, the group was reeling from battlefield defeats. But ISIS endured. The release of some ISIS prisoners due to the unanticipated and uncoordinated withdrawal increased the chances for ISIS's or a successor organization's renewal just as the release of Al-Qaeda prisoners, jailbreaks in Iraq, and Assad's release of jihadist terrorists in Syria were foundational to the resurgence of Al-Qaeda and the rise of ISIS and other terrorist organizations.[33] Isolating terrorists from sources of strength includes ensuring that committed jihadist terrorists are incarcerated under humane conditions until they are no longer a threat.

A long-term counterterrorism strategy must prioritize separating terrorists from the ideology that attracts people to their cause. In 2017 the Trump administration had high hopes for Saudi Arabia as Prince Mohammad bin Salman, also known as MBS, was named Crown Prince and began to exercise power. At thirty-one years old, MBS seemed like a reformer. He advocated for women having the right to work, drive cars, and travel without male chaperones. Even more significant were statements downplaying the importance of Wahhabism and Salafism (puritanical interpretations of Sunni Islam) and admitting to mistakes the Saudi kingdom had made in supporting radical ideologies.[34] Before King Salman switched the planned succession from his nephew Prince Muhammed bin Nayef (MBN), to MBS in May 2017, President Trump made Saudi Arabia his first overseas visit. The Saudis put on a tremendous show of hospitality, playing to the new president's ego and affinity for pageantry. But the trip was substantive as well. President Trump and King Salman oversaw the signing of an arms deal worth $110 billion at signing and $350 billion over ten years, alongside the signing of a number of agreements with American companies worth tens of billions of dollars. The two countries vowed to work with other Gulf states to dry up the funding for not only designated terrorist organizations, but also extremist organizations who brainwash young people and incite hatred of Jews, Christians, and any Muslims who do not adhere to their ideology.

Cooperation with Saudi Arabia was critical because, for nearly five decades, the kingdom had been the principal funder for mosques and schools that systematically extinguished empathy and removed obstacles to using violence against innocents. Indeed, until the group developed its own curriculum, ISIS used Saudi religious textbooks to preach intolerance and hatred of others.[35] Extremist and jihadist ideology is uniquely dangerous as new generations are taught to hate and rationalize the most horrible forms of violence. In 2019, the multinational Financial Action Task Force, an intergovernmental group devoted to fighting money laundering and terrorist financing, found that Saudi Arabia's efforts to combat terrorism financing were meager. During his speech at the 2017 Riyadh Summit, King Salman stated that "we have to stand united to fight the forces of extremism whatever their source." He added that Islam was a "religion of mercy, tolerance and coexistence" and that terrorists were misinterpreting its messages. Hopes rose that Saudi Arabia would become a leader in countering rather than fomenting Islamist ideologies foundational to jihadist terrorism.

It was past time for the United States to demand more from Saudi Arabia and other Gulf states in the effort to isolate terrorist organizations from sources of financial and ideological support. The terrible legacy of the Iranian Revolution and the export of Wahhabi and Salafi jihadist ideology set conditions for the horrors that would follow as Iran exported its revolutionary religious zeal to Shia proxies such as Hezbollah in Lebanon, militias in Iraq, Houthi rebels in Yemen, and a Shia proxy army fighting to keep Assad in power in Syria. Meanwhile, Saudi Arabia and other Gulf Arab States such as the UAE and Qatar, supported Sunni groups in Iraq, Syria, Yemen, and Libya. Support for the more radical elements of both sects had grown over the past four decades.[36] For too long, because of the importance of Saudi Arabia to the global economy, as highlighted during the 1973 oil embargo, U.S. leaders looked the other way as the kingdom aggressively exported a puritanical version of Islam. As President Donald Trump, with First Lady Melania Trump, stood alongside King Salman and Egyptian president Abdel Fattah al-Sisi in Riyadh at the opening of a center to counter Islamist ideology and placed their hands on a glowing orb that illuminated their faces, the odd scene was meant to communicate a common commitment to combating extremist ideology. If the opening at the center really did represent "a clear declaration that Muslim-majority countries must take the lead in combating radicalization" as President Trump stated in his remarks, the shift from supporting extremist ideologies to combating them could begin the most promising effort to separate jihadist terrorist organizations from sources of ideological support.[37]

Hopes rose, but then came disappointments that highlighted limits to the depth of the U.S.-Saudi relationship, especially the absence of common principles that bind the United States to other allies such as Canada, European countries, and Japan. To consolidate power after King Salman announced him as Crown Prince, MBS rounded up potential opponents and imprisoned many of them in Riyadh's Ritz-Carlton, holding them for months. In November 2017, he detained the prime minister of Lebanon, Saad Hariri, because MBS thought him too soft on Hezbollah—an illogical conclusion given that Hezbollah had assassinated the prime minister's father, former prime minister Rafic Hariri. The French secured Hariri's release and eventual reinstatement. Then, only a year later, in October 2018, MBS's ruthlessness and poor judgment became impossible to ignore when the journalist and legal U.S. resident Jamal Khashoggi was murdered in the Saudi Arabian consulate in Istanbul while attempting to obtain the documents he needed to get married. It did not seem possible that the murder—Khashoggi was gruesomely dismembered with a bone saw—was carried out without MBS's direct knowledge. Although Saudi Arabia initially tried to deny that a murder had even taken place, the evidence was overwhelming. The intelligence community unequivocally laid blame at Saudi officials' feet and Congress denounced the killing, but President Trump failed to condemn MBS or impose any meaningful costs on the Saudi regime.[38] Adding insult to injury, the kingdom fostered a cozy relationship with Vladimir Putin even as Russia continued to enable Saudi Arabia's nemesis, Iran, in Syria. King Salman visited Moscow before the Khashoggi murder and committed to purchases of Russian weapons, including the S-400 air defense system. And after the Khashoggi murder, MBS was filmed fist-bumping Vladimir Putin at the G-20 conference in Argentina. MBS's actions highlighted that when allies commit barbarous acts, silence does not purchase fealty.

In the Middle East, partners can be as vexing as adversaries, but sustained engagement and a willingness to sanction human rights abuses or support for extremist and terrorist organizations are foundational to a long-term strategy for the region. That is because the root cause of the Islamist contagion is a split within the world of Islam—between the intolerant, fundamentalist, and purist Salafi view of religion and the competing reformist view that Islam can and should be constantly evaluated and reinterpreted for modern times. The Muslim world will ultimately have to resolve that ideological competition. Our touchstones should be not lecturing or scolding, but rather, articulating and defending (not imposing) our own values and aligning ourselves with those whose views are most consonant with them.

* * *

SOME OF the region's most difficult problems are likely to remain intractable, but without U.S. engagement, they could become unmanageable. One of the most frustrating problems is, of course, the Israeli-Palestinian conflict, which I will touch on briefly here. America's ability to help Israel and the Palestinians progress toward an enduring peace agreement depends on our reputation for honesty and competence, as well as our ability to galvanize regional support and move toward normalization of relations between Israel and all its neighbors. Although American support for its ally Israel has been an element of continuity in U.S. policy since Israel announced its independence in 1948, Israelis' doubts about U.S. reliability grew based on what they perceived as a bad experience during the Obama administration. Trump administration support for Israel, such as approving the long-promised move of the embassy to Jerusalem, support for the permanent annexation of the Golan Heights (strategic terrain that was part of Syria before 1967), and the statement that Israeli West Bank settlements were legal, communicated support for Israel, but also removed incentives that might have been crucial in a future agreement. The February 2020 unveiling of a peace proposal in the White House with Israeli prime minister Benjamin Netanyahu in attendance was dead on arrival due to the lack of Palestinian participation in its development and the persistence of the most difficult obstacles to enduring peace, including Palestinians' claim to right of return to land lost since 1948; the associated fate of Israeli settlements in the West Bank; and the future of East Jerusalem, with the possibility of a portion of that holy and contested city becoming the capital of a future Palestinian state. Despite the proposal's detailed security provisions, many Israelis still fear that a peace settlement would allow security in the West Bank to devolve into the terrorist haven that exists in Gaza, which is controlled by groups committed to the destruction of what they call the "Zionist entity."

But the February 2020 "Peace to Prosperity" proposal may, at some point, help resurrect the possibility of a two-state solution. For the concept of "two states, Israel and Palestine," existing in peace "side by side within secure and recognized borders" to become a reality, the Israeli government and Palestinian Authority must be capable not only of agreeing to a deal, but also of enforcing it across their territories and among their people. The personalization and fragmentation of Israeli politics and the shift of political sentiment to the right, along with the growth of ultra-Orthodox populations, make Israeli approval of even the very favorable Peace to Prosperity vision problematic. And U.S. estrangement from the Palestinian Authority based on the Palestinian leadership's perception that the United States is not a fair interlocutor makes Palestinian assent to meaningful negotiations unlikely. Moreover, the Palestinian Authority's weakness and its vulnerability to radical spoilers such as Hamas are likely to remain insurmountable obstacles. If there is a ray of hope, it may be in efforts to galvanize real support from Arab states to lift the Palestinian populations in Jordan, the West Bank, and eventually Gaza out of poverty and convince the Palestinian people that their best hope for a secure and prosperous future lies in support for leaders who will pursue peace rather than perpetual violence. Breaking the cycle of sectarian violence in the region is connected to the Israeli-Palestinian conflict because the terrorist organizations (and their sponsors) that are parties to the civil wars in the Middle East both inflame and draw strength from the long struggle for control of the Holy Land.

⋆ ⋆ ⋆

THE STRATEGY to break the cycle of violence must include multilayered efforts to pressure and impose costs on those who perpetuate wars and state weakness. America's experience in the Middle East across the first two decades of the twenty-first century revealed lessons that might help guide its approach to the region in coming decades. It is clear that deposing tyrants in the region, whether through foreign invasion such as in Iraq in 2003 or through popular uprising such as during the "Arab Spring" in 2011, does not automatically usher in freedom and enlightened governance. It is also clear that U.S. disengagement from the region can unleash centripetal forces that generate not only more violence and human suffering in the region, but also threats that extend far beyond its geographic confines. The long-term problems of the Middle East require sustainable long-term engagement. The United States and like-minded partners should prioritize actions, initiatives, and programs that, over time, not only break the cycle of violence, but also restore hope through an evolution toward governance based on tolerance rather than hatred, representative government rather than autocracy, and a desire to join rather than reject the modern world. Although the obstacles to achieving peace and prosperity in the Middle East are daunting, there are signs of hope. Nascent reforms in Saudi Arabia and popular demands for an end to corrupt governance in Lebanon and Iraq provide new opportunities for collaboration in education, institutional development, and commerce.

Pursuing long-term opportunities, however, will require efforts to prevent ongoing crises from worsening and new ones from occurring. Supporting fragile states such as Lebanon and Jordan, which have borne the brunt of the humanitarian catastrophe associated with the Syrian Civil War, should remain a top priority. So should efforts to support governmental reform, education, and economic opportunity in Egypt, a country of 98 million people. The most important near-term effort, however, may be to counter Iran's effort to fuel sectarian violence to keep the Arab world perpetually weak, push America out of the region, threaten Israel, and extend Iranian influence to the Mediterranean Sea.

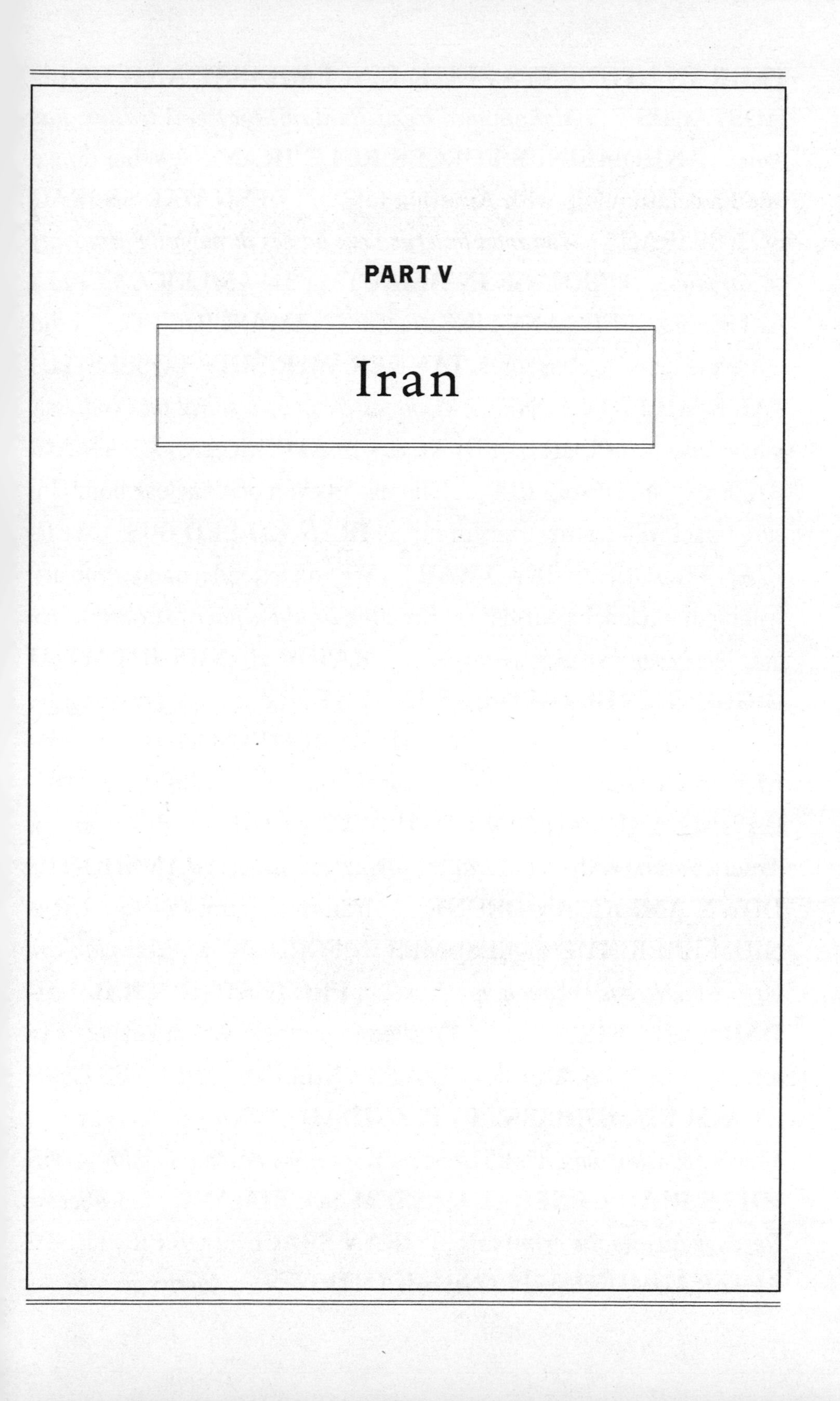

PART V

Iran

TEHRAN STUDENTS SEIZE U.S. EMBASSY AND HOLD HOSTAGES . . . *Mohammad Reza Pahlavi, that evil traitor, has gone* . . . **KHOMEINI'S FORCES RULE IRAN** . . . What do we need a relationship with America for? . . . OPEN WAR AS IRAQ BOMBS IRAN . . . *The road to Jerusalem passes through the Iraqi city of Karbala* . . . **CARNAGE IN BEIRUT** . . . 241 AMERICAN CASUALTIES EXCEEDS ANY SINGLE VIETNAM INCIDENT . . . I did not trade arms for hostages. **TANKER WAR SHIP LOSSES TOTAL 8 MILLION TONS** . . . *Good faith can be a spiral that endlessly moves on* . . . TRUCK BOMB KILLS 19 AMERICANS AT AN AIR BASE IN SAUDI ARABIA . . . The use of even one nuclear bomb inside Israel will destroy everything . . . **IRAN KILLED 608 AMERICAN TROOPS IN IRAQ WAR** . . . We will extend a hand if you are willing to unclench your fist . . . *Even the Iranians have to acknowledge that their economy is in shambles* . . . **MAJOR BANKS HIT WITH BIGGEST CYBERATTACKS IN HISTORY!** . . . this can still be resolved diplomatically . . . **DEAL REACHED ON IRAN NUCLEAR PROGRAM** . . . *America is the number one enemy* . . . **"DECAYING AND ROTTEN DEAL" WITH IRAN** . . . the people of America stand with you . . . *Iran is playing with fire* . . . **IRAN SHOOTS DOWN AMERICAN DRONE** . . . Iran made a mistake . . . URANIUM ENRICHMENT RESUMES IN FORDOW . . . *Hezbollah are terrorists! Here is Lebanon, not Iran!* . . . **PROTESTS ROCK BAGHDAD AND BEIRUT** . . . 1,500 killed in protests by Iranian security forces . . . AMERICAN CONTRACTOR KILLED . . . **PROTESTERS STORM U.S. EMBASSY IN BAGHDAD** . . . *Death to America!* . . . This is not a Warning, it is a Threat . . . *You cannot do a damn thing* . . . **US KILLS IRAN GENERAL QASSEM SOLEIMANI** . . . *A #Severe Revenge awaits the criminals* . . . IRAN BRACES FOR PROTESTS AFTER ADMITTING PLANE SHOOTDOWN . . . *Clerics get lost!* . . .

AZERBAIJAN
ARMENIA
AZER.
TURKEY
TURKMENISTAN
CASPIAN SEA
ĀZARBĀYJĀN-E GHARBI
ĀZARBĀYJĀN-E KHĀVARĪ
ARDEBĪL
Tabrīz
Ardabīl
Orūmīyeh
Rasht
GĪLĀN
Zanjān
ZANJĀN
KORDESTĀN
Sarī
MĀZANDARĀN
Mashhad
Sanandaj
Tehrān
TEHRĀN
Semnān
SEMNĀN
HAMADĀN
KERMĀNSHĀH
Hamadān
Qom
Kermānshāh
MARKAZĪ
Arāk
ISLAMIC REPUBLIC OF IRAN
KHORĀSĀN
Īlām
LORESTĀN
Khorramābād
EŞFAHĀN
Eşfahān
Baghdad
ĪLĀM
AFGHANISTAN
IRAQ
Shahr-e Kord
KHŪZESTĀN
CHAHARMAHĀL VA-BAKHTIYĀRI
Yazd
YAZD
Ahvāz
BOYERAHMAD VA-KOHGILUYEH
Yāsūj
Kermān
KUWAIT
Kuwait
Shīrāz
KERMĀN
Zahedān
FĀRS
PAKISTAN
Badar-e Būshehr
BŪSHEHR
SĪSTĀN VA BALŪCHESTĀN
PERSIAN GULF
HORMOZGĀN
Bandar-e 'Abbās
STRAIT OF HORMUZ
SAUDI ARABIA
Manama
BAHRAIN
OMAN
QATAR
Doha
UNITED ARAB EMIRATES
GULF OF OMAN
Abu Dhabi
OMAN
0 100 200 300 km
0 100 200 mi
44° 52° 56° 60° 64°
36° 32° 28° 24°
48°

CHAPTER 9

A Bad Deal: Iran's Forty-Year Proxy Wars and the Failure of Conciliation

America can't do a damn thing.

—KHOMEINI, 1979, AND KHAMENEI, 2020

THE JCPOA were initials likely to trigger an animated conversation with President Trump. Negotiated in a multinational forum comprising Iran, the five permanent members of the UN Security Council, and Germany, the Joint Comprehensive Plan of Action, better known as the 2015 Iran Nuclear Deal, was, to him, an example of an agreement in which the United States forfeited its bargaining power and gave away too much for too little.

I was sympathetic to the statement candidate Trump made many times during his presidential campaign characterizing the JCPOA as "the worst deal ever." I believed that the JCPOA had both strengthened an adversary and undermined U.S. interests because of fundamental flaws in two areas. First, there were practical problems. In an agreement designed to prevent Iran from threatening other nations with nuclear weapons, the omission of nonnuclear capabilities relevant to deployable nuclear weapons, such as missiles, and the inclusion of a "sunset clause," which would relax and then terminate restrictions on nuclear development after 2025, cut against the deal's purpose.[1] Second, the agreement was divorced from the very nature of an Iranian regime that was fundamentally untrustworthy and hostile to the United States. Inadequacies in the deal and the unwavering hostility of Iranian leaders made it likely that the JCPOA would succeed only in providing cover for a clandestine nuclear program while sanctions relief gave Iranian leaders more resources to fight Iran's proxy wars against the United States, Israel, and Arab states that opposed its ambition to extend its influence across the Middle East. Because monitoring and enforcement mechanisms were far from foolproof, the absence of evidence that Iran was breaking the agreement was not reassuring to those familiar with the regime's long record of hostility and duplicity.

Iranian leaders did not make it difficult for others to discern their intentions. President Obama declared that signing the "strongest nonproliferation agreement ever negotiated" would give the International Atomic Energy Agency (IAEA) "access where necessary, when necessary."[2] But before the ink on the agreement was dry, the spokesman for Iran's Atomic Energy Organization stated that "in the inked roadmap, no permission has been issued for the IAEA's access to any military centers and the nuclear scientists."[3] Iran's leaders constantly contradicted their diplomats, which should have raised doubts about trusting them to uphold any of the deal's stipulations.

The JCPOA was an extreme case of strategic narcissism based on wishful thinking—wishful thinking that led to self-delusion and, ultimately, the deception of the American people. It was wishful thinking to trust an openly hostile regime to adhere not only to the letter but also to the spirit of the deal. It was self-delusion to indulge in the conceit that lifting sanctions against Iran might evolve the regime into a responsible nation that no longer supported terrorist organizations. And it was deceptive to portray the flawed deal as unrelated to Iran's destructive behavior in the region and as the only alternative to war.

Far from persuading the Iranian regime to abandon its sponsorship of militias and terrorist organizations, the JCPOA had the opposite effect. The deal gave the regime a cash payment of $1.7 billion up front and a subsequent payout of approximately $100 billion in unfrozen assets.[4] Even more cash flowed to the regime after sanctions relief. Iran used the windfall to intensify its proxy wars and expand sectarian conflicts in the region.[5] In the words of former U.S. Central Command commander Gen. Joseph Votel, Iran grew "more aggressive in the days [after] the agreement."[6]

Although President Trump was eager to get out of the "terrible deal," I wanted to give him comprehensive options to address the broad range of challenges that the Iranian regime presented to U.S. security and prosperity. And it was important to consider the effects of pulling out. If other nations regarded the U.S. departure from the deal as unjustified, for example, efforts to impose costs on the regime might be diluted. Although the reimposition of U.S. sanctions alone would deliver a significant financial blow, we might stay in the deal and sanction Iran for behavior not covered in it, such as missile development and support for terrorists. Staying in the JCPOA despite common knowledge of President Trump's inclination to get out would create leverage for us to isolate the regime diplomatically as well as economically. The president might use that leverage to get others to support fixing the deal's flaws, insisting on robust inspections, and applying additional sanctions. Why not see how much we might accomplish before pulling out?

I was also concerned that pulling out could put the United States on the defensive and divert attention from the Iranian regime's criminality and brutality. Iran's smooth-talking, Western-educated foreign minister, Mohammad Javad Zarif, would undoubtedly attempt to portray Iran as a victim of a U.S. president whom some foreign leaders, especially in Europe, regarded as brash and impulsive. Corrupt Iranian clerics and officials would use the U.S. pullout to shift responsibility for Iran's failing economy away from themselves and toward the United States. Conversely, applying sanctions for nefarious Iranian activities abroad while staying in the flawed agreement might make it clear to the Iranian people that their leaders were the true authors of their problems, as they were squandering the nation's potential wealth on violence and destruction abroad.

It was for those reasons that I asked the president to give his cabinet time to develop options for an overall Iran strategy into which the JCPOA decisions fit, instead of viewing "stay or get out" in isolation. Toward that end, I asked our weapons of mass destruction senior director, Andrea Hall, and the senior director for Middle East affairs, Michael Bell, to intensify their work to frame the problems associated with Iran. They worked with Dina Powell, deputy national security advisor for strategy; Brian Hook, the director of policy planning at the Department of State, and all relevant departments and agencies to identify the challenges associated with the Islamic Republic and draft goals, objectives, and assumptions as the foundation for a fresh Iran strategy. In May 2017, I convened a Principals Committee meeting, composed of the relevant members of the cabinet, to review our collective framing of Iran's challenge to our national security and provide direction on the development of options. After the meeting, the president approved our assessment. All agreed that the fundamental problem was the Iranian regime's permanent hostility to the United States, Israel, its Arab neighbors, and the West.

But the president was impatient. And what made him more impatient was domestic legislation that the Republican-majority House of Representatives and Senate had passed in 2015, meant to force an expected Hillary Clinton administration to publicly and serially confront flaws in the nuclear deal. The Iran Nuclear Agreement Review Act of 2015 (INARA) required the administration to certify to Congress every ninety days that the agreement "meets United States non-proliferation objectives, does not jeopardize the common defense and security," and ensures that Iran's nuclear activities will not "constitute an unreasonable risk . . ."[7] It was a tall order, and one that cut directly against the president's assessment of the JCPOA.

The first INARA deadline arrived in April 2017, less than two months after I started as national security advisor. When I discovered that the secretary of state intended to send a perfunctory letter certifying that Iran was in compliance, I knew that the president would be incensed. Our team worked with the State Department and other agencies to propose alternative language that certified under the INARA legislation, but also stated clearly that Iranian behavior in the region, enabled by sanctions relief under the nuclear deal, threatened "common defense and security." Secretary of State Rex Tillerson, who at times seemed reflexively opposed to suggestions from the White House, rejected that option and sent the terse certification letter. After the president reacted as anticipated, I joined Secretary Tillerson in the Oval Office for a discussion. Tillerson subsequently amended the letter, noting that the nuclear deal "fails to achieve the objective of a non-nuclear Iran. It only delays their goal of becoming a nuclear state." He stated further that the Iran deal "represents the same failed approach of the past that brought us to the current imminent threat that we face from North Korea," telling reporters that "The evidence is clear: Iran's provocative actions threaten the United States, the region and the world."[8]

In anticipation of the two remaining INARA certifications in 2017, one in July and another in October, I asked our team to accelerate work on the Iran strategy. We referred to the series of discussions prior to those deadlines as the "gift that keeps on giving." It was our job to give the president options and, once he chose a course of action, to assist with the sensible execution of his decision. At each deadline, we were prepared to either get out or stay in—we had two sets of diplomatic and communications responses ready that reflected each option.[9] Though the decision changed after nearly every conversation, in the end, the president was persuaded to stay in the deal while asking other nations to join us in the imposition of non-JCPOA-related sanctions and help fix the deal's flaws, an endeavor that ultimately proved unfruitful. In July, simultaneous with the State Department's certification that Iran was conforming to the letter of the agreement, Treasury secretary Stephen Mnuchin announced sanctions on eighteen Iranian entities that supported terrorist organizations.[10]

Still, this action was not enough to placate critics who saw pulling out of the deal in much the same way that Obama administration officials had viewed signing it in the first place: as an end in and of itself. Partly in response to those critics, President Trump told the *Wall Street Journal* that he would be "surprised" if Iran were found compliant in ninety days.[11] Unfortunately, public discussion on Iran continued to focus almost exclusively on whether the United States would stay in or get out of the deal, rather than on the broad range of Iranian actions that undermined peace and security in the Middle East and beyond. Few people, either inside or outside the government, articulated how the decision would fit into an overall strategy not only to block Iran from developing nuclear weapons, but also to end its proxy wars.

I asked our National Security Council staff to develop Iran strategy options for presentation to the president prior to the next INARA deadline, in October. The president approved the new Iran strategy in early September, and we started work on a presidential speech meant to inform the American people and international audiences of the Iranian threat and our strategy to protect U.S. and allies' interests. Another painful conversation on INARA certification was impending, but the president could finally consider options in the context of a comprehensive approach to Iran.

In October, in addition to the binary choice of past certifications, we provided the president with a third choice on INARA: to decline to certify that the deal was in the national interest, but stay in the agreement conditionally as a way to incentivize other nations to address the deal's fundamental flaws and join the United States in sanctioning Iran's continued support for terrorists and militias. He approved that option. In his speech, Trump announced that "despite my strong inclination, I have not yet withdrawn the United States from the Iran nuclear deal. Instead, I have outlined two possible paths forward: either fix the deal's disastrous flaws, or the United States will withdraw."[12]

In retrospect, tying the speech on Iran strategy to the INARA decision was a mistake. Press coverage focused almost exclusively on that narrow issue and skipped over the significant shift in Iran strategy. We were running out of time to show how staying in the JCPOA, despite its flaws, was the best way to accomplish the new strategy's objectives. Mike Bell, Joel Rayburn, and Brian Hook, who would later become the State Department's lead on Iran policy, traveled to Europe to ask allies to support our efforts and keep the international conversation, and pressure, on Iran. Working with the European signatories, we attempted to marshal support for addressing missile development and restricting Iran's uranium enrichment permanently, rather than allowing those restrictions to expire in 2025 under the agreement, but I knew it was a long shot. In his last certification of the deal under INARA, the president said, "I am waiving the application of certain nuclear sanctions, but only in order to secure our European allies' agreement to fix the terrible flaws of the Iran nuclear deal. This is a last chance. In the absence of such an agreement, the United States will not again waive sanctions in order to stay in the Iran nuclear deal. And if at any time I judge that such an agreement is not within reach, I will withdraw from the deal immediately."[13]

We had created a window of opportunity for our allies to demonstrate the viability of staying in the deal while imposing costs on Iran for its destructive behavior in the Middle East. That window closed soon after I departed the White House. My last day as assistant to the president for national security affairs was April 9, 2018. A month later, the president withdrew from the JCPOA. The international reaction was as predicted: the conversation shifted from condemnation of Iran to exasperation with the United States. The following year, President Trump announced his intention to designate the IRGC as a foreign terrorist organization, recognizing that "Iran is not only a State Sponsor of Terrorism, but that the IRGC actively participates in, finances, and promotes terrorism as a tool of statecraft."[14] Israel and the Gulf states, those countries suffering directly from Iran's proxy wars, were supportive. Though the response from European allies was initially negative, I was confident they would recognize the importance of sanctioning Iran. It was only a matter of time before Iranian aggression clarified that the Iranian regime was the real problem—we could count on the mullahs in Tehran to demonstrate their hostility to the West.[15]

★ ★ ★

BY SUMMER 2019, Iran felt the pressure of the reimposed sanctions. The economy was contracting even faster than in 2018, when GDP fell from 3.7 percent growth per annum to 3.9 percent *contraction*. Crude oil exports, 2.3 million barrels a day in 2018, fell to 1.1 million barrels a day by March 2019. Inflation rose from 9 to 40 percent. Iranian leaders faced three fundamental choices. First, they could attempt to wait out President Donald Trump and work with other countries to avoid the sanctions. But the economic pressure was significant, and faced with doing business either with the United States or Iran, companies and investors were unsurprisingly choosing the United States. European efforts to circumvent the U.S. financial system in trade and investment were insufficient.[16] Second, the regime could enter into talks with the United States and other nations to renegotiate aspects of the deal and address their support for terrorist organizations and militias in exchange for sanctions alleviation. But the revolutionaries in Iran, particularly Supreme Leader Ali Khamenei and the IRGC, were not predisposed to conciliation. Lastly, they could increase adversarial activity against the United States, Europe, and Gulf states while violating the terms of the agreement to extort the United States and others to relieve sanctions.

Their choice became obvious on June 12. Prime Minister Abe was the first Japanese leader to visit Tehran in four decades. He met with Iranian president Hassan Rouhani in Sa'dabad Palace. Japan wanted to avoid an interruption of or reduction in the flow of oil from the Persian Gulf because, of all the industrialized nations, it had the weakest domestic production relative to its needs. After the sharp reduction in nuclear power generation following the Fukushima nuclear disaster in 2011, Japan's need for cheap oil grew. After reviewing a military honor guard, Abe and Rouhani sat on gold-framed furniture in a private meeting room decorated with flowers and the Iranian and Japanese flags. "If we witness some tensions, the root is the U.S. economic war against the Iranian nation. Any time this war stops, we will witness very positive developments in the region and the world," Mr. Rouhani said. It was the beginning of an attempt at extortion. During the subsequent meeting with the Supreme Leader, Abe delivered a message from President Trump. Khamenei refused to respond.[17]

As Rouhani and Khamenei hosted Abe, the Islamic Revolutionary Guard Corps Navy was tracking the movement of the Japanese oil tanker *Kokuka Courageous* in the Gulf of Oman on its way to the Indian Ocean. Just hours before Abe was to meet with Supreme Leader Khamenei, IRGC speedboats approached the tanker under cover of darkness and affixed limpet mines to it. After the boats sped away, an explosion ripped through the starboard side of the tanker, sending a shock wave toward the bridge and blowing a 1.5-meter-wide hole in the aft. Oil spilled from the hull into the ocean, but the ship's compartments limited the damage. Realizing that not all the mines exploded, the speedboats returned. By then, the crew of *Kokuka Courageous* had evacuated and the U.S. Navy had the tanker under surveillance, and a navy aircraft recorded IRGC operatives removing the unexploded mines.[18] The Japanese tanker was the second ship attacked that morning, as a limpet mine had damaged the Norwegian tanker *Front Altair* just an hour earlier. The IRGC clearly had timed the operations as an affront not only to Prime Minister Abe, but also to anyone who intended to bring a message of conciliation to the Islamic Republic. The attacks served as yet another corrective to those who preferred to separate negotiations with Iran from the nature of the regime and the ideology that drives its aggressive behavior.

As it became clear that Iran had chosen to escalate, the United States announced the deployment of additional military forces to the region. Less than a week after the tanker attacks, an Iranian missile shot down a remotely piloted U.S. surveillance aircraft over international waters. The United States was on the brink of retaliation until President Trump halted the planned strikes due to the estimated loss of Iranian life and his belief that such a response would be disproportionate to the provocation. While some applauded the decision as a way to play a longer game of diplomatic, economic, and financial pressure in which the United States had the advantage, the lack of a response emboldened Iranian leaders. It seemed as if President Trump were trying to give the Iranians an out. Trump told reporters that he found "it hard to believe it was intentional if you want to know the truth . . . I have a feeling . . . that it was a mistake by somebody."[19] The president's comments may have been well meaning but they were symptomatic of the tendency of American leaders to view the latest act of aggression in isolation, rather than in the context of a four-decade-long proxy war Iran continues to wage against us.

* * *

U.S. POLICY toward Iran across six U.S. administrations has suffered from a lack of strategic empathy and a failure to understand how historical memory, emotion, and ideology drive the Iranian regime's behavior. An effective Iran strategy requires strategic empathy, and that means rejecting both the flawed assumptions that underpinned a bad nuclear agreement and the ineffective, inconsistent strategies we have been employing to counter Iranian hostility since 1979.

Similar to the long-standing assumption that China's prosperity would lead to liberalization of its economy and government, President Obama hoped that "seeing the benefits of sanctions relief" would convince Iran to focus "more on the economy and its people." Obama's deputy national security advisor, Ben Rhodes, who promoted the deal to the American people based on a false choice between the JCPOA and war, suggested that the deal would cause "an evolution in Iranian behavior" as Iran became "more engaged with the international community."[20] The JCPOA was not the first case of American leaders believing that conciliatory actions, such as sanctions relief, would moderate Iranian leaders' behavior or cause them to prioritize interests over passion and ideology. Hope for warming relations led some to disregard the tendency of Iranian leaders, since the 1979 revolution, to engage with the United States when fearful or under duress. Such engagements used a veneer of sincerity to mask the leaders' true intentions of either avoiding consequences for terrorist acts or garnering more resources to fund their destructive operations.

In 1979, for example, President Jimmy Carter's administration did not recognize how deeply anti-Western sentiment drove the revolutionaries in Iran. Hoping to develop a relationship with Ayatollah Ruhollah Khomeini and preserve Iran as a Cold War bulwark against the Soviet Union, Carter administration officials closed their ears to the anti-American cheers of the revolution and averted their eyes from the reign of terror that the Ayatollah was inflicting on his people. While visiting Algiers on November 1, 1979, National Security Advisor Zbigniew Brzezinski sought out Iranian prime minister Mehdi Bazargan at a reception to tell him that the United States was open to a relationship with the new Islamic Republic. Iranian newspapers published photos of the two men shaking hands, which appeared alongside news that Mohammad Reza Pahlavi, the deposed Shah, had been admitted to the United States for medical treatment. Iranian revolutionaries put the two pieces of news together and assumed that the CIA and U.S. military were preparing to return the Shah to power. The Iranians in Algiers immediately ended the talks as outraged students in Tehran seized the U.S. embassy and took fifty-two Americans hostage. It was the beginning of a 444-day crisis that would dominate the rest of Jimmy Carter's presidency. On January 20, 1981, minutes after Ronald Reagan was sworn in as president, the Iranian government released the hostages, a supposed conciliatory action of goodwill.[21] As a first-year cadet at West Point, I was part of a cordon of cheering cadets who lined Thayer Road to welcome the hostages back to American soil as six green-and-white army buses took them through the campus on the way to a three-day respite with their families at the Hotel Thayer.

The regime had released the hostages under duress. Iraq's invasion of Iran in September 1980 had increased the cost of Iran's diplomatic and economic isolation. The first stages of the destructive war depleted Iranian weapons and munitions stocks. Because the United States built the Iranian military when the two nations were allies, Iranian officials had no option but to turn for assistance to the nation they called "the Great Satan."

During Ronald Reagan's second term as president, and just two years after the October 1983 Iranian-sponsored attack against a marine barracks in Lebanon that killed 241 servicemen, U.S. officials offered missiles in exchange for the release of U.S. hostages taken in Lebanon. After the Iranians got the munitions they wanted, an Iran-backed terrorist group in Lebanon took three more Americans hostage. Radical revolutionaries in Iran exposed the embarrassing arms-for-hostages scandal, miring the Reagan administration in controversy for its final two years. [22]

Despite having been tarnished by this scandal as vice president, President George H. W. Bush had high hopes for engaging Iran and, in particular, for gaining the release of nine American hostages held in Lebanon. The new U.S. president offered an olive branch in his inaugural address, and stated that Iranian assistance in releasing the hostages could transform the relationship: "Goodwill begets goodwill. Good faith can be a spiral that endlessly moves on."[23] The Bush administration requested that a UN intermediary travel to Tehran to test Iranian willingness to engage.

The cycle of rising expectations and dashed hopes continued. In 1989, after Ayatollah Khomeini died, Ayatollah Ali Khamenei became Supreme Leader and Akbar Hashemi Rafsanjani, who had negotiated the arms-for-hostages deal, became President. Rafsanjani, who was a businessman and politician as well as a cleric, forged a strong relationship with the *bazaar*, or Iranian mercantile class. The war had made a shambles of the Iranian economy. Infrastructure was damaged and dilapidated. Oil production suffered. The Bush administration offered more than words to explore improved relations, releasing the $567 million frozen by Washington after the Tehran embassy attack in 1979.[24] But Rafsanjani had neither the power nor the inclination to dispense goodwill. He and the merchant class were far weaker than Khamenei and his conservative clerics, security service, and IRGC allies.

All the while, the IRGC was supplying terrorist cells in Europe with weapons to attack their political enemies and Western interests. In 1989, Iranian agents murdered prominent Kurdish-Iranian resistance leader Abdul Rahman Ghassemlou in Vienna; that same year, Khomeini issued a *fatwa* (a judgment rendered by an Islamic jurist) directing the assassination of the novelist Salman Rushdie, whose book *The Satanic Verses* contained passages that the Supreme Leader deemed blasphemous; the next year, Iranian "diplomats" shot Kazem Rajavi, brother of People's Mujahedin of Iran cofounder Massoud Rajavi, in Geneva; in 1991, Iranian assassins killed the Shah's last prime minister, Shapour Bakhtiar, in Paris after failing a decade earlier; and in 1992, Iranian agents murdered three prominent Kurdish Iranian leaders and their interpreter in a Greek restaurant in Berlin.[25]

As President Bush offered his olive branch and Europe expanded economic relations with Iran, the Iranian-supported Lebanese terrorist group Hezbollah was going global. Worldwide attacks included a 1989 failed bombing in London in an attempt to assassinate Rushdie, a 1992 bombing of the Israeli embassy in Argentina that killed twenty-nine people, a 1994 bombing of a Jewish community center in Argentina that killed eighty-five people,[26] and the bombing of Alas Chiricanas Flight 901 on its way from Colón to Panama City, Panama, that killed all twenty-one of its passengers. In 1996, Hezbollah bombed the Khobar Towers complex in Saudi Arabia, killing nineteen U.S. Air Force personnel. Rafsanjani's successor, Mohammed Khatami, categorically denied, as Iranian leaders always do, that Iran supports terrorist operations overseas. He held out hope for reform in Iran, describing an internal political competition in which "one political tendency" that "firmly believes in the prevalence of logic and the rule of law" grapples with "another tendency that believes it is entitled to go beyond the law."[27] Maybe this new Iranian president who called for a "dialogue between civilizations" could end Iran's use of terrorism. Once Iran's responsibility for the Khobar Towers bombing became clear, the possibility of improved relations convinced Bill Clinton to forgo retaliation.

Strategic narcissism endured, in the form of continuing reluctance to confront Iranian aggression. As the George W. Bush administration commenced at the beginning of 2001, there was still hope for improved relations based on the perceived strength of moderates and reformers inside the Islamic Republic. The tragedy of the 9/11 attacks seemed to present an opportunity to work together against common enemies such as Al-Qaeda and the Taliban. After the U.S. invasion of Afghanistan, Iranian and U.S. diplomats discussed the formation of the new Afghan government, but that cooperation was limited and short-lived.[28] In his January 2002 State of the Union address, President Bush included Iran, along with Iraq and North Korea, as part of an "axis of evil." The Iranians suspended diplomatic contacts. Even more significant than the offense Iran took from the speech, the combination of Iranian economic weakness, growing international awareness of Iran's nuclear weapons program, and Iran's intensification of its proxy wars would shape the next phase of the contentious U.S.-Iranian relationship.

So-called moderates in Iran were moderate mainly in American and Western imaginations, but rarely at home. In December 2001, former president Rafsanjani, the man who had served as the vessel for Western dreams of Iranian moderation prior to Khatami, spoke from the podium at Tehran University to deliver the government's official weekly sermon. He declared, "If one day, the Islamic world is also equipped with weapons like those that Israel possesses now, then the imperialists' strategy will reach a standstill because the use of even one nuclear bomb inside Israel will destroy everything." In August 2002, an Iranian exile group revealed the existence of a secret facility in Natanz capable of enriching uranium for use in nuclear weapons as well as civilian nuclear power reactors.[29] The Iranian bomb was meant to be the ultimate weapon in the Islamic Republic's proxy wars to push the United States out of the Middle East, dominate its Arab neighbors, and destroy Israel.

Even as it initiated programs to encourage a change in the nature of the Iranian regime through Persian-language broadcasts and support to civil society groups, the Bush administration pursued cooperation with Iran against Al-Qaeda, an organization that seemed to hate Iranian Shia Muslims as much as American Christians and Jews. In 2003, after the United States and British militaries accomplished in weeks what the Iranian military could not accomplish in eight years, deposing Saddam, the Iranians engaged in discussions, fearful that they might be next on the Bush administration's regime change agenda. But instead of cooperating with the United States against what seemed to be a common enemy, the IRGC and Iran's security apparatus gave Al-Qaeda leaders safe haven and helped them target the United States and Arab monarchies.[30] Iran also used Al-Qaeda to jump-start a sectarian civil war in Iraq that allowed Iran to settle old scores with former Baathists, build powerful proxy forces, and infiltrate its agents into Iraqi institutions. The U.S. post-invasion failure to consolidate gains in Iraq opened the door for the IRGC and the MOIS, the Islamic Republic's domestic and foreign spy service, comprising many of the Shah's former SAVAK secret police). Iranian operatives and intelligence agents moved freely across unguarded borders. As the United States and Coalition forces struggled with a growing insurgency, Iranian fear of America's conventional military prowess dissipated. The IRGC and their allied militias in Iraq added American soldiers to their list of targets as they intensified their proxy war against the United States.

These militias began killing and maiming American servicemen and women with Iranian-manufactured roadside bombs called explosively formed penetrators (EFPs). EFPs are as lethal as they are simple. A metal or PVC pipe packed with explosives and capped with a curved copper or steel disc is detonated. The explosion transforms the disc into a high-velocity molten slug capable of penetrating vehicles' armor protection.[31] EFPs required precision manufacturing in Iran. To transport them to battlefields abroad, the regime developed complex and innovative smuggling networks and techniques.

But Washington was slow to respond to Iran's escalation of the war, despite the urging of some civilian and military officials to confront Iranian aggression. Similar to the Obama and Trump administrations' self-delusion that the Taliban was separate from Al-Qaeda in Afghanistan and Pakistan, the Bush administration indulged an implausible theory that Iran's leaders might simply be unaware that their agents in Iraq were killing hundreds of American soldiers. When asked about the irrefutable evidence that the deadly EFPs were coming from Iran, President Bush said, "What we don't know is whether or not the head leaders of Iran ordered the Qods [Quds] Force to do what they did."[32] Two days earlier, Chairman of the Joint Chiefs of Staff Peter Pace stated that the weapons shipments from Iran did not mean that "the Iranian government per se, for sure, is directly involved in doing this." I read those statements when I was in Iraq as commander of the Third Armored Cavalry Regiment. It was not ambiguous to me or to our troopers; the preponderance of our casualties in the area south of Baghdad was due to Iranian-made EFPs provided through a network run by the Revolutionary Guard Quds Force. Those who conducted the attacks were trained and directed by the IRGC Quds Force. It was implausible that Iranian leaders were not responsible for killing more than six hundred U.S. soldiers, over 17 percent of all U.S. deaths in Iraq from 2003 to 2011.[33] Conciliatory action, even to the point of developing and promoting a cover story for Teheran, led neither to a reduction in Iran's destructive activity nor to a stronger position for reformers. Instead, the lack of a strong response emboldened the revolutionaries.

The years 2005 to 2013 were ones of confrontation under President Mahmoud Ahmadinejad. He and the conservatives were ascendant, as were oil revenues, and the regime intensified not only rhetoric, but also actions against Israel, the United States, Saudi Arabia, and the United Kingdom. On July 12, 2006, Hezbollah kidnapped two Israeli soldiers and ignited a war that went far beyond what Hezbollah secretary-general Hassan Nasrallah expected. Twelve hundred Lebanese, including more than 270 Hezbollah fighters, died along with 158 Israelis. In the wake of the war, Ahmadinejad vastly increased support for Hezbollah and for the Palestinian terrorist groups Hamas and Palestinian Islamic Jihad. On January 20, 2007, Tehran's proxies in Iraq became more audacious in direct action against U.S. forces. Qais al-Khazali led the militant group Asa'ib Ahl al-Haq in an attack on the Provincial Joint Coordination Center in Karbala. The militants wore U.S. uniforms to sneak past Iraqi guards. They killed one U.S. soldier and took four more hostage, all of whom were later murdered in cold blood. The Quds Force even planned a bold assassination and terrorist attack in the United States. On October 11, 2011, U.S. government officials foiled an assassination attempt on the Saudi ambassador to the United States, Adel al-Jubeir. Their planned attack at the ambassador's favorite restaurant in Washington would have killed many bystanders. A month and a half later, on November 29, 2011, Iranian protestors stormed and overwhelmed the British embassy in Tehran, chanting "Death to England" and ransacking the premises and its sensitive contents. This unrest came after Great Britain announced new sanctions on the Iranian regime, and the movements appeared to have been state sponsored.[34]

But Iran was paying a price for the intensification of its proxy wars and its pursuit of nuclear weapons. In Iraq, despite the reluctance in Washington to attribute attacks to leaders in Teheran, the U.S. military responded to the mounting violence from Shia militias. In January 2007, U.S. special operations forces raided the Iranian consulate in Erbil, which the United States suspected of being an IRGC base. Two months later, in March, Coalition special operations forces also attacked a terrorist cell in Basra that had been responsible for the deaths of the five U.S. soldiers in the January attack in Karbala. Among the captured were al-Khazali, the leader of that attack; his brother; and Mullah Ali Mussa Daqduq, a Lebanese Hezbollah advisor who had been working with the Iranians to create an Iraqi version of Lebanese Hezbollah. The Coalition also supported Prime Minister Nouri al-Maliki's 2007 "Charge of the Knights" offensive against the Mahdi Army in southern Iraq after Maliki uncovered a plot to replace him with someone who would be completely beholden to

Iran, such as Ibrahim al-Jaafari or Ahmed Chalabi. The U.S. military's heightened response caught the Shia militias and the IRGC by surprise. Quds Force commander Qasem Soleimani was particularly concerned with the capture of five Quds Force officers from the Erbil consulate, and fearing similar future incidents, he scaled back IRGC operations and personnel in Iraq.[35]

Beginning in 2005, the Bush administration and European allies increased pressure on Iran in the form of economic sanctions and reportedly clandestine operations in response to Iran's pursuit of nuclear weapons.[36] Pressure continued in the Obama administration, and beginning in 2012, the Iranian economy contracted severely. Then, in 2014, oil prices fell, compounding the effect of tightening sanctions and bringing Iran's economy to the brink of collapse.[37] Iran was losing externally as well as internally. Iranian ally Bashar al-Assad seemed to be a dead man walking in Syria.

Iran needed a way out. In 2013, new leaders President Hassan Rouhani and Foreign Minister Javad Zarif began a charm offensive, even feigning a friendlier approach to Israel. In contrast to President Ahmadinejad's denial of the Holocaust, Foreign Minister Zarif described the Nazi genocidal campaign against the Jews as a "horrifying tragedy" and even suggested that if Israel and the Palestinians reached a peace agreement, Tehran might recognize Israel.[38] Once again, Western leaders were ready to believe that Iran might really, this time, moderate its behavior in response to a conciliatory gesture.

But Iran was intensifying its proxy wars with increased support for the Syrian regime, Hezbollah, Iraqi militias, and the Houthi militia in Yemen. In February 2014, for example, Iran sent hundreds of "military specialists," Quds Force commanders, and IRGC troops to Syria to boost the Assad regime. The Fatemiyoun Division, an Afghan Shia militia established and commanded by the IRGC, grew to an estimated twenty thousand fighters. Over the course of 2014, Iran also quietly increased its support for the Taliban in Afghanistan, both materially and through manpower.[39] Meanwhile, the mullahs were advancing the nuclear program. As Iranian leaders negotiated with Western diplomats, Supreme

Leader Khamenei announced that Iran would pursue 190,000 centrifuges rather than the 10,000 that negotiators were discussing. Nonetheless, the Obama administration, like those before it, chose to alleviate the pressure on the Iranian regime based on hopes of what conciliation might bring. When Iranian protesters reached out to Western democracies during the 2009 Green Movement, the administration issued a tepid statement to avoid upsetting the regime and foreclosing on the possibility of improved relations. The decision not to enforce the "red line" in Syria against the use of chemical weapons to murder civilians was, in part, a concession to Tehran. The administration even portrayed the Islamic Republic of Iran as a partner in the effort to remove chemical weapons from the Syrian regime's arsenal.[40]

The administration's high hopes for the nascent Iranian Nuclear Deal led it to scale back what had been a promising effort to constrain Iran's aggression. From 2008 to 2016, Project Cassandra disrupted Iran's ability to fund its proxies abroad, including Lebanese Hezbollah's international terrorist network. But as Treasury official Katherine Bauer later recalled, "the investigations were tamped down for fear of rocking the boat with Iran and jeopardizing the nuclear deal."[41]

And once the deal went into effect, the Obama administration was determined to avoid confrontations that might undo the agreement. As American money flowed into Iran and Iranian exports tripled, funding for terrorist organizations and IRGC operations across the region soared. Hezbollah received an additional $700 million per year; another $100 million went to various Palestinian militant and terrorist groups. The JCPOA strengthened the Iranian regime psychologically as well as financially. In contrast to the language in the deal's preamble stipulating that signatories would "implement this JCPOA in good faith and in a constructive atmosphere" and "refrain from any action inconsistent with the letter, spirit, intent" of the agreement, the IRGC intensified operations in Syria, Iraq, Lebanon, Yemen, and eastern Saudi Arabia. For example, in October 2015, only months after the signing of the JCPOA, hundreds of Iranian troops arrived in Syria over a ten-day period to bolster offensive operations in Idlib and Hama. The IRGC also continued a series of ballistic missile tests in violation of UN Security Council resolutions, testing fourteen missiles from the signing of the agreement to February 2017, including a long-range ballistic missile under the guise of a satellite launch. Although a number of those launches failed, the Iranians were improving. In response to a June 2017 terrorist attack in Tehran, and to demonstrate its new capabilities, Iran launched six missiles from its territory over Iraq to strike an ISIS-controlled area in Dayr al-Zawr, Syria.[42]

The Obama administration took conciliation with Iran to a new level. Just prior to Iran signing the agreement in the summer of 2015, the U.S. State Department flew pallets of euros and Swiss francs into Geneva, where trams loaded them onto Iranian cargo planes headed for Tehran. That same day, Iran released four Americans who had been, in effect, hostages. It was an operation reminiscent of the arms-for-hostages arrangement under the Reagan administration. Iran's leaders regarded the thinly veiled cash-for-hostages payment as a sign of weakness rather than the metaphorical "outstretched hand" of conciliation that President Obama offered in his June 2009 speech in Cairo. The lie that the cash payment and the hostage release were disconnected encouraged Iran's long practice of using hostages for coercion to extort favorable terms, and the revolutionaries in Tehran portrayed the ransom payment as an admission of American guilt and weakness. Hossein Nejat, deputy intelligence director of the IRGC, stated that ransom payments demonstrated that "the Americans themselves say they have no power to attack Iran."[43] In the months that followed the payoff, in addition to multiple missile launches, the regime boasted about its nuclear stockpiles, awarded a medal to an IRGC commander with American blood on his hands, and seized two U.S. Navy vessels, arresting ten sailors and parading them in front of cameras before releasing them fifteen hours later. Iran even failed to refrain from hostage taking, detaining Princeton graduate student Xiyue Wang in 2016 while he was conducting research on the Qajar dynasty and learning Farsi for a PhD in Eurasian history. As in the past, conciliation had led to Iranian escalation, not moderation.

* * *

DESPITE THE United States' repeated hopes for moderate leaders and reformist government, Iran's internal political dynamics make real reform highly unlikely. Modest efforts over the years to liberalize the press, boost the economy, and reduce extrajudicial killings have typically been stifled by a coalition of conservatives and Supreme Leader Khamenei.

Despite their veneer of piety, the conservative mullahs who dominate the Guardian Council and exert influence across the government maintain their power through moral and financial corruption. Much of this is done through the use of *bonyads*, religious foundations that provide cover for extensive patronage networks from which the ayatollahs and government officials profit. *Bonyads* control businesses, receive government contracts, launder money, operate without any external audits, and pay no taxes. The Supreme Leader appoints the heads of the *bonyads*. Many are children of influential mullahs. The regime's largest *bonyad*, Astan Quds Razavi, controls more than one hundred businesses in diverse fields ranging from car manufacturing to agriculture to oil and gas to financial services.

Corruption extends beyond the clerical order into the security services and Revolutionary Guard. The IRGC profits from smuggling contraband and trafficking drugs. Corrupt networks also stifle political reforms to maintain their control of the government and the economy. For example, the constitutionally mandated Guardian Council, a body of clerics and lawyers, rigged the Majlis parliamentary elections in 2004 and marginalized reformers to ensure that the populist mayor of Tehran, Mahmoud Ahmadinejad, defeated former president Rafsanjani in the 2005 presidential election.[44] The most notorious theft of an election was the presidential election in 2009. Most often, the Guardian

Council simply excluded reformist candidates from elections as it did in February 2020 when it denied over seven thousand potential Majlis candidates the opportunity to run for office.[45]

Paradoxically, the West's conciliatory policies have often reinforced the revolutionaries' efforts to stifle reform. For example, in 2004 and 2005, European Union negotiators overlooked the election irregularities that led to conservative victories in the Majlis and Ahmadinejad's election. Avoiding confrontation, they thought, might lead to cooperation on Iran's nuclear program. Instead, Ahmadinejad ushered in a period of aggression abroad while the revolutionaries continued to consolidate power internally.

Aggressive actions abroad derive from and depend on the revolutionaries' strong grip on power domestically. High oil revenues allowed Ahmadinejad to empower and enrich the IRGC expeditionary organization responsible for carrying out terrorist operations abroad, the Quds Force. To show their gratitude, the increasingly powerful combination of the IRGC and the MOIS stole the 2009 election when Ahmadinejad faced a popular reform candidate, Mir-Hossein Mousavi. Mousavi's Green Movement threatened to morph into a Green Revolution to overturn the corrupt clerical order. When people took to the streets in the largest protests since 1979, the IRGC and the Basij, a paramilitary organization mobilized for security, brutally suppressed them.[46]

The JCPOA has proved a windfall for Supreme Leader Khamenei, the *bonyads*, and the IRGC, allowing them to extend their patronage networks and intensify proxy wars in the region. Integrating Iran into the global economy was, in theory, supposed to strengthen the private sector, loosen the government's grip on the commercial sector, empower moderates, and, over time, produce a less hostile Iranian government. But rather than opening up the Iranian market and liberalizing the country, the sanctions relief strengthened the revolutionaries, especially the *bonyads* and the IRGC. Like the *bonyads*, the IRGC is central to Iran's economic system. It gained a high degree of economic influence during the Iran-Iraq War and controls 20 to 40 percent of the Iranian economy. In the first eighteen months after the JCPOA payment of $1.7 billion, at least 90 of the 110 commercial agreements and approximately $80 billion in outside investment went to state-controlled companies.[47]

* * *

THE BELIEF that sanctions relief would change not only the behavior but also the very nature of the regime was based on the narcissistic assumption that U.S. actions were the principal source of Iranian attitudes and behaviors. The Iranian political structure was not well understood or was ignored. Wendy Sherman, the lead negotiator for the Iran Nuclear Deal, suggested that "to make a meaningful deal, we need to see our adversaries not as eternal enemies or as dispensable ones, but as virtual partners."[48] The counterproductive Iran policy was the result of self-delusion, a lack of Iran expertise, and the associated misunderstanding of history and underappreciation for the emotions and ideology of Iranian leaders.

A superficial understanding of history is often more misleading than complete ignorance. The Obama administration accepted the founding myth of the Iranian Revolution—that the 1953 coup that toppled Iranian prime minister Mohammad Mosaddeq and consolidated the rule of the Shah was externally planned and executed. For the Iranian revolutionaries, the coup myth reinforced their meta-narrative of victimization at the hands of Western colonialists. For the Obama administration, the story reinforced its tendency to see the United States as the key determinant of Iran's actions. Never mind that the Shah had the legal right to dismiss his prime minister and that Mosaddeq's rejection of that dismissal was in fact unconstitutional and illegal. Although the prime minister, lionized by the revolutionaries in Iran and New Left historians in the United States, was indeed an honorable patriot, the Mosaddeq myth overlooked its protagonist's obstinance and how his inflexibility crippled the Iranian economy, opening the door for radicals on both sides of the political spectrum. Never mind that the monarchy and the Shah were still popular and that Mosaddeq was a monarchist. Although British and U.S. intelligence agencies did conspire against Mosaddeq, a trove of documents released in 2017 demonstrated that their efforts would have failed without the support of domestic actors. In the end, the Shah's coalition proved stronger than Mosaddeq's narrow coalition of unhappy intellectuals and leftist politicians.[49]

But the simplistic history of the coup appealed to those sympathetic to the New Left's interpretation of history, in which the modern ills of the world are attributed mainly to capitalist imperialism and an overly powerful United States. The standard interpretation of the coup in U.S. universities is, in part, a late by-product of opposition to the Vietnam War.

The flawed interpretation of the Mosaddeq coup contributed to a predisposition toward atonement for America's alleged sins as the first step toward improved relations. In 2019, the University of Texas at Austin's *Texas National Security Review*, for example, released an issue examining *why* the Eisenhower administration chose to overthrow the Mosaddeq government, taking as a foregone conclusion that it had actually done so. Or consider the February 2019 headline on NPR's website that read, "How the CIA Overthrew Iran's Democracy in 4 Days." In 2009, during the Cairo speech on American relations with the Muslim world, President Obama noted that, "In the middle of the Cold War, the United States played a role in the overthrow of a democratically elected Iranian government."[50] Although he went on to remind his audience that Iran had also committed its share of misdeeds against Americans, the oblique reference to the Mosaddeq coup was meant as an admission of guilt that might lead to better relations. Regardless of the reality in 1953, it is important to recognize the Mosaddeq myth as both a cause and a symptom of the deep resentment, sense of victimhood, and thirst for vengeance that drive and rationalize the Iranian revolutionaries' most egregious acts.[51] The United States, however, should not abet an abuse of history by a clerical order whose forebears were far more responsible for Mosaddeq's demise than the CIA.

* * *

THE ASSUMPTION that the Islamic Republic, once welcomed into the international community, would evolve into a force for stability in the Middle East grew out of the narcissistic tendency to view outside actions as more important than internal dynamics in determining the regime's behavior. Although the Iranian government was not hostile because of historical wrongs that sanctions relief could right, some actually welcomed Iranian regional hegemony as a potential source of peace and stability in the region reminiscent of the Persian empires of the sixth century BC through the nineteenth century AD. But this was an ahistorical fantasy.

Officials in the Obama administration focused on selling the deal rather than subjecting it to scrutiny. As a setup for the sales pitch, Ben Rhodes stated that those who were against the JCPOA were for the Iraq War. "Wrong then, wrong now became our mantra," he recalled. The sales point not only was a red herring, but also demonstrated how opposition to the Iraq War dominated other policy initiatives of that administration whether relevant or not. The red herring of the Iraq War led to a false dilemma, an informal fallacy in which something is falsely claimed to be in an "either/or" situation when, in fact, there is at least one additional option. Rhodes was proud of posing the false dilemma between either supporting the JCPOA or going to war with Iran. President Obama liked it, too, describing it as their "best argument" for the agreement.[52] In this instance, the option omitted was to continue to pressure the regime with sanctions at a time when it was feeling the pressure. Long before the JCPOA, Iranian leaders, nervous about their corrupt regime's ability to forestall opposition to their autocratic rule, began to talk about the "China model," in which economic growth, jump-started by sanctions relief, would placate the unhappy Iranian population.[53] Sanctions relief under the JCPOA gave the regime an injection of cash to arrest economic deterioration and devaluation. But by strengthening a repressive regime and the IRGC, the deal weakened rather than strengthened reformers. Even worse, the conciliatory atmosphere employed to pursue a flawed nuclear agreement and American assurances of the Iranian leadership's trustworthiness actually disempowered the Iranian people as actors and potential forces for moderation. Rhodes gave instructions to make sure that the agreement was portrayed only as addressing the nuclear issue because "we don't want to let the critics muddy the nuclear issue with the other issues."[54] By "other issues," he meant the regime's tyrannical repression of its own population, its support for terrorists, and the perpetuation of violence in the Middle East.

* * *

LATER, THOSE determined to preserve the JCPOA argued that pulling out was shortsighted. The JCPOA itself, however, was also shortsighted because it divorced Iran's nuclear program from not only the regime's behavior, but also its very nature. It produced a bad political outcome disguised as successful diplomacy by giving up on core American values, by siding with a repressive regime against its own people and the peoples of the region. However, the assumptions and illusions that underpinned the JCPOA were not unique to that agreement or the Obama administration. Across six administrations, goodwill never begot goodwill with Iran. Conciliation has never brought moderation or a shift in the regime's permanent hostility to the United States, Israel, Europe, and the Arab monarchies. While the JCPOA was presented as a major turn in American policy, in fact it was consistent with a long history of errors and illusions. Rectifying failed policies of the past required a better understanding of the Iranian regime and, in particular, how ideology and emotions drive and constrain its behavior.

CHAPTER 10

Forcing a Choice

Anyone who will say that religion is separate from politics is a fool; he does not know Islam or politics.

—KHOMEINI

IT WAS harder than it should have been to implement the Iran policy announced in President Trump's October 2017 speech. Any policy shift is difficult across all departments and agencies of the U.S. government, especially when the change in direction is significant, as with Iran. And some of the friction in implementation was due to lingering sympathy for a conciliatory policy. I believed we had to force Iran to choose between either acting as a responsible nation and enjoying the corresponding benefits of such behavior or suffering sanctions and isolation as it continued to wage its destabilizing proxy wars. But some continued to prioritize avoiding confrontation with Iran over forcing that choice, clinging to the forlorn belief that one day Iranian leaders would, as President Obama had hoped, unclench their fists.[1]

During the summer and fall of 2019, about a year after the United States withdrew from the Iran Nuclear Deal, and as Iranian oil exports and the value of its currency hit historic lows, Iran attacked Saudi, Emirati, Japanese, and Norwegian oil tankers; conducted drone strikes on Saudi Arabian oil facilities; and shot down a U.S. drone.[2] Those actions should have confirmed to the United States that until there is an evolution in the nature of the regime, Iranian leaders will not abandon their proxy wars. Iranian aggression might also have disabused U.S. allies still supporting the JCPOA of the premise that the West can, through engagement and economic enticements, convince the regime to forgo aggression. What became known as the Gulf Crisis of 2019 reinforced the notion that the ideology of the revolution drives Iran's external actions. Iran's revolutionaries do not want to be conciliated, and Iran believes it can continue to have it both ways: use extra-state violence to achieve its objectives and still be treated internationally like a responsible nation. It was probably encouraged in this by the international reaction to its hostile acts, which assigned blame more to the U.S. withdrawal from the Iran Nuclear Deal than to the decisions made by Supreme Leader Ayatollah Khamenei, the Guardian Council, or IRGC commanders. "America made Iran do it" was the subtext behind too much news analysis.

But the regime's "Great Satan," "Death to Israel," and "Death to America" language is not mere bluster. Hostility toward the United States, Israel, and the West is foundational to Iran's revolutionary ideology, but has historical roots that well predate the revolution. Iran's leaders deeply resent colonial and foreign powers who are seen as the cause of the Persian empire's collapse and their country's loss of sovereignty in the nineteenth and early twentieth centuries. Due to its strategic geography and oil reserves, Iran was an arena of competition in the "Great Game" between Britain and Russia for power and influence across Central Asia. But Iran was also an active player in that game. In the 1930s, for example, authoritarian ruler Reza Shah sought to consolidate his power and cultivated relationships with fascists in Germany, Italy, and Turkey. Early Axis victories in World War II seemed to create an opportunity to expel the British he resented for profiting inequitably from Iran's oil wealth, but he bet on the wrong side. After the Soviets invaded from the north and the British from the south, Reza Shah abdicated to his son, Mohammad Reza Shah Pahlavi, who would go on to rule Iran for thirty-seven years. Iran declared war on the Axis in 1943.[3] The Cold War contest between the Soviet Union and the West introduced the new Shah to an updated version of the Great Game, with the

United States replacing Britain as the primary external influence on Iran's politics and economy. Resentment toward America and Britain for undermining sovereignty, especially during the overthrow of Mosaddeq in 1953, remains a principal emotional determinant of Iranian foreign and military policy nearly seventy years later.

In the years leading up to 1979, opponents of the Shah, such as Ayatollah Ruhollah Khomeini, gained popularity for expressing anti-American sentiment. After being exiled from Iran in 1964 and expelled from Iraq in 1978, Khomeini moved to France, where he took full advantage of freedoms he would later suppress in Iran to spew anti-American and anti-Western propaganda. In his cottage in the Paris suburb of Neauphle-le-Château, surrounded by telephones and cassette tape recorders, he conducted more than 450 interviews, portraying the Shah as a puppet of the United States. When the Ayatollah returned to Iran on February 1, 1979, the crowds that greeted him chanted anti-Western and anti-Israel slogans. The 444-day-long hostage crisis also gave Khomeini and his fellow revolutionaries the opportunity to use anti-Americanism to consolidate their grip on power. It is a technique the regime uses to this day.[4] In November 2019, when protests erupted around Iran in response to a government announcement rationing petrol and raising gas prices by up to 300 percent,[5] the IRGC called a rally to direct popular discontent away from the regime and toward the United States. The commander of the IRGC, speaking in front of a crowd, blamed the United States for the recent protests, stating, "We are in a great world war and at this moment you are defeating the arrogant powers. The war that was started recently in our streets was an international plot."[6]

While understanding what drives and constrains Iranian leaders is critical to U.S. policy, so is an appreciation for the broad range of beliefs and perspectives held by the Iranian people themselves. In 1998, when proposing a "dialogue between civilizations," Iranian president Mohammad Khatami observed that one should "know the civilization with which you want to maintain a dialogue."[7] He was making a case for strategic empathy. The Iranian people's attitudes toward the United States and the West are neither uniform nor immutable, which is why a long period of friendship between the American and the Iranian people preceded the revolution. Across public and private sectors, discussions with Iranian counterparts can foster understanding of the positive as well as the negative aspects of Iran's and America's intertwined history. One might share the story of Howard Baskerville, a young American who taught English, history, and geometry in Tabriz and went on to become a martyr of the 1905–1911 Persian Constitutional Revolution. Quoted as stating that "the only difference between me and these people is my place of birth, and this is not a big difference," Baskerville died in command of 150 young constitutionalists while trying to break a siege of tsarist forces and bring food to the people of Tabriz.[8] We should recognize the Iranian people's diversity of viewpoints and take advantage of communities that possess an affinity for American and Western literature, film, music, and the performing arts. Many of Iran's communities possess cultural and religious identities incompatible with the Marxist and Islamist fundamentalist ideology of the regime.[9]

While the Iranian regime cannot be changed from the outside, engagement with the Iranian people can help constrain the regime's use of demagoguery to justify external aggression and internal repression. Penetrating or circumventing the regime's restrictions to build relationships with Iranians can help counter the regime's narrative. Although many foreign citizens of Iranian descent have been unlawfully imprisoned in Iran, communication and visits between the Iranian diaspora in the West and their friends and family is an effective catalyst for countering regime disinformation.[10] Soon, technological means of bypassing Iran's censors will be available. Satellite-based internet and other empowering technologies will make it harder for the Iranian regime to block communications and access to information.

Dialogue might also increase social pressure on the regime by reducing Iranian leaders' ability to blame "the Great Satan" (the United States), "the Little Satan" (Israel), and others for the tragedy of the modern Islamic Republic. The United States and other nations should not take credit for the failing Iranian economy. Credit should go to those Iranian leaders whose corruption and militarism are preventing normal economic engagement and Iranian prosperity. Iran is a tragedy not only because of the devastation and suffering it has caused, but also because of its leaders' failure to take advantage of the tremendous potential of its people and natural resources. During Ahmadinejad's presidency, the kleptocratic regime wasted an estimated $800 billion in oil wealth.[11] Historically, during periods of instability, the nation's corrupt leaders hoarded even more wealth to bolster their positions of privilege.

The United States and other nations can also do more to expose the hypocrisy in the regime's flawed ideology and claims to virtue. Parallels are easy to draw between the corrupt mullahs of Iran and the tyrannical Communist totalitarian regimes of the Cold War era. The statement that the Iranian revolutionaries issued to explain mass executions in the early days of the 1979 civil war echoed the language of the Bolsheviks who rationalized the Red Terror in 1918: "To destroy and kill evil is part of the truth and that the purging of society of those persons means paving the way for a unified society in which classes will not exist."[12] The Supreme Leader's theocratic authority through the concept of *velayat e faqih*, or "rule of the jurist," is not universally accepted in Shiism. Many see the corrupt authoritarian system cloaked in religion as heretical. Investigative journalists and analysts should expose the money wasted on Iran's proxy wars, and the vast wealth of government officials and clerics associated with the *bonyads*, to show the Iranian people how their wealth is squandered. In public statements, foreign leaders should be careful to distinguish the Iranian regime from the Iranian people. Failure to do so only allows the regime to continue to deflect criticism away from its own failures to take advantage of the country's tremendous gifts, including its educated population, geostrategic location, and natural resources.

* * *

DESPITE THEIR failing economy, Iranian leaders feel emboldened because they perceive the United States and Europe as divided and weak in resolve. The IRGC's successes are due not only to its unscrupulousness and talent for deception, but also to the lack of a sustained response. Khamenei saw the divisiveness and contention in the United States after the 2016 presidential election as a sign of political and moral decline, stating that the United States "is becoming hollow from the inside like what termites do." Like the Chinese Communist Party leadership, he had regarded the 2008 financial crisis as an indication of weakness, stating that the U.S. economy "had declined astonishingly in recent decades" and "American power has declined in the area of politics as well."[13] The lack of a U.S. military response to Iranian attacks in 2019 fit what Iranian leaders view as a pattern of weak reactions to Iranian provocations. The European Union clung to a conciliatory policy even as Iran conducted escalatory attacks, working to circumvent reimposed sanctions after the U.S. withdrawal from the JCPOA.[14] It is entirely possible that Iranian leaders concluded that neither the United States nor European nations possess the will to see through a military confrontation with Iran. In November 2019, IRGC commander Salami boasted to the United States that "you have experienced our power in the battlefield and received a powerful slap across your face and could not respond . . . If you cross our redlines, we will annihilate you."[15]

The U.S. strike that killed Qasem Soleimani and Abu Mahdi al-Muhandis in Baghdad on January 3, 2020, must have come as a surprise. Just prior to the strike, the Supreme Leader, referring to the prospect of U.S. retaliation for Iranian proxy attacks on U.S. bases and the U.S. embassy in Iraq, had taunted President Trump, saying, "You can't do anything."[16] The regime had clearly been conditioned to believe so.

Absent demonstrated American resolve to impose physical and financial costs on Iran, Iran's proxy wars will intensify. The ability of the Islamic Republic to modernize its military and wage its proxy wars is partially dependent on the success of the overall Iranian economy. From 2008 to 2018, Iran spent nearly $140 billion on its military and combat operations abroad. Between 2017 and 2019, the United States sanctioned approximately one thousand Iranian individuals and organizations. In 2018, the rial declined fourfold against major currencies, and oil exports, which generate most of the regime's income, dropped to 1 million barrels a day from a high of 2.5 million. Sanctions, a decrease in GDP, and high inflation resulted in a 10 percent reduction in military spending. In spite of these economic constraints, Iran continued to draw on foreign currency reserves to fund its proxy wars.[17]

Until the regime ends its hostility to the United States, Israel, the West, and the Arab world, the United States and its partners should improve defenses against Iranian capabilities. Iran's proxy wars have grown more dangerous because they have expanded both geographically and in the number of participants. Other nations have joined in the conflicts to protect their interests in Syria, Yemen, and in the waters of the Persian Gulf and the Bab-el-Mandeb strait. From 2012 to 2020, Iran's use of the Levantine air and land bridge to increase the threat to Israel prompted the Israel Defense Forces to intervene in Syria.[18] In 2019, Israel increased strikes in Lebanon and Gaza to disrupt Hezbollah, Palestinian Islamic Jihad, and Hamas capabilities and leadership while Iran continued to support those organizations with weapons and cash. Between 2015 and 2020, Iranian proxies in Yemen launched more than 250 missiles into Saudi Arabia and the United Arab Emirates and attempted dozens of attacks on ships in the Bab-el-Mandeb.[19] As a result, Saudi Arabia threatened to retaliate directly against Iran. In a November 2017 interview, Prince Mohammad bin Salman called Supreme Leader Khamenei "the new Hitler of the Middle East" after stating in May that the kingdom would make sure any future struggle between the two countries "is waged in Iran."[20]

The proxy wars are also more dangerous because of the destructive weaponry Iran provides its militias. Hezbollah and Houthi rebels have both demonstrated an ability to hit ships with guided missiles. Iran's use of cyber and drone attacks, as well as ballistic missiles, serves as a warning to nations in the region and beyond. Increased U.S. and European defense cooperation with Gulf states in areas such as missile defense, air defense, and long-range fires might convince Iran's leaders that they cannot accomplish objectives through the use of force. As Iran's efforts spread geographically, its span of control and logistics became stretched. A multilateral effort should exploit existing weaknesses in the IRGC and proxy network, including its broad geographic scope and overstretched logistic capabilities.

Meanwhile, defensive measures in Arab nations must extend beyond military means. Political and governance reforms to address grievances and meet the needs of non-Sunni citizens will reduce Iranian influence and the numbers willing to aid and abet Iranian subversion, especially in countries with Shia majorities, such as Iraq and Bahrain, and in those with significant Shia minorities, such as Kuwait, Lebanon, and Saudi Arabia.

In particular, strengthening governance in Lebanon and weakening Lebanese Hezbollah should be a top priority. Lebanon was the site of Iran's first proxy war. The Revolutionary Guard took advantage of a brutal multi-sectarian civil war that began in 1975 by portraying themselves as patrons of the poor Shia community. They began forming and training Shia militias that coalesced into Hezbollah after Israel invaded in 1982. Members of Hezbollah and other Iranian-controlled militias went on to launch a series of devastating terrorist attacks, including the 1983 bombing of the U.S. embassy in Beirut. In the 1970s, anti-Shah Iranians who would later become leaders in the Iranian Revolutionary Guard received training and support at Lebanese Palestine Liberation Organization camps. After the revolution, the newly formed IRGC maintained its militant relationships.

Over the years, Iran strengthened the bond between Hezbollah and its Shia base through the provision of social services and the political power Hezbollah exercises in the Lebanese government.[21] Hezbollah was first legitimized as a defender of Lebanese Shia by its ongoing resistance to the Israeli occupation of Lebanese territory and later by initiating war with Israeli forces during Israel's 2006 invasion of Lebanon. Sunni jihadist sectarian violence has allowed Hezbollah to continue to portray itself as a protector of Shia Muslims and their sacred sites, such as by fighting ISIS in Syria since 2013. Indeed, Hezbollah has a reach far beyond Lebanon, with strategic goals in Syria and Israel, and a demonstrated ability to conduct terrorist attacks in Europe, South America, and the greater Middle East. Iran uses Hezbollah to provide an "Arab face" to its subversive efforts across the region.[22]

The U.S. government must do far more to target Hezbollah and other Iranian proxies, using its full range of financial, military, and law enforcement authorities. We should monitor and sanction Hezbollah-linked companies and *bonyads*, support countervailing forces in Lebanon, and cooperate with the European Union and regional partners.[23] Exposing Hezbollah's corruption and use as a tool of the Iranian regime can help galvanize the Lebanese people against the organization. It is important to understand the depth of the challenge, as Hezbollah is militarily strong, and Lebanon is politically fragile. Weakening Hezbollah requires a powerful appeal to the Lebanese people. The suffering that Hezbollah inflicted on the Syrian population and the costs the Lebanese people bore on behalf of the Iranian regime might boost anti-Hezbollah sentiment. More than 7,000 Hezbollah fighters were injured and 1,139 killed in Syria between 2011 and 2019, including more than 600 from heavily Shia southern Lebanon.[24]

In October 2019, the Lebanese people's frustrations with their dysfunctional government overflowed. Much of their ire was directed at Hezbollah for defending the government and the corrupt status quo, tarnishing the image of the party as above the fray. By November, hundreds of antigovernment protestors were chanting "Hezbollah are terrorists!" and "Here is Lebanon, not Iran!" in a significant break from the past.[25] In early 2020, the Lebanese financial system was in free fall and its government had collapsed. As Haider al-Abadi once told me, sectarianism and corruption go together.

Lebanon remains Iran's primary front in its proxy war against Israel. Even as its economy falters, Iran will continue to prioritize its aggression toward the "little Satan." It is possible that, despite efforts to deter another war, Iran will use Hezbollah to precipitate a crisis with Israel as it did in 2006. That war proved inconclusive, but Hezbollah claimed "divine victory" over the Israel Defense Forces. Despite the casualties suffered, the group promoted a resilient survival narrative in which its leaders boasted of their courage and appealed to anti-Israel sympathies. Hezbollah remains committed to Israel's "total annihilation," as do Hamas and Palestinian Islamic Jihad, two other organizations Iran sponsors. Since 2017, Iran has funded both groups, amounting to a combined $100 million annually. Just as Iran increased Hamas's rocket supply during the 2008 Gaza War, Khamenei offered Hamas weapons and increased funding in response to Israeli airstrikes in July 2019. There is no reason to doubt Iran when it simultaneously emboldens these forces and makes threats, such as when the deputy commander of the IRGC warned Israel in 2018, "Listen! Any war that occurs will result in your annihilation."[26]

* * *

UNLIKE IN China, where the ideology of "communism light" and the narrative of national rejuvenation are meant to preserve the party, the Iranian state exists to spread its ideology. The leaders of the Quds Force, the element of the IRGC that directs unconventional warfare and intelligence activities, believe they are protecting the "purity of the revolution." After suffering more than a million casualties and losing nearly $645 billion during the Iran-Iraq War, those leaders committed to extraterritorial operations under the theory that the best defense was a good offense.[27] The IRGC oriented its "forward defense" strategy on two principal enemies: Saudi Arabia and Israel.

Israel has responded much more forcefully to Iran's threats than have Western nations. In 2019, as Iran attempted to complete its Levantine land bridge across Iraq and Syria and place a proxy army on the border of Israel, IDF strikes targeted nodes in the Iranian network in Syria, Lebanon, and reportedly even Iraq. Israeli prime minister Benjamin Netanyahu hinted that this offense is only the beginning, stating that "Iran has no immunity, anywhere."[28]

Obstacles may remain insurmountable, but mediation between Israel and the Palestinians as well as between Israel and its neighbors would diminish Iran's ability to portray itself as a patron of the Palestinians as it pursues its objective of destroying Israel. Many factors concerning the prospects for peace between Israel and the Palestinians depend on the Palestinian people, such as whether an alternative to Hamas will develop in the Gaza Strip or if the Palestinian Authority will evolve in a way that produces not only a renewed desire to negotiate final status, but also the ability to enforce an agreement. Other factors depend on the Israeli people, including whether the highly personalized Israeli political landscape allows the sincere pursuit and eventual approval of an agreement. Progress also depends on the United States' ability to pressure Israel and the Palestinian Authority or mediate between the two, an ability dependent on whether both sides view the United States as an honest broker.

★ ★ ★

ALTHOUGH SOME analysts have described the relationship between Saudi Arabia and Iran as a cold war, it is actually more dangerous and destructive than a cold war because the escalating political and religious struggle drives the cycle of sectarian violence across the region. In 1987, the threat of the Iranian Revolution and its ideology to the Saudi royal family struck home during the Hajj, the annual pilgrimage to Mecca that all Muslims are expected to make at least once during their lifetime. Iranian Shiite pilgrims gathered for a political demonstration, chanting "Death to America! Death to the Soviet Union! Death to Israel!" Subsequent clashes with Saudi riot police left four hundred dead.[29] In response, Khomeini asserted that "these vile and ungodly Wahhabis, are like daggers which have always pierced the heart of the Muslims from the back." In 1991, the two countries reinstated diplomatic relations, but efforts to improve the relationship failed. Tensions grew as Saudi Arabia became concerned over Iranian influence in Iraq and Yemen, and relations were again suspended in 2016, when Iranians stormed the Saudi embassy in Tehran following the execution of Shia cleric Sheikh Nimr al-Nimr in Riyadh on terrorism charges.[30] Efforts to mediate between Saudi Arabia and Iran started again in early 2020. Although prospects are dim, the alternative is the grim continuation of sectarian violence.[31]

There are continuities across the four-decade-long conflict, but new technologies are generating new dangers. For example, the 2019 attack on Saudi Arabia's oil infrastructure was reminiscent of the 1987 failed commando attack on Saudi Arabian and Kuwaiti oil fields using a flotilla of fast boats, an attack interdicted by American helicopters. The 2019 attack was successful as it employed armed drones in an unprecedented swarm attack. Another successful attack, in 2012, was carried out through cyberspace as Iranian hackers shut down thirty thousand computers and ten thousand servers belonging to Saudi Aramco, causing system damage that took five months to repair.[32] Iranian development of ballistic missiles, a nuclear program, and chemical weapons would be unacceptable to Saudi Arabia, Israel, and other nations. Thus, a preventive war to deny Iran these destructive capabilities is a growing possibility, and given its potential to escalate into a devastating conflict, reducing Saudi Arabian and Iranian tensions is vital.

The United States and others should not take sides in the Shia-Sunni competition but can encourage authorities within those sects to isolate extremists who advocate for violence and fuel sectarian civil war. Gulf states such as Saudi Arabia, the UAE, and Qatar must stop private and government support for jihadists, as those organizations allow Iran to claim that its support for the Houthis in Yemen, various militias in Iraq, and the proxy army fighting in Syria is a legitimate counterterrorism effort rather than an attempt to extend Iranian influence across the region. Enduring political accommodations among Sunni, Shia, and Kurdish populations in Syria and Iraq and between the Zaidi Shiites and Sunnis in Yemen are important steps toward curbing Iranian designs on the region.

Exposing Iran's support for jihadist terrorist organizations hostile to both Shiism and its own people might place pressure on the Supreme Leader to end his cynical efforts to keep the Arab world perpetually weak. Iran, for example, has harbored Al-Qaeda leaders and eased the movement of Sunni jihadist terrorists. In a letter found in Osama bin Laden's compound, a senior Al-Qaeda official reported in 2007 that Iran had "offered to some of Saudi brothers . . . to support them with money and arms and everything they need, and offered them training in Hezbollah camps in Lebanon, in return for striking American interests in Saudi Arabia and the Gulf." Later that year, bin Laden chastised Abu Musab al-Zarqawi, the leader of Al-Qaeda in Iraq, for threatening Iran: "For as you are aware, Iran is our main artery for funds, personnel, and communication, as well as the matter of hostages."[33] All should condemn jihadist terrorist attacks against Iran such as the 2010 suicide attack on the Chabahar mosque that killed thirty-nine Iranians, the 2017 killings of twelve at the Iranian Consultative Assembly (parliament) and Khomeini's tomb, and the 2018 murder of twenty-five Iranians at a military parade in Ahvaz. But it is also important to point out that, like Pakistan, Iran is vulnerable to those attacks because the regime's reliance on religious oppression drives sectarian violence within the country.

Pakistan provides a stark warning. Iran could become, like Pakistan today, a nuclear armed state in which terrorists already enjoy a support base. The greatest threat to humanity in the coming decades may lie at the nexus between terrorists and the most destructive weapons on earth.

★ ★ ★

THAT IS why blocking Iran's path to a nuclear weapon should remain a top priority. Aside from the potential transfer of such weapons to terrorists, it is also likely that Saudi Arabia and other nations will conclude that they, too, need nuclear weapons to deter Iran. The breakdown of nonproliferation in the Middle East would increase the potential for apocalyptic war in a region already enmeshed in persistent political and religious conflict. Iran's leaders' messianic ideology and their cult of martyrdom raise doubts about the ability to effectively deter a nuclear-armed Iran, as they may be willing to risk massive casualties among their own people. The billboards that Iranian teenagers walked by as they went to their deaths during the Iran-Iraq War stated tellingly, "The sword does not bring victory; it is the blood that brings it."[34]

The JCPOA could not address the underlying problem: the Iranian regime's hostility to the United States, Israel, the Arab monarchies, and the West. The agreement actually allowed Iran to have it both ways. The regime benefited economically and used those resources to intensify its proxy war. It is past time to force Iran's leaders to choose between economic ruin and isolation or an agreement that combines a peace treaty to end its proxy wars and strong verification of promises not to pursue nuclear weapons, missiles, or other weapons of mass destruction.[35]

Combined with the effect of sanctions, it was the credible threat of a military strike against Iran's nuclear program that moved Iran to enter multiple rounds of nuclear negotiations between 2006 and 2015.[36] But after the U.S. departure from the JCPOA, an Iranian government dominated by the revolutionaries is unlikely to enter another agreement in the near future. As the United States and others attempt to influence a shift in the nature of the Iranian regime, it would be prudent to implement all available measures to delay and disrupt its program. Integrated intelligence, law enforcement, and cyber efforts should have support across the political spectrum as they did after President George W. Bush briefed President-elect Barack Obama about the range of U.S. efforts against Iran's nuclear program, telling him, "We want you to succeed."[37]

As pressure on Iran mounts, the United States and other nations must be prepared for escalation. Past patterns of escalation are instructive. In June 2010, a number of computer viruses, including a particularly elegant cyber malware called Stuxnet wrecked approximately one third of the centrifuges at Iran's Natanz enrichment facility. Cybersecurity experts judge the virus to have been the creation of Israeli and U.S. scientists.[38] Then, in November 2011, a large explosion occurred during an Iranian missile test, killing seventeen members of the Revolutionary Guard, including Maj. Gen. Hassan Moghaddam, director of Iran's missile program. As Israel prepared for military strikes against Iranian nuclear and missile facilities, Israeli leaders appear to have expanded their efforts through clandestine attacks. Between 2010 and 2012, five Iranian nuclear scientists were assassinated. Motorcyclists pulled up next to their cars and attached "sticky bombs" to their car doors before speeding away.[39] Iran then sought retribution with multiple assassination attempts, including the brazen 2011 plot against the Saudi Arabian ambassador to the United States, Adel al-Jubeir, at Washington's Café Milano. Contingency plans should identify the range of potential Iranian escalations and identify the actions that could be taken now to prevent them or mitigate their effects.

* * *

IN THE fall of 2017, I asked our NSC staff to coordinate across the government to develop such contingency plans. As in the wars in Afghanistan and Iraq, the future course of events in Iran's proxy wars depends on continuous interaction with a determined adversary. U.S. strategy toward Iran should be flexible and attempt to anticipate Iranian reactions and initiatives. Some potential Iranian actions are not difficult to predict, as we have seen most of it before across the four-decade-long proxy war: mining of the Persian Gulf and the Bab-el-Mandeb; shore-to-ship missiles fired at U.S. or other nations' naval or commercial vessels; rockets fired into Arab partner nations from Yemen or into Israel from southern Lebanon or Gaza; assassinations, kidnappings, and hostage taking; attacks on U.S. forces in the region; bombings of U.S.

military facilities; attacks on Saudi Arabian oil infrastructure. Even the relatively new options available to Iran, such as drone strikes and cyber attacks, are somewhat predictable. It would have been negligent not to prepare for these potential actions. With time to think, it is possible to understand better and mitigate risk as well as identify and take advantage of opportunities. Importantly, it is also possible to work with partners to craft multinational responses to Iranian aggression.

In 2020 it was clear that Supreme Leader Khamenei, after realizing that the regime could no longer enjoy the benefits of foreign investment and international trade while continuing wars of terror and subversion, chose to intensify Iran's proxy wars and violate portions of the JCPOA in an effort to extort concessions. Given the regime's behavior over the last forty years, the choice should not have been surprising. The revolutionaries' pride and resentment, sustained by their ideological cocktail of Marxism and Shia millenarianism, made concessions impossible. Due to the combination of sanctions, falling oil exports and prices, and the regime's corrupt practices, the economic pressure was too severe to wait out Donald Trump.

Although the Supreme Leader's and the IRGC's escalating attacks at the end of 2019 were predictable, the U.S. killing of Soleimani and Muhandis must have surprised Iranian leaders. As the regime encouraged large-scale protests in Iraq and Iran and whipped up anti-U.S. sentiment, it was also predictable that Iran would have to respond. Khamenei vowed to exact "severe revenge." On January 7, the IRGC fired sixteen ballistic missiles at two Iraqi bases hosting U.S. forces.[40] There were no fatalities, although soldiers suffered brain injuries in the attack.

The reaction among U.S. allies on the death of Soleimani and Muhandis was mixed. Some called the strike an escalation, but those comments did not account for the forty-year-long proxy war that Iran had waged against the United States or give due consideration to Iran's escalation of that war on its own terms without fear of significant retribution. Soleimani was not only a purveyor of death and suffering outside Iran's borders, but also a scourge on the Iranian people. Under his leadership, the Quds Force squandered Iran's wealth while earning the country's status as a pariah terrorist sponsor that deserves isolation and sanction. Less than two months before Soleimani's death, the deadliest protests since the 1979 revolution spread to twenty-nine of the thirty-one Iranian provinces. Protesters shouted, "Death to Khamenei" and "Death to the dictator" as they ransacked state-run banks owned by corrupt leaders. The regime unleashed the state security forces on its own citizens. It is estimated that more than three hundred people were killed, two thousand wounded, and seven thousand arrested.[41] It is likely that Khamenei hoped that Soleimani's killing might divert the people's anger away from him and the corrupt order. It was not to be.

Public anger toward the regime erupted again just eight days after Soleimani's death, over the Iranian military's shooting down of a commercial airliner with 176 people on board. "They tell us the lie that it is America, but our enemy is right here," a crowd shouted as protesters referred to the IRGC as akin to ISIS.[42] Although it is possible that financial ruin and calls for an end to the regime could chasten Iranian leaders and make them reluctant to intensify proxy attacks or resume their nuclear weapons programs, it is more likely that they will order a continuation of their proxy wars. Iran's leaders will almost certainly continue to use external conflict as a way to divert the public's anger away from them . . . and toward "Great Satan" and the "Little Satan."

* * *

AN INCREASINGLY desperate regime could use combinations of old tactics and new capabilities to escalate conflict. The regime could decide to inflict mass casualties with chemical agents or a dirty bomb, which combines explosives with radiological materials. The IRGC would no doubt attempt to conduct those attacks through proxies, but everyone would know the return address. What seems even more likely, however, is that Iran would find ways to attack U.S. and European interests beyond the Middle East, including cyber attacks against critical infrastructure.

And those cyber attacks could be larger in scope and more effective than in the past. In 2012, the same year malware attacks hit Saudi Aramco, Iranian hackers targeted U.S. financial institutions with 176 days of distributed denial-of-service attacks similar to those that Russia used to attack Estonia's system in 2007. Targeted banks were temporarily paralyzed. In 2013, in what was likely a rehearsal for cyber attacks on U.S. infrastructure, Iranian hackers broke into the control system of the Bowman Avenue Dam, in Rye Brook, New York.[43] Cyber attacks will become more likely if Iran's revolutionaries conclude that they have little to lose.

A conflict initiated by an Iranian cyber attack would likely continue after U.S., Saudi Arabian, Israeli, or multinational retaliation. Iran could escalate further with rockets and missiles from proxy forces or from its own territory. Iran is increasing its ability to strike targets in Israel, the Gulf states, and the waterways of the Persian Gulf and Bab-el-Mandeb with precision rockets and missiles. Therefore, the deployment of additional integrated missile defenses, air defense, surveillance, and strike-and-counter drone capabilities to prevent, defend against, or respond to an Iranian attack is prudent. As would be the removal of U.S. forces in the region that are vulnerable to attack and are not contributing materially to important missions. The 1983 bombings of French paratroopers and U.S. Marines in Lebanon and the 1996 bombing of mainly U.S. Airmen in Saudi Arabia should serve as warnings.

Still, despite the best efforts to anticipate Iranian actions and reactions, a conflict could easily produce unintended consequences on both sides. Consider how Iranian riots in Mecca during the 1987 Hajj led to military clashes that involved Iran, Saudi Arabia, Kuwait, and the United States, resulting in the deaths of innocent civilians as well as combatants. Mohsen Rezaei, the head of the Revolutionary Guard, responded to the riots and the Saudi response by ordering a commando attack on Saudi Arabian oil fields. Since the Kuwaiti royal family had asked the United States to protect Gulf shipping from Iranian interdiction as the Iran-Iraq War raged on Kuwait's doorstep, American helicopters were in place patrolling the Gulf. Those helicopters destroyed IRGC Navy boats, forcing the rest of the attackers to withdraw. Chastened by the failure, the Revolutionary Guard next struck two oil tankers near a dock outside Kuwait City with Chinese-made Silkworm missiles. One was a reflagged U.S. vessel; seventeen crewmen and the American captain were injured. The U.S. Navy responded by shelling two Revolutionary Guard bases located on oil platforms that had been used to stage attacks on shipping. In a reprisal attack four months later, the IRGC Navy laid mines, and the frigate USS *Samuel B. Roberts* struck one; it blew a hole in its hull, injuring ten sailors. The U.S. responded with attacks on two Iranian frigates and Revolutionary Guard bases. Iranian attacks on neutral ships dropped, but tensions remained high. On July 3, 1988, USS *Vincennes*, while engaged with Iranian boats, mistook Iran Air Flight 655 for an Iranian F-14 and shot it down over the Strait of Hormuz, killing the 290 innocent passengers and crew on board, including 66 children. The United States did not admit fault, and President Reagan's lamentable decision to award the captain of the *Vincennes* with a medal deepened the legacy of mutual distrust and enmity between the United States and the Iranian regime.[11] The series of events demonstrated how interaction between Iran and the objects of its proxy wars can lead to escalation and unintended, tragic consequences.

* * *

THE IRGC and mullahs in Tehran are in a weakened position. The country's infrastructure is deteriorating. The corruption of the *bonyads* and the IRGC-controlled companies are a further drain on the economy. Iranians with the means and opportunity are leaving; the country is experiencing a massive brain drain. Approximately 150,000 educated Iranians emigrate every year, costing the country up to $150 billion annually. Pressure on the regime to focus on nation building at home instead of destruction abroad may mount, as it did during the widespread demonstrations of 2018, 2019, and early 2020. It was not an unprecedented reaction to the diversion of resources to the military. During the oil boom of 1973–74, vast expenditures on military hardware instead of investments in industry, agriculture, and education led to resentment of the Shah's military establishment.[45]

Revolution in Iran can be sudden and violent. The Iranian regime today has created conditions that are analogous to 1979. The Shah fell, in part, because the economy was collapsing, corruption was rife, military spending was excessive, and efforts to develop political alternatives to his rule were stifled. The Shah thought he had escaped historical dangers of porous borders, hostile neighbors, and internal divisions. He had not. There are earlier historical precedents for the regime's problems. In the seventeenth and early eighteenth centuries, the rulers of the declining Safavid dynasty governed their empire through a system that balanced theirs and their military's power against that of the clergy. The clergy sometimes shifted its allegiance between the regime and the merchant class. Today, the Supreme Leader finds himself in a position redolent of that of his monarchical predecessors. As the economy worsens, clerics and Iranian citizens increase their criticisms; Khamenei has responded by tightening his grip on Shiism's holy city of Qom, which, in turn, strengthens the clerics of its Iraqi competitor Najaf, who adhere to the quietist tradition and oppose the rule of the jurisprudent, which underpins Khamenei's power. Khamenei has to prevent internal opposition, defend against those he has provoked, and continue the pursuit of his messianic vision to export the revolution. Paradoxically, he and his fellow revolutionaries may have created political, economic, social, and military conditions similar to those that led to the demise of both the government they overthrew and the empire they want desperately to restore.

The tension between religious tradition and secular modernity is also not new. The Shah's suppression of the Shia ulema (scholars of Islamic sacred law and theology) contributed to his fall. The revolutionaries' brutal repression of republicans who prefer secular representative government to theocratic authoritarianism may also generate growing internal opposition. The Guardian Council's denial of approximately seven thousand candidates for parliamentary elections in 2020 made clear that the revolutionaries remained unwilling to grant political space to the reformers. The Shah was unable to reconcile tensions between the traditional and the modern, the religious and the secular, the rural and the urban. The Supreme Leader faces the same dilemma.

The Iranian people may tire of and reject the rule of jurisprudence. The concept is not inherent to Iranian culture. There are signs that Shia clerics of the quietist tradition in the Iraqi city of Najaf and the Iranian city of Qom are increasingly critical of clerical rule, and those criticisms are inspiring others. As Shiism's preeminent Marja', Grand Ayatollah Ali al-Sistani, entered his ninth decade and Ayatollah Khamenei was well into his eighth, it was not clear how their successors might influence clerical rule in Iran.

It is possible that the Iranian regime can evolve such that it ceases its permanent hostility to the United States, Israel, its Arab neighbors, and the West. Although, since 1979, the Iranian regime has proven consistently hostile and the revolutionaries are ascendant, the Iranian regime is not a monolith. Also, the IRGC and the Iranian regime are particularly vulnerable to a concerted multinational effort to force them to choose between continuing their murderous proxy wars or behaving like a responsible nation. The United States and other nations can encourage them to choose the latter if we implement a long-term strategy to defend against the Iranian regime's aggression, and force Iranian leaders to make a choice.

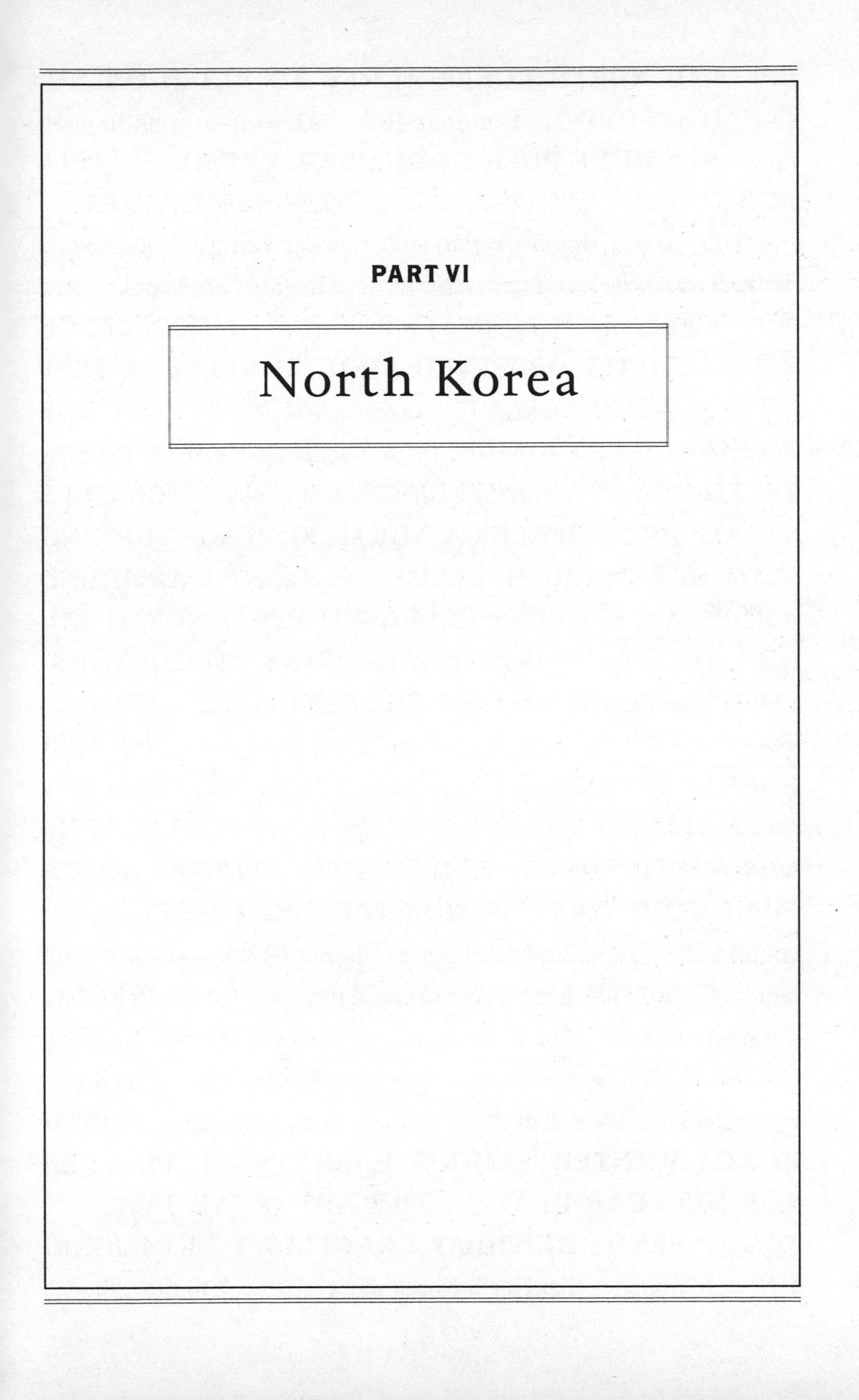

PART VI

North Korea

U.S. AND NORTH KOREA AGREE TO BUILD ON NUCLEAR ACCORD . . . Kim Jong-Il doesn't seem to want to open up . . . **SUNSHINE POLICY WARMS OLD RIVALS** . . . *POLITICS SANK ACCORD ON MISSILES WITH NORTH KOREA* . . . *North Korea is a regime arming with missiles and weapons of mass destruction, while starving its citizens* . . . The talks are hopeless and North Korea can't be negotiated with . . . **U.S. INTELLIGENCE SAYS ACTIVITY OBSERVED AROUND NUCLEAR TEST SITES** . . . *NORTH KOREA TO CLOSE REACTOR IN EXCHANGE FOR AID* . . . **KIM JONG-IL** . . . *NORTH KOREAN NUCLEAR TEST DRAWS ANGER, INCLUDING FROM CHINA* . . . **OBAMA'S STRATEGIC PATIENCE ON NORTH KOREA IS TURNING INTO STRATEGIC NEGLECT** . . . **OTTO WARMBIER, AMERICAN STUDENT RELEASED FROM NORTH KOREA, DIES** . . . **NORTH KOREA VOWS "THOUSANDS-FOLD" REVENGE ON U.S. OVER SANCTIONS** . . . They will be met with fire and fury like the world has never seen . . . **NORTH KOREA FIRES MISSILE OVER JAPAN** . . . *Rocket Man is on a suicide mission for himself and for his regime* . . . **A PLOT TO MURDER NORTH KOREA'S EXILED SON** . . . **HAWAII TO RESUME COLD WAR–ERA NUCLEAR SIREN TESTS** . . . The country has fired 23 missiles perfecting its technology with each launch . . . 2017 has been a year of rapid progress for North Korea's missile program . . . I too have a Nuclear Button, but it is a much bigger one . . . *They wanted the sanctions lifted in their entirety* . . . *Sometimes you have to walk and this was one of those times* . . . **NORTH KOREA WANTED FAMOUS BASKETBALL PLAYERS FOR NUCLEAR DEAL** . . . **THE ART OF NO DEAL** . . . **TRUMP SENDS BIRTHDAY GREETING TO KIM JONG-UN** . . . My hope is that North Korea will come back to the table . . .

RUSSIA
CHINA
SAKHALIN IS.
Khabarovsk
Yuzhno-Sakhalinsk
Harbin
KURIL IS.
HOKKAIDŌ
Sapporo
Kushiro
Vladivostok
Hakodate
NORTH KOREA
SEA OF JAPAN
Pyongyang
Wonsan
Sakata
Sendai
Seoul
HONSHŪ
YELLOW SEA
SOUTH KOREA
Nagano
JAPAN
Gifu
Tokyo
Busan
Kyoto
Nagoya
Hiroshima
Ōsaka
Fukuoka
Matsuyame
Kōchi
SHIKOKU
Nagasaki
KYŪSHŪ
Miyazaki
NORTH PACIFIC OCEAN
EAST CHINA SEA
AMAMI
OKINAWA
Naha
RYŪKYŪ ISANDS
0 200 400 600 km
0 200 400 mi
120° 125° 130° 135° 140° 145° 150°
50° 45° 40° 35° 30° 25°

CHAPTER 11

The Definition of Insanity

If the American imperialists provoke us a bit, we will not hesitate to slap them with a preemptive nuclear strike. The United States must choose! It's up to you whether the nation called the United States exists on this planet or not.

—NORTH KOREAN PROPAGANDA VIDEO, *LAST CHANCE*, 2016

FORT LESLEY J. McNair, a quiet place on the south side of the busy U.S. capital, lies at the confluence of the Anacostia and Potomac Rivers. It is home to the National Defense University, including the National War College and a range of military education and research activities. While I was away in South Asia in April 2017, my wife, Katie, with the help of our daughters, moved into one of the World War I–era general officer quarters on the banks of the Washington Channel, which flows into and out of the Tidal Basin. The homes have views of the Jefferson Memorial and the Washington Monument. Sunsets behind Hains Point, a peninsula between the channel and the Potomac River, are spectacular. Our home was the perfect setting to host administration colleagues and foreign counterparts. Because conversations there were relaxed, they tended to be more creative and productive than those in the West Wing. Katie and our enlisted aide, Sgt. First Class Juan Sanchez, always made our guests feel welcome.

Our first guest was my South Korean counterpart, Ambassador Chung Eui-yong. Ambassador Chung and his assistant, Park Jang-ho, joined Matt Pottinger and me in June. The timing was only weeks after President Moon Jae-in of South Korea's left-wing

Minjoo (Democratic) Party won a special election, brought forward nine months early by the downfall of President Park Geun-hye and ending nine years of conservative Liberty Party rule. Our relationship would be important. The Democratic People's Republic of North Korea (DPRK, or North Korea) was one of the most pressing national security challenges. Chung and I both felt a sense of urgency due to that country's nuclear and missile programs. We needed to ensure that the relationship between the one-month-old South Korean administration and the four-month-old Trump administration got off to a good start.

The growing threats from North Korea had become palpable, but they were not new. As we faced those threats, it would be necessary for the United States and the Republic of Korea (ROK, or South Korea) to be aligned. Over the years, the approaches that our two nations took toward North Korea, however, had all too often been divergent.

* * *

CHUNG KNEW the history of U.S.-ROK relations well. After his election to the National Assembly, Chung worked on the U.S.-ROK trade agreement and observed interactions between U.S. president George W. Bush and South Korean president Roh Moo-hyun. The two presidents were close; Bush would speak at Roh's funeral after the latter's tragic suicide in 2009. But the two presidents' conflicting approaches to North Korea created daylight between the allies, daylight that Pyongyang was all too happy to exploit.

The stakes were high in 2002. During a visit to Pyongyang by U.S. Assistant Secretary of State James A. Kelly, a North Korean official did not deny that the North was secretly working to produce highly enriched uranium as a new source of fuel for nuclear weapons, despite the 1994 Agreed Framework, the treaty under which North Korea was supposed to freeze its nuclear weapons program and welcome international inspectors into its facilities. In return, North Korea would get energy aid from the

United States, including shipments of oil and the construction of two light-water nuclear reactors. But implementation of a weak agreement that had not been approved by the U.S. Senate was problematic from the outset.[1]

In response to the disclosure of the North's uranium-enrichment program, Bush discontinued aid to the country he had described as part of an "axis of evil." At the same time, the South Korean government under Roh pursued Roh's version of a "Sunshine Policy" that sought reconciliation through peaceful cooperation and opening up to the North. So, when, in 2003, the Bush administration pursued the Six-Party Talks (which included the United States, China, Russia, Japan, and the two Koreas) to achieve the verifiable denuclearization of the Korean Peninsula, the economic leverage was toothless due to Seoul's assistance to Pyongyang. Economic opening was supposed to induce North Korea to give up its nuclear weapons, but why would Pyongyang give them up when the Sunshine Policy was delivering economic benefits for free?[2]

The Sunshine Policy maintained its allure, even though what appeared to be early successes were manufactured. In 2000, Roh's predecessor, Kim Dae-jung, a politician who had made an unlikely comeback after being condemned to death for his role in the antigovernment Gwangju Uprising of twenty years earlier, received the Nobel Peace Prize for visiting Kim Jong-il in Pyongyang and "paving the way for a brighter future for all Koreans and other peace-loving peoples of the world." In pursuit of the historic summit, however, Kim Dae-jung's administration had secretly paid the dictator $500 million in cash.[3] After the payoff was exposed, Kim's opponents quipped that it was the most expensive Nobel Prize in history. Just a few weeks after the summit, North and South Korean athletes marched jointly in the opening ceremonies of the 2000 Summer Olympics in Sydney, Australia. Some commentators got caught up in the emotion and predicted imminent unification. But as with Kim Dae-jung's summit, there was more to the story. North Korea had demanded and secured secret payments from Seoul.[4] North Korea also demanded that athletes from the South not outnumber those from the North.

As a result, many South Korean athletes had to sit out the ceremony. It would not be the last time that a sporting/cultural event would raise hopes for sudden change in the North-South relationship and the achievement of an enduring peace on the peninsula. Inter-Korean business projects and exchanges also followed the summit, including the Mount Kumgang Tourist Region, the Kaesong Industrial Zone (KIZ), and Kaesong city tours. Those projects proved lucrative for cash-starved Pyongyang, but did not lead to the anticipated opening or to gradual reform in the North. The programs, especially the KIZ, did allow limited interaction between South and North Koreans, interaction that challenged the North's propaganda that the people of South Korea were suffering from poverty under a miserably incompetent government. Economic projects allowed South Korean companies to employ cheap North Korean labor while providing the North with much-needed foreign currency as DPRK workers' salaries went directly to the North Korean regime.[5]

In 2007, despite the incompatible policies of South Korea and the United States, a prolonged stalemate in the Six-Party Talks ended, and the parties signed a tentative disarmament agreement. In June 2008, North Korea destroyed cooling towers at the Yongbyon Nuclear Scientific Research Center and even allowed foreign journalists and diplomats to witness the demolition. In response, the United States removed North Korea from the State Sponsors of Terrorism list and returned $25 million to Pyongyang that the United States had convinced Macao authorities to freeze in an account that North Korea used for money laundering. The agreement, as with the 1994 Agreed Framework, however, created only an illusion of progress. Four months later, North Korea reneged on verification measures and expelled IAEA inspectors from Yongbyon grounds.[6] The Six-Party Talks died and never revived.

After 2008, South Korea's Sunshine Policy disappeared behind a storm of North Korean aggression. As he replaced Roh, President Lee Myung-bak and his conservative administration believed that a decade of massive aid had neither improved the lives of destitute North Koreans nor induced any change in Pyongyang's reckless behavior. By then, the North had abandoned even the appearance of cooperation and had resumed provocations, including a long-range missile test in April 2009 and its second underground nuclear test on May 25, U.S. Memorial Day.[7] Amid the tests, the Kim regime took two U.S. journalists, Euna Lee and Laura Ling, hostage after they crossed into North Korea without visas. Even though they worked for former vice president Al Gore's Current TV, a North Korean kangaroo court sentenced them to twelve years of hard labor. Despite the release of the journalist-hostages after President Bill Clinton visited Pyongyang in August, provocations continued. In March 2010, a North Korean midget submarine sank the South Korean naval vessel ROKS *Cheonan*, killing forty-six sailors. Eight months later, the Korean People's Army fired 170 artillery shells onto South Korea's Yeonpyeong Island, killing four people and injuring nineteen.[8] Later that year, the regime revealed to visiting Stanford University metallurgist Siegfried Hecker the apparently fully operational uranium enrichment facility at Yongbyon. North Korean leaders had vehemently denied the facility's existence for nearly a decade.[9] Behind the series of well-timed provocations lay North Korea's effort to bolster the military qualifications of successor-in-waiting Kim Jong-un to consolidate power and initiate negotiations from a position of strength.[10]

Yet Washington remained weak in its response by continuing to engage the regime for potential talks, under the misguided assumption that reconciliatory diplomacy could generate a fundamental shift in Pyongyang's policy. After the sinking of the *Cheonan*, former president Jimmy Carter visited Pyongyang, where he called for a new dialogue; he returned to the United States with a detained American. Two months later, North Korea carried out the artillery barrage against Yeonpyeong Island. Still, Carter maintained in an op-ed that the attacks were meant to "remind the world that they deserve respect in negotiations."[11]

The former president warned the Obama administration that without direct talks, "North Koreans [would] take whatever actions they consider necessary to defend themselves."[12]

The Obama administration judged the status quo as preferable to actions that might escalate to military conflict. As he hosted ROK president Lee in Washington during a state visit in October 2011, President Obama declared that "If the North abandons its quest for nuclear weapons and moves toward denuclearization, it will enjoy greater security and opportunity for its people. That's the choice that North Korea faces." The Obama administration hoped that a policy of "strategic patience" might devalue the North's provocations as the United States ignored rather than responded to Pyongyang's efforts to get attention and extort concessions. The administration assumed that the Kim regime was, in the words of the North Korea expert Victor Cha, an impossible state that would ultimately collapse due to its brutality, corruption, and dysfunction.[13] Besides, the ailing dictator Kim Jong-il would soon be replaced by a relatively unknown twenty-seven-year-old who seemed an unlikely dictator.

As Chung and I spoke over dinner at my home six years later, it was clear that Obama's strategy of strategic patience, like Roh's Sunshine Policy, had failed. In their first and only meeting, President Barack Obama told President-elect Donald Trump that the DPRK had become his most pressing problem.[14] Kim Jong-un, who had taken on the moniker of the Great Successor, had consolidated power in brutal fashion, executing anyone deemed a potential challenger to his authority. One might even say that Kim Jong-un was pursuing his version of a policy of strategic impatience. The year 2016 marked the fifth year of the "Great Successor's" rule, and he did not like being ignored. He accelerated the North's nuclear and missile programs; both were progressing faster than most anticipated. Optimism that the North Korean regime could not be sustained by its third-generation dictator faded. The United States and South Korea were at a crossroads of rethinking their North Korea policies.

Chung and I agreed that, as we implemented a new strategy,

the United States and South Korea needed to avoid working at cross-purposes. I suggested that we pledge to reject two flawed assumptions that undercut previous policies toward North Korea: first, the Sunshine Policy's notion that an opening up to North Korea would change the nature of the regime; and second, the fundamental premise of the strategic patience policy, that the regime was unsustainable and on the brink of collapse or, at least, that it would collapse before the emergence of a nuclear-armed North Korea—something that presented an unacceptable risk to the United States and its allies.

Matt Pottinger and I summarized the strategy that he had started working on prior to the inauguration. The United States would work with others to apply unwavering, integrated, and multinational pressure on the Kim regime. We thought that the alignment of South Korean, U.S., and Japanese policies toward North Korea was the starting point for garnering broad international support for denuclearization. We needed a realistic strategy designed to convince Pyongyang that its nuclear and missile programs were a danger rather than an asset to the Kim regime.

* * *

CHUNG AND I agreed that aligning our efforts would be easier said than done. The Moon government was entering office at a tumultuous time. In 2016, a political scandal broke in Seoul that ultimately led to the impeachment, removal, prosecution, and imprisonment of Moon's predecessor, President Park Geun-hye.[15] As the first liberal president in a decade, many believed that Moon, who had served as Roh's chief of staff, would resurrect the Sunshine Policy, which the South Korean press could not resist labeling a "Moonshine Policy" toward the North. I was frank with Chung that moonshine and what we were calling "maximum pressure" would not mix well.

I told Ambassador Chung that misalignment in our approaches could generate a perfect storm. Like warm air from a low-pressure system hitting the cool air from a high-pressure system, the deep skepticism of overseas military commitments among Trump-supporting "economic nationalists" could collide with the wariness toward dependence on America among Moon-supporting leftists. That collision might not only undermine the approach to North Korea, but also do irreparable damage to the alliance. Some Trump supporters were isolationists. Some of Moon's supporters were sympathetic to New Left interpretations of history that blamed U.S. "capitalist imperialism" for problems on the Korean Peninsula and beyond. Although the Korean Peninsula lies at the far reaches of American power, it is central to debates over America's role in the world.

We had about thirty thousand troops stationed in South Korea, and Chung knew that Trump was not the first U.S. president to question the need for U.S. forces there. While campaigning for president in 1976, the year that two American soldiers were axed to death by North Korean soldiers in the demilitarized zone for attempting to cut down a poplar tree, candidate Jimmy Carter often stated his desire to bring American troops home.[16] Carter was frustrated with the corruption and human rights abuses of the general turned politician Park Chung-hee, who served five consecutive terms as president until he was assassinated by his intelligence chief in 1979. South Korea subsequently strengthened its democratic institutions and experienced extraordinary economic success during the two Park administrations and the thirty-four years that separated them. Yet that success provided some U.S. skeptics with a new rationale for American withdrawal: South Korea was rich and strong enough to defend itself. The argument that U.S. forces are vital for preventing another major conflict that could be even more destructive than the Korean War of 1950–1953 leaves these skeptics unconvinced, in part because it is impossible to prove a negative. History, however, can give warning in the form of potential consequences.

As I got to know Ambassador Chung and to learn about his background, I became aware that he was four years old when North Korea invaded. He lived in Seoul, the South's capital city that changed hands four times during the war. His earliest memories included entire city blocks in rubble; his mother forcing him and his siblings to wear makeshift helmets she sewed out of seat cushions, to protect them from shrapnel; and his home full of patients screaming in pain as they waited for his father, a doctor, to treat their wounds. He remembers macabre scenes while playing on streets surrounded by partially covered corpses. And he remembers great hardships; he walked more than two hundred kilometers in freezing cold weather to escape the Chinese army's assault on the capital in 1951. He remembers attending his first day of school in a "classroom" with no walls or ceiling; its only structure was a makeshift chalkboard. Ambassador Chung was aware that my father, Herbert McMaster, served as an infantryman in the Korean War, and he often expressed his gratitude for not only my father's service but also the service and sacrifice of so many others. The Korean War cost the lives of close to 37,000 Americans, 200,000 South Korean and UN soldiers, 400,000 North Korean soldiers, 600,000 Chinese troops, and a million and a half civilians. All told, approximately 3 million people died as a direct result of a war, after which neither side could claim victory. Statistics can be numbing, but Chung understood the war through distant memories of horrors that we were determined to prevent from happening again. Chung and I agreed that it was cheaper to prevent a war than to fight one. We also agreed that the Korean War had been preventable.

At the end of World War II, Korea was freed from Japanese control by the Allies. President Harry Truman sent American soldiers to Korea to prevent the Soviets from occupying its entirety, and the United States and the Soviet Union would each occupy half of the peninsula over the next few years. When the formal UN-endorsed trusteeships over the two Koreas expired in 1948, the two superpowers failed to agree on how a united Korean state would be governed. After a general election in the South, Syngman Rhee, who held a doctorate in politics from Princeton University, took over from the U.S. military government. Meanwhile, the Soviets established the Democratic People's Republic of North Korea (DPRK) and chose as its leader Kim Il-sung, an ambitious young Communist guerrilla leader who had impressed his Soviet patrons while serving in the Soviet Army in the Far East.[17] From the start, the ideologically incompatible two Koreas were each trying to undermine the other's government, each wanting reunification under its control.[18]

Despite the obvious prospects for war on the peninsula, the first secretary of the new U.S. Department of Defense, James Forrestal, was skeptical about maintaining U.S. forces there. He saw the deployment as an unnecessary "drain on Army resources," describing duty there as "a source of unceasing complaint from parents of the enlisted men who were unhappy, dissatisfied, and bored."[19] Stalin, emboldened by the drawdown of American forces, gave Kim Il-sung the green light to prepare for a large-scale invasion of South Korea.[20] The war began at four o'clock on the morning of June 25, 1950, as six North Korean infantry divisions reinforced with tanks poured across the Thirty-Eighth Parallel.

It should not have been a surprise. A February 1949 CIA top-secret study predicted that the U.S. troop withdrawal "would probably in time be followed by an invasion" and that "continued presence in Korea of a moderate U.S. force would not only discourage the threatened invasion but also would help sustain the will and ability of the South Koreans to resist any future invasion."[21]

I described to Chung my concern over a new strain of American isolationism based on a wariness of foreign entanglements. This strain traces back to our founding. In his 1801 inaugural address, Thomas Jefferson listed "peace, commerce, and honest friendship with all nations, entangling alliances with none" as "essential principles of our government." Twenty-first-century skeptics of U.S. military engagement abroad and especially of the wars in Afghanistan, Iraq, and Syria are fond of quoting former president John Quincy Adams that America "goes not abroad in search of monsters to destroy," while leaving out the context of the fledgling nation's incomplete task of western expansion, preoccupation with conflicts on the frontier with Native Americans, and lack of financial and military power.[22] Profound expressions of isolationist sentiment were manifest in America's rejection of membership in the League of Nations after World War I and its reluctance to intervene directly in World War II until after Japan attacked Pearl Harbor in December 1941. Pearl Harbor delivered a severe blow not only to the U.S. naval fleet but also to the isolationist movement in the United States. I warned Chung about the revival of American isolationism, as an activist element of President Trump's political base rallied around building a border wall, defending the United States at the ocean's edge, and ending protracted overseas commitments. At the very least, the president's supporters wanted others (especially allies) to pull their own weight. Many viewed allies as free riding on security provided by Uncle Sam while the United States got little in return. Shouldering a fair share of the burden was not a new American concern, but the voices of those who held it were growing louder.

As we spoke, I realized that Chung was the right person in the leftist Moon government to help prevent the perfect storm. His quiet, calm confidence stemmed from his long experience as a diplomat. He had served as South Korea's ambassador to the United Nations and Israel as well as the deputy minister for trade and minister of economic affairs in Washington. As a member of the South Korean National Assembly, he chaired the Foreign Relations Committee. He looked much younger than his seventy-one years, despite a grueling travel schedule that would have exhausted a much younger man. Chung could also help generate international support for the North Korea strategy as he was well respected across Asia and in Moscow and Beijing. We would have to work hard to stay aligned due not only to the incompatibility of Trump's and Moon's domestic supporters, but also because China's leaders would do their best to divide us.

* * *

AFTER DINNER, we moved out onto the patio, where I conveyed my impressions from President Trump's summit with Xi Jinping at Mar-a-Lago of two months earlier. I thought that China was likely to use DPRK tensions opportunistically to gain dominant influence in Northeast Asia. If the Kim regime collapsed, the South would dominate a unified Korea, given its much larger population (51 million compared to 25 million) and an economy estimated to be as much as eighty-eight times the North's size.[23] If the Kim regime was ultimately doomed to failure, the best way for China to prevent strong U.S. influence from extending north to the Yalu River was to drive a wedge between Washington and Seoul. Pushing the United States military off the peninsula would leave South Korea vulnerable to Chinese co-option and coercion, the objectives of which would be getting Seoul to align more closely with Beijing than with Washington and isolating China's most powerful regional competitor, Japan.

China's strategic priority of pushing the United States out of Northeast Asia explains why American efforts to get China to do more on North Korea were so often met with Chinese officials' assertions of moral equivalency between North Korea, South Korea, and the United States. After even the most egregious DPRK acts of aggression, China would invariably call on all parties to reduce tensions. To absolve China from responsibility for North Korea's aggression, Chinese leaders would persistently state that it was a problem between the United States and North Korea.

Beijing tries to obscure the obvious coercive power it has over the Kim regime. At Mar-a-Lago, President Trump told our Chinese counterparts that they could solve the problem of North Korea if they wanted to. He was correct. Over 90 percent of North Korea's trade is with China, and virtually all the North's fuel and oil imports come across the Chinese border.[24] It is impossible to fire a missile without fuel. To appear as an honest broker, China, after DPRK provocations such as nuclear and missile tests, would invariably suggest a "freeze for freeze," meaning that North Korea would cease testing in exchange for suspension of ROK-U.S. alliance activities such as joint training exercises. The problem was that each "freeze" reinforced the narrative that North Korea's actions were defensive while locking in their more advanced capabilities as the new normal. At Mar-a-Lago, however, Pottinger and I discerned a subtle shift in Chinese leaders' language on North Korea that might have communicated a willingness to soften their cynical manipulation of DPRK's behavior to advance their long-term goal of hegemonic influence in Asia.

Chinese leaders need to recognize that a nuclear-armed North Korea is bad for China and the world. North Korea's nuclear weapons would not only pose a direct threat to China, but also incite other nations to consider building their own nuclear capability to deter the Kim regime. Those nations would certainly include Japan and South Korea, but it would not be difficult to imagine similar discussions in Taiwan or Vietnam. Just as the United States played a role in discouraging both South Korea and Japan from pursuing nuclear weapons programs, it was past time for China to do something similar with North Korea.

We did not expect Chinese leaders to have an epiphany and commit fully to denuclearization, but the initial response was, at least, not disappointing. Xi, whose relationship with Kim Jong-un was nonexistent at the time—the two leaders had not yet met in person—dropped the moral equivalency language. There was no invocation of a "freeze for freeze." Xi seemed to understand that complete, verifiable, irreversible denuclearization was the only acceptable outcome. Chung and I discussed how, despite those positive indicators, China would continue to find ways to undermine the U.S.-ROK alliance and use historical ROK-Japan animosities to isolate Japan. China shares with Korea a deep resentment associated with Imperial Japan's brutal behavior across Northeast Asia. In addition to occupying Korea for thirty-five years, Japan invaded Manchuria, Shanghai, and Nanking. Especially brutal was the infamous Nanking Massacre, a campaign of killing, torture, and rape from December 1937 to January 1938 that left as many as two hundred thousand to three hundred thousand civilians dead. China takes advantage of animosity for Japan within the South Korean public and President Moon's party, given this history of shared horrors.

* * *

STRATEGIC EMPATHY applies to allies as well as adversaries. I tried to understand Chung's and President Moon's perspective on our alliance as well as on the problem of North Korea. Ambassador Chung was older than the majority of officials in the Moon government and had lived through not only the horrors of the Korean War but also the long history of North Korean aggression after the cease-fire in 1953. He graduated from Seoul National University with a degree in diplomacy in 1968 as North Korea's United Front strategy to destabilize the South Korean government and undermine the U.S.-ROK alliance intensified. That same year, during his last semester, thirty-one North Korean commandos infiltrated Seoul intending to storm the Blue House, official residence of the South Korean head of state, and kill everyone inside. They were intercepted, and all but two died in the firefight in downtown Seoul. That same month, DPRK submarine chasers and torpedo boats attacked the intelligence ship USS *Pueblo* near the North Korean coast, killing one sailor and capturing the ship and eighty-three sailors, who were abused and held in captivity for over a year. In October, 120 North Korean commandos attacked from the sea, landing on the east coast and occupying villages in an effort to instigate a Communist revolution. The invaders tried first to recruit the villagers to their cause with tales of the people's paradise in the North and then murdered those who were visibly unconvinced of how happy they could be under the Kim regime. It took two months for the South Korean military to hunt down the unpersuasive thugs.[25] North Korean attacks continued into 1969. On April 15, two People's Liberation Army Air Force MiG interceptors shot down a defenseless U.S. Navy surveillance aircraft ninety-five miles off the east coast of the DPRK, killing thirty-one Americans.

Moon came of age well after Chung, in the 1980s, as part of a new, politically active generation nicknamed the 386 Generation because they were in their thirties, had attended university in the eighties, and were born in the sixties.[26] Also, the Intel 386 microprocessor was the prevalent computer chip then, in the 1990s. For the older generation, the Korean War and South Korea's recovery were their formative life events, in which the U.S. defended then assisted in their recovery. This led to very strong anti–North Korean and pro-U.S. viewpoints. For the 386ers, their formative experience was the Gwangju Uprising and the perceived U.S. assistance or acquiescence in allowing South Korean forces to violently put down the anti-government demonstrations that eventually led to Chun Doo-hwan's ascendance to the presidency in 1979. This led to strong anti-U.S. feelings among the 386ers.

The 386ers drove the pro-democracy movement that in 1987 ended decades of authoritarian rule in Seoul. They jump-started an era of democratic reform and spectacular economic growth that remains one of the greatest successes in the history of democracy and capitalism. Once judged by some as incapable of popular rule, South Korea became one of the most vibrant democracies in Asia.

I shared with Chung my concern about some of President Moon's campaign rhetoric. Moon risked resurrecting the underlying flawed assumption of the Sunshine Policy: that an opening up to North Korea through political concessions and unconditional aid would lead to a gradual change in the regime similar to what occurred in China under Deng Xiaoping or in Vietnam after economic reforms in 1986. It would be challenging to generate maximum pressure against the North if South Korea and other nations believed there was a solution that did not require tough economic measures, a united diplomatic effort to isolate Pyongyang, and military preparation for a worst-case scenario. Any South Korean effort at conciliation with the North would not only dissipate pressure on Pyongyang, but also encourage Kim to continue rather than curtail provocations in the hope of extorting concessions.

By 2017, it was also clear that the assumptions that underpinned the strategy of strategic patience were false. The third-generation dictator of the Kim family, "the Great Successor" Kim Jong-un, would not transform North Korea into a responsible state. And he had proved capable of continuing the brutal repression of the North Korean people. During the transition of power between

Kim Jong-il and Kim Jong-un in December 2011, it was easy to find dark humor in the nature of the Kim dictatorship. Spectators sobbed hysterically during the over-the-top funeral procession. The man who led the procession seemed comical: a twenty-seven-year-old educated in Switzerland who kept an odd haircut so he might resemble his beloved grandfather Kim Il-sung. Even Chinese Communist Party leaders, not known for their sense of humor, referred to the Great Successor uncharitably as "Kim Fatty the Third."[27] The newest dictator seemed to take his family's penchant for fabricating stories of their own brilliance, prowess, benevolence, and infallibility to new, even more laughable levels. In a manual for middle school teachers, Kim appeared as a child prodigy who started driving at the age of three and composed numerous musical scores.[28] That aside, Kim Jong-un was deadly serious and consolidated power with a ruthlessness befitting his family's long history of inhumanity.

Just two months before my meeting with Chung, the Great Successor's estranged half-brother, Kim Jong-nam (KJN), suffered a gruesome death in Malaysia. KJN was a gambler, playboy, and heavy drinker who enjoyed the good life in Asia's most vibrant cities. Kim Jong-un would never trust his half-brother because his mere existence, backed by China, meant that he was a plausible replacement. KJN's fatal mistake, however, was to criticize his half-brother. On February 13, 2017, the ordinary-looking, balding forty-five-year-old walked through the bustling Kuala Lumpur International Airport terminal on his way to check in for a flight to Macau. As he stood at the kiosk, two women approached. The first, a young Indonesian woman, came up behind him, covered his eyes, and then wiped her hands down his face and over his mouth. Then a Vietnamese woman repeated the action. Both were purportedly told they were participating in a television prank. As the women ran off, KJN began to experience the symptoms of exposure to the internationally banned VX nerve agent. His muscles started to contract uncontrollably. He was escorted to the airport's medical clinic and died in the ambulance on the way to the hospital after approximately fifteen minutes of agony.[29]

The assassination underscored Kim's determination to preempt opposition. The Institute for National Security Strategy, a South Korean think tank, estimated that in his first five years as dictator, Kim personally ordered the executions of at least 340 people, 140 of whom were senior military, government, or party officials.[30] Executions in North Korea were not unprecedented, but including one's own family was a new twist. KJN was not the first victim in Kim's family. A few years earlier, Kim Jong-un had ordered the execution of his uncle by marriage, Jang Song-thaek. Jang, who was thought to be the real power behind the young dictator, was accused of various offenses, including treason and graft.[31] He was blown apart by antiaircraft cannons in front of military cadets. The scene must have pleased Kim, as he used the same method two years later to execute the chief of the Korean People's Army's General Staff, General Hyon Yong-chol, whose offenses included treason compounded by falling asleep in meetings with the Great Successor.[32] Death by antiaircraft cannon may not be the most economical means of execution, but like the use of a nerve agent on Kim Jong-nam in a busy airport, it sent a dramatic message to anyone who might want to challenge the young dictator's authority.

* * *

NO ONE was surprised by Kim's tightening grip on power. Most knew that the death of Kim Jong-il would not change the DPRK regime's ability to control every aspect of people's lives. Outside information is forbidden. Every citizen is monitored to detect any sign of dissent. The Kim family also uses North Korea's *songbun* political class system, which categorizes the population into three classes: loyal, wavering, and hostile. While the 40 percent of the population stuck in the undesirable or hostile category have no hope for advancement, those at the top can be threatened with plummeting to the bottom.[33] The party uses local informants, who report even the slightest hint of disloyalty. The accused find themselves in North Korea's massive penal labor colonies, in which an estimated two hundred thousand "counter-revolutionaries" are consigned to reeducation, hard labor, starvation, beatings, and torture. Women are raped, forced to have abortions, and sometimes have to watch as their newborn babies are killed in front of them. Some are forced to kill their babies or be killed themselves. The United Nations and an international tribunal of human rights judges concluded that Kim should be tried for crimes against humanity.[34] North Korea's nuclear and missile programs appeared even more dangerous in the context of the regime's brutality. Strategic patience was no longer a viable strategy.

The fact that North Korea's nuclear and missile programs were advancing more rapidly than most had imagined was another reason to lose our patience. On January 6, 2016, North Korea announced that it had conducted a fourth nuclear weapon test, claiming to have detonated a hydrogen bomb for the first time, although experts were dubious given the seismic evidence. A month later, in defiance of UN prohibitions, North Korea launched a long-range ballistic missile carrying what it claimed was an earth observation satellite. Eight intermediate-range missile tests followed in the next eight months. Although seven of those tests failed, it was clear that Kim Jong-un had prioritized the program and that its scientists were learning from those failures. And just two months before the 2016 U.S. presidential election, North Korea conducted a fifth nuclear weapon test of what it claimed to be a nuclear warhead.[35]

It was those multiple tests in 2016 that prompted the Obama administration and the Park government in South Korea to deploy the Terminal High Altitude Area Defense (THAAD) transportable missile system to South Korea. The system intercepts incoming short- to intermediate-range ballistic missiles during their terminal phase of flight. It seemed like a logical response to a growing threat to South Korea and to the approximately 23,000

U.S. troops and 130,000 American civilians living there.[36] But in the summer of 2017, THAAD threatened to generate that perfect storm between Donald Trump's and Moon Jae-in's political bases.

THAAD was contentious with President Trump and his political base because of its cost and the perception that it was another case of the United States defending a faraway place at American taxpayers' expense. Skeptics of U.S. military presence overseas did not care that the system was wholly owned and operated by the U.S. Army, nor that it was a cheaper missile defense solution than the alternative of multiple Patriot air defense batteries. They were especially doubtful that countries like South Korea could not afford their own defenses.

President Moon's party had its own reservations about THAAD. China had already punished South Korea economically for the missile system, including with financial sanctions, restrictions on South Korean corporations in China, and a sharp reduction in tourism. During his campaign, Moon said that THAAD required more study as well as parliamentary approval. He had also expressed disappointment over a preelection early-morning emplacement of two rockets and radar on the former golf course designated as the site for THAAD as well as the expedited arrival of the remaining four missiles in South Korea. Moon felt that the rushed deployment was to get the missiles in place without his new government's consent. Some South Korean critics fell for the baseless claim of the Chinese Communist Party leadership that THAAD would conduct radar surveillance deep into Chinese territory and shoot down Chinese rather than North Korean missiles. The THAAD deployment was especially contentious among South Koreans skeptical of the U.S. military presence, who, despite the North's provocations, tended to blame America for everything, including China's punitive economic actions against the South for hosting the missile system.[37]

President Trump and other American leaders were likely to view any South Korean hesitation to deploy THAAD as a sign of weakness in deference to China as well as a sign of ingratitude for U.S. assistance, even as the threat from North Korea grew.

Issues with THAAD were being conflated with the perception that the United States–Korea Free Trade Agreement (KORUS) was a bad deal for the United States. Prior to the South Korean presidential election, President Trump had threatened to pull out of KORUS and force South Korea to pay for THAAD.

Chung brought to our dinner a proposal to delay the deployment of the remaining missiles and other components in the THAAD battery, to allow for more analysis of the issue, parliamentary approval, and the completion of an environmental study. He drew on a napkin the portion of the converted golf course in which the initial two missiles were deployed as the others were readied for emplacement. I told Chung that his proposal could lead to disaster. President Trump, a real estate developer, would likely have a visceral reaction to even the suggestion of an environmental study. He would also see it for what it was: a delaying tactic or, even worse, an effort to bargain away an alliance capability to placate China. The delay of the THAAD deployment would also allow alliance skeptics on the right in the United States to portray the Moon government as ungrateful and strengthen those in the United States who questioned not only the THAAD deployment, but also KORUS and even the presence of U.S. forces on the Korean Peninsula. I told him that I was not an alarmist, but that the delay of THAAD could be the first step toward the rending of the alliance that had prevented war for more than six decades.

I slid the napkin toward myself and drew the remaining four missiles into the smaller area with the other two. I asked Chung why all the missiles could not go into the smaller space to prevent further delay. The environmental study could then be done deliberately, after which the missiles could be spread out over the larger space. After Chung said that he would try to make it work, I passed the napkin to Pottinger and told Chung that if he and President Moon made the THAAD deployment happen quickly, we would frame the napkin for posterity.

Pottinger and I thought that the dinner with Chung was successful. Not only did we begin to develop what would become a strong friendship, but we also resolved that we would not approach the North Korea problem consistent with the definition of insanity mentioned previously: doing the same thing over and over again while expecting a different result. We resolved to stay aligned as we endeavored to break the historical pattern of North Korean provocation, extortion, and negotiation that had worked so well for the Kim family dictators in the past. And we resolved not to take our alliance for granted.

⋆ ⋆ ⋆

IN THE summer of 2017, I thought that we were off to a good start in replacing the strategy of strategic patience with a strategy of maximum pressure. President Trump had approved the maximum pressure strategy in March. All in attendance at that National Security Council meeting had agreed that we should assume that the North Korean regime was likely neither to collapse nor to reform. We resolved not to repeat the failed pattern of past efforts and emphasized the importance of getting other nations to make that same resolution.

Coordination with South Korea was going well, and the potential dangers to our relationship were receding. At the end of June, President Moon visited Washington. He and President Trump agreed on the strategy of maximum pressure to achieve the denuclearization of North Korea. But we were under no illusions that this would be easy to achieve. We were testing a thesis that the United States and other nations could force Kim Jong-un to envision a future in which he continued to rule in an increasingly prosperous North, and thus conclude that he and his regime were safer without nuclear weapons than they were with them.

After his visit to Washington, President Moon reversed his earlier decision to wait for an environmental assessment and approved the plan drawn on the napkin at my house. I gladly received President Trump's ire for South Korea's not paying for a missile system our army owned and our soldiers operated. Meanwhile, U.S. trade representative Robert Lighthizer, National Economic

Council director Gary Cohn, and I made the case that renegotiating KORUS was better for the American people than withdrawing from the deal, both from an economic and a national security perspective.

But as maximum pressure increased, so did North Korea's attempts to force a return to the failed cycle of the past. The DPRK would conduct seventeen missile tests throughout 2017, including the test of an intercontinental ballistic missile on July 4, America's Independence Day. These tests helped solidify support for our strategy as the best alternative to what would be a dangerous and costly military confrontation or premature negotiations that would lock in the status quo as the new normal while North Korea continued to mature its nuclear program. Ambassador Nikki Haley, the U.S. permanent representative to the United Nations, masterfully negotiated four new UN Security Council resolutions that helped place significant economic pressure on North Korea. The administration worked with Congress to impose additional sanctions. The Department of State, the Department of Treasury, and the Department of Justice redoubled efforts to enforce those sanctions and disrupt North Korea's organized crime and cybercrime activities.

Because of our flawed assumptions and misaligned policies, the North Korean regime had never felt diplomatic, economic, financial, or military pressure sufficient to convince its leaders that denuclearization was in their interests. On the contrary, the cycle of North Korean provocation, feigned conciliation, negotiation, extortion, concession, promulgation of a weak agreement, and the inevitable violation of that agreement actually encouraged the North's aggression. As pressure mounted in 2017 and as nations around the world, including China, began to enforce sanctions, North Korea redoubled its efforts to get back to the negotiating table. On September 3, over Labor Day weekend in the United States, North Korea conducted a nuclear test that it again claimed to be a hydrogen bomb. The blast was estimated to have destructive power equivalent to 140 kilotons of TNT, ten times as powerful as Little Boy, the atomic bomb dropped on Hiroshima in 1945.[38]

We had work to do. Implementation would depend on maintaining unity of effort internally and with multinational partners. But despite promising, unified early efforts, it proved difficult to keep everyone committed. Some officials who helped develop the campaign of maximum pressure started using the term *peaceful pressure* instead. It was easy to lapse into strategic narcissism and, in particular, to do what one might prefer to do rather than what the situation demanded. Pottinger, the NSC staff, and I worked hard to prevent divergent and inconsistent approaches to North Korea across our departments and agencies. At one point, in October 2017, as the State Department reverted to form by seeking several channels of communication with Pyongyang despite the president's guidance to let Pyongyang first feel the consequences of its actions, the president tweeted to Secretary of State Rex Tillerson that he was "wasting his time trying to negotiate with Little Rocket Man."[39]

As we prepared for President Trump's first State of the Union address, one of three moving speeches in which he criticized North Korean human rights violations, we were reminded of the importance of our work on the North Korea challenge. I hosted the family of Otto Warmbier in my office with Pottinger and our director for Korea, Allison Hooker, before they met President Trump in the Oval Office. Otto had been a student at the University of Virginia when, on his way to study abroad in Asia, he joined a tour to North Korea. The regime's thugs arrested him and charged him with crimes against the state for allegedly trying to steal a propaganda poster. He was sentenced to fifteen years of hard labor. He was tortured nearly to death and repatriated to the United States just before he died. Meeting Otto's parents, Fred and Cindy, and his brother, Austin, and sister, Greta, as well as North Korean escapees, prior to the State of the Union address strengthened my belief in the importance of placing the nature of the Kim regime, its warped ideology, and its profound inhumanity at the foundation of our strategy. It was unfortunate that after the Singapore Summit in 2018, President Trump ceased criticism of North Korean human rights, downplayed Kim Jong-un's knowledge of Otto's treatment, and described Kim as "honorable" and a person who "loves his people."

CHAPTER 12

Making Him Safer Without Them

Although Chairman Kim Jong-un has good personal feelings about President Trump, they are, in the true sense of the word, "personal" . . . There will never be such negotiations as that in Vietnam, in which we proposed exchanging a core nuclear facility of the country for the lift of some UN sanctions in a bid to lessen the sufferings of the peaceable people even a bit.

—**KIM JONG-UN'S SPOKESPERSON, KIM KYE-GWAN, JANUARY 11, 2020**[1]

IN MID-MARCH 2018, I hosted Ambassador Chung and Japanese national security advisor Yachi Shotaro in San Francisco, the city that has served as America's gateway to the Indo-Pacific region since the mid-nineteenth century. It was our third and final trilateral meeting in less than a year. The city and the particular venue for our meetings, the Marines' Memorial Club Hotel, near Union Square, were well suited to our purpose. Since the height of the California gold rush, San Francisco has been home to vibrant Indo-Pacific expatriate communities, including Chinese, Taiwanese, Japanese, Vietnamese, Indian, and Korean immigrants. Since the 1990s, the booming Silicon Valley economy attracted a new wave of highly educated expatriates from the region, who helped make the city a global hub of technological and commercial innovation. The artifacts on display within the Marines' Memorial Club Hotel commemorated the role of the United States in determining the historical trajectory of the Indo-Pacific in the twentieth century. Founded in 1946 as a living memorial for

U.S. servicemen and women, the hotel featured many images of combat in the Pacific during World War II, from Imperial Japan's attack on Pearl Harbor, Hawaii, on December 7, 1941, to the dropping of atomic bombs on Hiroshima and Nagasaki on August 6 and 9, 1945; and those of the Korean War, from North Korea's invasion of the South on June 25, 1950, to the signing of the armistice on July 27, 1953. These images were reminders of the costs and horrors of war. They were also reminders of the achievements of an emerging superpower, a former enemy, and a nation victimized by that former enemy in the wake of those wars in building a better future.

During our first meeting at the White House in March 2017, Yachi had expressed concerns over what an "America First" foreign policy would mean for the Japanese-U.S. relationship. I assured him that although President Trump would insist on reciprocity, especially in the areas of trade, market access, and defense burden sharing, all in the administration understood that a strong United States–Japan alliance was essential to realizing the vision of a "free and open Indo-Pacific." Yachi and I became close friends. I was grateful for the opportunity to work with one of Japan's most experienced and respected diplomats. The growing threat to Japan from North Korea's missiles and nuclear program was his top priority.

Yachi, born one year before the end of World War II, was a true believer in a strong U.S.-Japanese alliance. He was an infant when six million Japanese soldiers and civilians returned home to find their country ravaged by a sustained bombing campaign that culminated in America's use of the most destructive weapon in human history to end the costliest war in human history. Yachi grew up at a time when the Japanese people wondered how their country, the size of the state of Montana and with a population of over one hundred million people yet few natural resources, could ever recover from the devastation. But the day after Japan's surrender in 1945, America extended the hand of friendship to it, and the Japanese people proved to be resilient and determined to rebuild and reshape their nation. By Yachi's eighth birthday, when the Allied occupation ended, the Japanese economy had almost recovered to prewar levels of production. It was only the beginning of an astonishing success story.[2] It was also the beginning of what would become a strong, enduring alliance between Japan and the United States. Although the alliance was strained at times, such as in 1960, during the demonstrations against the renewal of the U.S.-Japanese Treaty of Mutual Cooperation and Security (or Anpo, as it is abbreviated in Japanese), the relationship between the former foes not only benefited their citizens, but also contributed to a remarkable economic expansion that lifted tens of millions out of poverty in East Asia. Abe Shinzo, Japan's longest-serving prime minister (and the grandson of a prime minister and son of a foreign minister) described the Japan–United States alliance as one that has given the world hope. He asked rhetorically, "What should we call this, if not a miracle of history? Enemies that had fought each other so fiercely have become friends bonded in spirit."[3]

I intended our meetings in San Francisco to serve as an implicit homage to what is known as the San Francisco System of U.S. alliances in East Asia. After World War II, that hub-and-spoke system of bilateral alliances included a range of political, economic, and military commitments with Japan, South Korea, the Philippines, Thailand, and Australia.[4] U.S. officials appreciated the mutually beneficial nature of those alliances, labeling Japan the "cornerstone" and South Korea the "linchpin" for security and prosperity in Northeast Asia. Relations between the cornerstone and the linchpin remained tense, however, and Prime Minister Abe's grandfather Nobusuke Kishi inflamed those tensions when, in the late 1950s, as prime minister, he promoted postwar nationalist revisionism with acts such as dedicating a monument to Gen. Hideki Tojo and six other military leaders convicted and sentenced to death by the Tokyo War Crimes Tribunal.

Like Yachi, Chung had witnessed extraordinary changes as South Korea emerged from the decades of war and brutal occupation. Against all odds—destroyed infrastructure, a denuded countryside, illiteracy, and corrupt governance—the South Korean people created a thriving democracy and the fifth-largest economy in Asia. Between 1960 and 2020, the South Korean economy increased by a factor of 350, life expectancy rose from fifty-four to eighty-two years and the country saw the most dramatic rise in standard of living in modern history.[5]

The Korean miracle, however, ended at the armistice line that divided North and South. In the second decade of the twenty-first century, only forty miles north of Seoul's bustling streets, fewer than half of North Korea's population have access to electricity, and residential plumbing remains a luxury exclusive to the upper class.[6] Among children under the age of five, nearly 30 percent suffer stunted growth due to malnutrition.[7] Meanwhile the North Korean regime distributes goods to the privileged few based on perceived loyalty. Those who demonstrate the highest loyalty live in relative comfort in Pyongyang, while others are condemned to lives of destitution and even starvation. The regime continues to prioritize its military, weapons programs, and even the building of monuments to the Kim family dictators over the welfare of its people.

The South Korean and Japanese people, despite their tremendous success and their common commitment to democratic governance and rule of law, are divided by difficult historical memories that create tensions in their relationship. Those tensions sometimes crept into the tenor of the discussions between two men who began their lives in the midst of devastating wars. I was determined to encourage a positive relationship between our allies and between Chung and Yachi. Tensions between Seoul and Tokyo would only benefit common adversaries. Beijing spoke of the U.S. alliance system in Asia as an irrelevant relic of the Cold War. The corollary to the Chinese narrative was that South Korea and Japan should resign themselves to China's growing power.[8] A rift between South Korea and Japan could allow China to drive a wedge between the United States and both allies and allow Beijing to pose as beneficent mediator even as it pursued primacy in Northeast Asia. Lack of unity among our three nations would also diminish both Beijing's incentive to support denuclearization and Pyongyang's fears that its nuclear and missile programs were driving the three nations together.

The images and artifacts in the Marines' Memorial Club Hotel underscored the importance of remembering our history. I hoped that these artifacts might allow us to reflect on our achievements and the shared values fundamental to those achievements. Moving beyond a painful past was necessary to overcome challenges of the present and build a better future.

★ ★ ★

MATT POTTINGER and I met separately with Chung and Yachi on Saturday afternoon. NSC director for Korea, Allison Hooker, joined us for the South Korea meeting, and our NSC director for Japan, Eric Johnson, attended the Japan meeting. Pottinger and I then joined Chung and Yachi and their "plus ones" for dinner together at the Leatherneck Steakhouse, on the hotel's top floor. The spectacular view of the San Francisco skyline, along with good steaks and California pinot noir, helped dissipate some of the tension between Chung and Yachi that had accumulated since the last meeting. South Korea's renewed calls for atonement and compensation for the Japanese occupation's crimes from 1910 to 1945, including its use of Korean "comfort women" as wartime sex slaves and forced labor in Japanese industries, had elicited a defensive response from Japanese leaders, who felt that they had atoned already for the crimes of previous generations. Matt could always be counted on for a humorous story to break the ice, and the dinner allowed us to catch up on each other's families and build strong personal relationships important to the work we had before us. Chung and Yachi were statesmen. They respected each other and were able to transcend the latest tumult in the South Korea–Japan relationship. We held our trilateral meeting the next morning, after which I would depart so Chung and Yachi could have time together without me.

★ ★ ★

THE TRILATERAL meeting had three main purposes: first, to foster a common understanding of the precise nature of North Korea's nuclear threat; second, to agree on principles we deemed essential to ensuring that the Kim regime no longer posed a grave danger to our security; and third, to identify mutually reinforcing actions our leaders and our governments could take to advance our collective efforts.

It was important that Chung, Yachi, and I agree on the North Korean regime's motivations and intentions, because differences of opinion over the best approach to North Korea stemmed from divergent understandings of why Kim Jong-un wants nuclear weapons and advanced missiles. For example, those who argued that the least risky and least costly course of action would be to accept North Korea as a nuclear power and then deter its use of nuclear weapons assumed Pyongyang wanted the most destructive weapons on earth mainly for defensive purposes. As David Lai and Alyssa Blair wrote in August 2017, "facing continued hostility from the United States and its allies, Japan and South Korea, North Korea felt extreme concern about its national survival; as a result, it viewed nuclear weapons as a necessity."[9] Lai, professor at the U.S. Army War College, asserted that "in practice a North Korean nuclear capability to attack America would not threaten U.S. security" because "the North is looking for a deterrent to U.S. military action." I disagree. The assumption that Kim wants nuclear weapons only for deterrence is based on mirror imaging of an adversary that is not "like us" and on simplistic historical analogies to nuclear deterrence against the Soviet Union during the Cold War. Chung, Yachi, and I agreed that we had to base our approach on the possibility that Kim's family dictatorship wanted these weapons for more than defensive purposes.

Chairman Kim is the third in a succession of ultranationalist leaders whose legitimacy rests on the promise of "final victory."[10] It was important to at least consider the Kim regime's own explanation for investing and sacrificing so much in pursuit of nuclear weapons and missiles. Kim Jong-un and his father both spoke of their planned nuclear arsenal as a "treasured sword" designed to cleave the alliance between the United States and South Korea and make the United States think twice about ever coming to South Korea's aid in time of war. Because the United States would likely determine that the security of South Korea was not worth a nuclear holocaust on its own territory, nuclear weapons would help push U.S. forces off the peninsula as the first step toward "red-colored unification" (적화통일), or "final victory," after which South Korea would submit to Kim family rule.[11] After he assumed power, Kim Jong-un reportedly directed the military to come up with a new war plan so that the Korean People's Army could occupy Seoul in three days and the peninsula in seven. The North Korean missile launches in 2016 and 2017 were meant to exercise the war plan, which included practicing nuclear airbursts over disembarkation airfields and ports in South Korea and U.S. bases in Japan.[12]

Moreover, if North Korea was concerned mainly with deterring South Korea and the United States, it did not need nukes. North Korea has a tremendous conventional deterrent capability with more than 21,000 artillery and rocket systems able to bombard the city of Seoul, which lies only thirty-one miles from the DMZ.[13] And what was North Korea so eager to deter? Every act of aggression and violence against the United States, South Korea, and Japan since the invasion of South Korea in June 1950 was initiated by the North. North Korea directly attacked the South Korean leadership during the 1968 commando attacks of the Blue House and the failed assassination of President Chun Doo-hwan in Rangoon, Burma, in 1983, to set conditions for the ultimate objective of red-colored unification. And the North has not hesitated to employ terrorism against the innocent. In 1987, less than a year before the 1988 Seoul Olympic Games, two North Korean agents exploded Korean Air Flight 858, killing 113 South Korean civilians. One agent, who survived a suicide attempt, confessed later that the attack had been ordered by Kim Jong-il. North Korea has also abducted thousands of South Koreans and as many as a hundred Japanese citizens since the armistice.[14] Testimonies from North Korean defectors suggest varying motivations, ranging from finding native language teachers for its spies to a long-term breeding project that sought to train the abductees' children as secret agents.[15]

Rather than deter conflict, North Korea would likely become more aggressive and prone to initiate a war once the Kim regime had its desired weapons. Under the cover of nuclear capabilities, Kim Jong-un could increase physical attacks on South Korea and cyber attacks globally.[16] Pyongyang would almost certainly try to extort payoffs and concessions from South Korea, Japan, the United States, and others.[17] And if North Korea develops an intercontinental ballistic missile capable of carrying a nuclear warhead to the United States, it is easy to imagine Pyongyang issuing an ultimatum to Washington demanding that U.S. forces depart the peninsula. But even if Kim Jong-un really did want nuclear weapons for the less ambitious purpose of forestalling efforts to end Kim family rule, there are other reasons a nuclear-armed North Korea is a grave danger for which deterrence is an inadequate solution.

As with Iran and the Middle East, accepting and deterring a nuclear-armed North Korea would create strong incentives for the further proliferation of nuclear weapons in the region. If North Korea was capable of striking the United States with a nuclear weapon, thus raising further doubts about the willingness of the United States to keep South Korea and Japan under the protection of its "nuclear umbrella," it could be only a matter of time before Japan and South Korea concluded they needed their own nuclear weapons. And soon enough, other countries across Asia and beyond might conclude they needed them, too.

Another factor is the North Korean regime's record of selling every weapon it has ever possessed, including its nuclear and missile technologies. At the end of 2006, Israel Defense Forces intelligence chief, Gen. Amos Yadlin, concluded that a cube-shaped structure in the Syrian desert was a nuclear reactor intended to produce plutonium for military purposes.[18] A few months later, Mossad (Israeli intelligence) agents broke into the Vienna hotel room of Ibrahim Othman, the head of the Syrian Atomic Energy

Commission. The images they downloaded from his computer included photos of North Korean scientists and workers at the site. At 10:30 p.m. on September 5, 2007, Israel launched Operation Outside the Box as Israel Defense Forces (IDF) fighter jets streaked across the Syrian desert only one hundred meters above ground level. The fighters dropped seventeen tons of explosives on the facility, destroying it completely. Israel did not admit to the attack until 2018. And Syria, perhaps reluctant to divulge both the impotence of its defenses and the nature of its nuclear weapons programs, acknowledged only an intrusion of its airspace. Ten North Korean scientists are believed to have perished in the strike.[19]

North Korea has smuggled weapons to Houthi rebels in Yemen, militias in Libya, and armed forces in Sudan, all in violation of UN sanctions. The DPRK not only sold its nuclear program to Syria, but also assisted it with the production of chemical weapons used to mass murder civilians during the Syrian Civil War.[20] North Korea has also shared its missile and nuclear technology with Iran. In return, Iran financed the Al-Kibar nuclear reactor in Syria and brokered a range of North Korean arms sales to Bashar al-Assad.[21] Shifting its criminal and weapons-smuggling networks to the marketing and sale of nuclear weapons and facilities would be a simple task for Pyongyang. It is not unreasonable to envision North Korea selling nuclear devices to the highest bidder, even if that bidder is a terrorist organization. As with Iran, decisions concerning North Korea's nuclear program should not be separated from the nature of its depraved leadership. Yachi, Chung, and I agreed that the stakes were high for our security and the security of all nations.

* * *

WE DEVELOPED three first principles. First, we would convince other nations to support the strategy of maximum pressure and resist the temptation to accede to weak initial agreements just to get to the negotiating table. In the past, weak deals, such as "freeze for freeze," in which the United States and South Korea suspended military exercises in return for flimsy North Korean promises to suspend testing of nuclear weapons and missiles, gave North Korea what it wanted, alleviating sanctions and delivering payoffs that reduced pressure on the regime. These kinds of agreements should be nonstarters.[22]

Second, we could not view diplomacy and the development of military options as separate, sequential efforts. Successful diplomacy would depend on demonstrated will and capability to employ force against North Korea if necessary. Consistent with negotiation and mediation theory taught in business schools, we must make Pyongyang's "best alternative to a negotiated agreement," or BATNA, look bleak while making denuclearization look attractive based on economic benefits and assurances that the Kim regime would remain intact.

Third and most important, we would resist efforts to lift sanctions prematurely or to reward the DPRK government just for talking. Sanctions on the regime would remain in place until there was irreversible momentum toward denuclearization. I believe that adhering to those principles gave us the best chance to convince Kim that his family's decades-old playbook no longer worked; he could no longer keep his nuclear weapons and missile programs going while extorting concessions. But events since our previous meeting in August 2017 were already testing our ability to maintain this nascent strategy.

* * *

IN LATE 2017, after the September nuclear test by North Korea, the effort to apply maximum pressure was beginning to take effect. The UN Security Council had approved unprecedented sanctions. American, South Korean, and Japanese armed forces trained and prepared for contingencies. Chairman Kim was isolated. In December, the United States and Canada announced that they would host nations from around the world in Vancouver in January to show solidarity against North Korea's nuclear program. The Trump administration sanctioned more North Korean entities in eighteen months than the Obama administration did in eight years. Kim must have decided that it was time to make an overture and alleviate the pressure. His first move was to respond to President Moon's invitation for an opening to the North.

On January 9, 2018, when North Korean officials met with South Korean officials at the DMZ for the first time since 2015, they announced that they would accept President Moon's invitation to participate in the upcoming Winter Olympics in Pyeongchang, South Korea. During the games in the following months, fawning coverage of the North Korean athletes, its cheerleaders, the joint North Korea–South Korea women's hockey team, and especially the lead of the North Korean delegation, Kim Jong-un's younger sister Kim Yo-jong, seemed to give Kim what he wanted. South Korea seated the U.S. delegation, Vice President and Mrs. Pence and First Daughter Ivanka Trump, directly in front of the North Korean delegation, which included Kim Yong-nam, the nearly ninety-year-old nominal head of state.[23] President Moon no doubt hoped that the two delegations might break the ice and engage in conversation, but neither party spoke to the other. The good feelings that the Olympics generated, however, boosted support for a more conciliatory approach to Pyongyang, especially in Seoul. Old hopes were rekindled. Maybe participation in the Pyeongchang Winter Olympics was the start of Pyongyang's transformation into something like the government in Hanoi.[24]

Moon, eager to pursue inter-Korean dialogue and reduce tensions on the peninsula, followed up the Olympics with a meeting with Kim at the demilitarized zone on April 27, 2018. The two leaders held hands as they dramatically stepped across the line dividing the DMZ and then held a conversation at the "Truce Village," Panmunjom. The ill winds on the Korean Peninsula seemed to be receding.

After the Moon-Kim meeting, Chung and South Korea's intelligence chief, Suh Hoon, traveled to the United States to brief me and Gina Haspel, acting director of the CIA, on the results of the meeting. I arranged for them to also brief members of the president's cabinet in advance of an Oval Office meeting with President Trump. The main purpose of their trip was to communicate Kim Jong-un's willingness to meet with President Trump, convey President Moon's request that President Trump reciprocate, and to inform us of President Moon's plans for sustained engagement with the North.

I was skeptical of a Trump-Kim summit because it was likely to alleviate diplomatic and economic pressure on Kim. Other nations, especially China and Russia, would probably become slack on sanctions enforcement. Moreover, the maximum pressure strategy was still in its nascent stages; some of the sanctions would not take full effect until the end of 2019. I believed that this was why Kim Jong-un had agreed to participate in the Olympics, met with Moon, and reacted positively to Moon's proposal of a meeting with Trump. Kim needed a way to alleviate pressure, bolster his reputation, and break out of his isolation. Plus, engaging Seoul and Washington would make him more attractive to other world leaders. Chairman Xi, for example, would no doubt fear missing out on Pyongyang's dialogue with us and the South Koreans, a dialogue that could produce an outcome inconsistent with China's interests.

But I also knew that President Trump would say yes to a summit. He would find a historic first meeting between a North Korean leader and a U.S. president irresistible. His confidence in using relationships to solve problems, and in his own negotiating skills, would render ineffective any argument to wait until Pyongyang began to feel the effects of the maximum pressure campaign. Given the certainty of that outcome, our team worked with the State Department and others across the government to make the most out of a forthcoming summit while keeping the pressure campaign intact.

Despite my misgivings, I also believed that a summit with Kim would present an opportunity. Trump was unconventional, and the North Korea challenge had proven immutable to conventional approaches. A summit would drive a process that was more top down than bottom up, and that seemed positive. Bottom-up, protracted negotiations with North Korean officials who had no real decision-making authority and who were fully vested in the status quo, had in the past proven frustrating and futile. Although we knew that Kim Jong-un had a tremendous capacity for brutality, we did not know how he would respond to President Trump's argument that denuclearization was in Kim's interest. His background was different from that of his father and grandfather. He was quirky to say the least. He took his father's interest in the National Basketball Association and, in particular, the famous Chicago Bulls championship teams of the 1990s to the extreme. Kim's odd alliance with that team's eccentric, much-pierced, and excessively tattooed defensive specialist Dennis Rodman (who had also appeared on Trump's reality TV show, *The Apprentice*) revealed his tendency toward immoderation and maybe also toward making unexpected decisions.[25] Moreover, in his first five years as "Great Successor," Kim had created an unprecedented class of wealthy entrepreneurs, military officers, and party officials known as *donju* (돈주), loosely translated as "masters of money," whose extensive access to smuggled goods and foreign currency helped the economy remain stable as sanctions tightened.[26] Kim may prove reluctant to dash the rising expectations of the donju. Allison Hooker, who had traveled to North Korea several times and understood the country as well as anyone in the U.S. government, agreed that there was at least a slim chance of a breakthrough.

Our meeting with Yachi and Chung focused on how to take full advantage of the opportunities presented by the inter-Korean dialogue and the Trump-Kim Summit while minimizing risks and adhering to our principles. Keeping Korea, Japan, and the United States aligned would communicate resolve to North Korea and the world. After listening to Chung and Yachi, I summarized their comments around three points of agreement. First, while improvements in inter-Korean relations should reduce tensions, we remained committed to keeping sanctions in place until there was irreversible and verifiable progress toward denuclearization.

Second, we would emphasize to Kim and others that obligations under UN Security Council resolutions were for North Korean action, not for U.S. or South Korean concessions. In particular, we should reject talk of a freeze for freeze or other preliminary agreements that would fail to address the problem and reduce pressure on the North. Finally, we would ask all nations, including China and Russia, to encourage Kim to take advantage of the opportunity for enduring peace and prosperity. I still had profound reservations about the summit, but Kim was enjoying the temporary alignment of an unconventional U.S. president willing to take risks and a South Korean president willing to pursue a fundamentally different relationship between North and South.

I departed the White House the following month, on April 9, 2018. My regrets at leaving the job as national security advisor were threefold, and all reflected in that earlier meeting in San Francisco. I would miss working with dedicated foreign counterparts such as Yachi and Chung. I would also miss my colleagues on the NSC staff, including Matt Pottinger, Allison Hooker, and Eric Johnson. And I would regret leaving unfinished our work on crucial challenges to our freedom and security. But I also realized that the toxic environment in Washington, in the administration, and the White House had hobbled my ability to make a positive contribution to the president and our nation.

By the time of the San Francisco meeting, I had already spoken with the president about the best time for me to transition to my successor. Yachi was aware of my departure, as the White House chief of staff's office had already leaked rumors of it to the press. After our meeting, I sat with my friend in front of the large fireplace in the Marines' Memorial Club library. When he expressed his and Prime Minister Abe's hope that I would continue serving, I answered obliquely and told him how much I appreciated the opportunity to serve with him. I knew that there was no guarantee that maximum pressure would achieve denuclearization, but I would finish my tour of duty as national security advisor hoping that the strategy would survive and that we would not succumb to the tendency of returning to the failed pattern of past efforts.

* * *

TWO YEARS after my last meeting with Yachi and Chung, the threat from North Korea's missile and nuclear programs continued to grow despite President Trump's best effort to achieve a breakthrough. Although the first Trump-Kim Summit, in Singapore, ended on a positive note, with an ambiguous DPRK commitment to denuclearization of the Korean Peninsula, the course of events that followed revealed that Kim's and Trump's definitions of denuclearization were incompatible. In the summer of 2018, perhaps as a gesture of good faith, President Trump even gave Kim something for nothing, postponing what he described in a tweet as "ridiculous and expensive" joint U.S.–South Korean military exercises.[27] But Kim may have interpreted Trump's gesture and other conciliatory actions and words—such as Trump's overruling new Treasury Department sanctions on North Korea, or stating that he no longer wanted to use the term *maximum pressure* because "we're getting along," or stating that he did not want to impose more sanctions because of his relationship with Kim Jong-un—as an indication that Trump was returning to previous ineffective tactics like a freeze for freeze and premature alleviation of sanctions.[28] In early 2019, President Trump reminded North Korea that there was a much better future awaiting should Kim denuclearize. A presidential tweet predicted that the hermit kingdom could become a "great Economic Powerhouse." While the investment and real estate boom that would follow North Korea's opening sounded good from a Western perspective, to Kim the prospect of opening North Korea to the world could mean the beginning of the end of his family dynasty. In the same tweet, Trump praised Kim as a "capable leader."[29]

President Trump was trying to separate his relationship with Kim from the negotiations. At a rally in West Virginia soon after the Singapore Summit, Trump summarized their first meeting. He told his enthusiastic supporters that "I was really being tough, and so was he. And we would go back and forth. And then we fell in love. No, really. He wrote me beautiful letters."[30] President

Trump even excused Kim for personal responsibility in Otto Warmbier's death, stating that he would "take him at his word" that Kim knew nothing of the abuse and fatal injuries inflicted on the college student in his prison.[31] Such excuses and professions of affection seemed to render Kim reluctant to criticize Trump personally, but they were insufficient to achieve a breakthrough with the North Korean leader.

North Korea was adept at cultivating hope. In May 2018, to clear the way for the Singapore summit, Kim released three American hostages, Tony Kim, Kim Hak-song, and Kim Dong-chul, to Secretary of State Mike Pompeo. After the June 2018 summit, Kim dropped his most incendiary rhetoric, as did the DPRK's state-controlled media. Meanwhile, Kim and Moon continued their inter-Korean dialogue. In September 2018, Moon and his wife, Kim Jung-sook, visited Pyongyang for three days, signing a "Pyongyang Joint Declaration" promising civilian exchanges, economic cooperation, family reunions, and the destruction of two missile facilities. During that historic visit, Moon gave a powerful speech in which he reported that he and the Great Successor had "agreed on concrete measures to completely eliminate the fear of war and the risk of armed conflicts on the Korean Peninsula" and instead "turn our beautiful territory from Baekdu Mountain to Halla Mountain into a land of permanent peace, free from nuclear weapons and nuclear threats, and to bequeath it to our future generations."[32] Moon, like Trump, promised prosperity as he shared a vision of "three Economic Belts" to connect the Koreas to each other and to their neighbors. He tried to create additional momentum and goodwill by removing land mines from the DMZ and excluding military aircraft from portions of the border region. He did his best to really, this time, jump-start the gradual transformation of the regime and erode its hostility. Sadly, as in the past, Moon, like Trump, was unable to achieve a breakthrough.

As noted earlier, strategic narcissism works both ways. The second Trump-Kim Summit, in Hanoi, Vietnam, in February 2019, exposed misunderstandings on both sides. One of Kim's officials had watched the TV dramas *The West Wing* and *Madam Secretary* to understand how U.S. administrations made policy decisions.[33] Kim Jong-un, perhaps advised poorly by Chinese officials as well as his television-watching aides, seemed to believe that President Trump was so eager for a foreign policy "win" in advance of the 2020 presidential election and so weakened by the Republican Party's defeat in the 2018 midterm elections and the ongoing Mueller investigation that he would grant sanctions relief in exchange for the symbolic destruction of a used-up nuclear facility at Yongbyon.[34] And President Trump, perhaps too confident in his own persuasive abilities and the irresistibility of economic incentives, may have overestimated Kim's ability to abandon the regime's Juche ideology, the ideas promulgated by Kim Il-sung that assert the self-reliance of the North Korean people, the primacy of their interests, and their purity. Juche celebrates deprivation as a sign of the North Korean people's virtues and superiority. President Trump may have also underestimated the dictator's willingness to discard the opportunity to improve the lives of North Koreans.

After the failure of the Hanoi summit, some argued that Trump had missed his shot at a deal with the DPRK because he refused to make concessions to Kim. But to have done so would have replicated the pattern of the 1994 Agreed Framework, during which sanctions were eased in exchange for the suspension of missile tests and energy assistance was promised in exchange for halting activities at Yongbyon. Remaining true to the principle of not lifting sanctions prematurely or rewarding the DPRK merely for talking kept alive the possibility of convincing Kim, maybe, at some point in the future, that he was safer without nuclear weapons than with them.

In 2019, the United States and South Korea both tried to keep the door open to the North, but Kim kept pushing it closed. President Trump continued to emphasize his positive personal relationship with Chairman Kim and even did a last-minute pop-in to the DMZ to meet Kim after a visit to the G20 Summit in Osaka, Japan, in June. Meanwhile, the regime picked up its aggressive rhetoric, calling National Security Advisor John Bolton a "human defect" and lashing out at Secretary of State Mike Pompeo for "fabricating stories like a fiction writer."[35] President Moon continued to offer humanitarian aid and cooperative efforts, such as joint quarantine of pigs to protect against swine fever. At the end of 2019, Kim Jong-un hosted the Fifth Plenary Meeting of the Seventh Central Committee. He seemed to reject flatly the vision of a prosperous, denuclearized North Korea, stating that "we further hardened our resolution never to barter the security and dignity of the state and the safety of its future for anything else."[36] It appeared that President Trump's love had gone unrequited as Kim described the Trump administration's engagement with him as "double-dealing behavior of the brigandish state in trying to completely strangle and stifle the DPRK through its provocative political, military and economic maneuvers." He threatened to "shift to a shocking action to pay back for the pains that our people had to suffer." In other words, back to the old cycle. And in contrast to his 2019 New Year's address, in which he referred to the economy thirty-nine times and predicted that 2019 would be "full of hope," while extending greetings to the "compatriots in the south and abroad who shared our will in writing a new history of reconciliation, unity, peace and prosperity,"[37] Kim seemed in 2020 to predict more deprivation as "the DPRK-U.S. standoff has now finally been compressed to that between self-reliance and sanctions." Still, in January 2020, President Trump sent Kim a warm birthday letter.[38] Kim avoided insulting Trump personally, but his spokesperson accused the United States of deceiving North Korea over the past eighteen months of negotiations.

Meanwhile, North Korea showed no signs of slowing its nuclear or missile programs. Soon after the Hanoi summit, it tested a "tactical guided weapon" of an unknown quality and fired several short-range missiles into the Sea of Japan.[39] In 2019, North Korea fired twenty-six missiles, the most violations of UN resolutions by the DPRK in a single year. However, Kim refrained from long-range missile tests or another nuclear test. But as prospects for improved relations faded, defensive mechanisms kicked in. Kim dismissed the humanist Moon as "officious" and "double-dealing" and did his best to dash hopes of opening to the South by stating that the South Koreans should "mind their own business."[40] After the novelty of their initial affections wore off, Kim may have felt stuck in his relationship with President Trump, and doubted whether the professions of love and promises of security for his regime were true.

President Moon's and President Trump's bold diplomatic foray might have been the beginning of a brighter and better future for the people of North Korea while removing a grave threat to the world. Having borne witness to a dramatic shift in a geopolitical landscape that many did not think possible in October 1989 along the East-West German border, I had permitted myself to at least hope. But Kim could not yet transcend his deep resentment of the South's success nor his reluctance to open up to the world lest his people's access to the truth expose Juche ideology as a fraud and reveal the Kim dynasty as neither superior nor virtuous.

* * *

IN 2020, as Kim Jong-un entered his ninth year as dictator and Donald Trump prepared to run for reelection, the strategy of maximum pressure was still intact but had not yet been achieved. Sanctions have been imperfectly enforced, with both China and Russia calling for sanctions relief. Both countries have also tried to renege on commitments to return North Korean "guest workers," the estimated one hundred thousand North Korean citizens who work in Russia and China under conditions that border on slavery. Reports of Chinese evasion of sanctions increased, including evidence that China was providing components critical to North Korea's production of transporter erector launchers (TELs) for its missiles, while ship-to-ship transfers of fuel imports as well as illicit coal exports grew.[41] After it was clear that there would be no sudden breakthrough on denuclearization, it was time to increase pressure, including tougher sanctions enforcement, the exposure of human rights violations, cyber actions, and information operations.

The United States and its allies should penalize those nations that are failing to enforce sanctions and take a range of actions to improve enforcement. Secondary sanctions on financial institutions that facilitate illicit commerce with North Korea—as is alleged in the case of two of China's largest banks—could be particularly effective. Thirty representatives of North Korean banks are stationed overseas in China, Russia, Libya, Syria, and the United Arab Emirates to help evade sanctions. North Korea also uses diplomatic privileges and property to generate more of the hard currency it needs for its weapons programs. A sustained campaign of fines, sanctions, and law enforcement actions could collapse North Korea's evasion of sanctions. Cyber interdiction should complement these efforts; as should offensive cyber actions against DPRK state-sponsored cyber criminals, many of whom operate outside the DPRK. For example, defectors have testified that teams of North Korean hackers receive training and carry out cyber attacks in Shenyang, China.[42] UN sanctions on North Korean overseas laborers should be enforced. The United States and like-minded countries should sanction countries and commercial entities that help the Kim regime evade sanctions and continue its nuclear weapons program through a form of slave labor.

Diplomatic efforts should focus on getting other nations not only to enforce sanctions, but also to go beyond those sanctions and do their part to impose greater cost on Pyongyang for continuing its nuclear and missile programs. For example, the U.S. State Department and its diplomats abroad have been effective in encouraging others to take action against North Korea's extensive organized crime network.[43] Those efforts should be intensified and adapted continuously as North Korea finds new ways to evade sanctions and engages in novel illicit activities such as cybercrime.

Some argue that sanctions have not been effective, but sanctions against North Korea have never been fully enforced. North

Korea's nuclear and missile programs are dependent on sanctions erosion, as none of the major components are manufactured in North Korea. Sanctions authorized in 2017, if enforced, would generate unprecedented pressure on the North. For example, forcing the return of North Korean guest workers from China and Russia would constrain further the regime's access to hard currency and force tradeoffs between spending on its nuclear and missile programs and spending to improve the lives of North Koreans.

After promising an improved economy in his 2019 New Year's speech, Kim returned to prioritization of military capabilities over quality of life, telling his people in 2020 that "it is our firm revolutionary faith to defend the country's dignity and defeat imperialism through self-prosperity even though we tighten our belts."[44] After a period of rising expectations, not only the North's vast peasant class, but also the privileged class in Pyongyang may begin to question the wisdom and effectiveness of the Great Successor. It seems likely that the severe restriction in trade associated with the coronavirus in early 2020 was bound to not only restrict the economy, but prove to be an unintended and unfortunate means of enforcing sanctions on North Korea.

U.S. diplomats should also work with other nations and international organizations to expose and sanction North Korea's human rights abuses, including its abuse of overseas laborers. A UN special Commission of Inquiry on Human Rights in the DPRK concluded in 2014 that "the gravity, scale, and nature of these violations reveal a State that does not have any parallel in the contemporary world."[45] In the ensuing years, the regime's brutality has been undiminished. Some may argue that pressuring Pyongyang on human rights will diminish the likelihood of negotiations. But any negotiations, as seen during the Hanoi summit, are premature if Kim has not yet concluded that he could be better off without nuclear weapons.

U.S., South Korean, and Japanese militaries and the militaries of other nations play an important role in this pressure campaign. We should seek legal justification based on "reasonable grounds" to interdict and search North Korea–linked vessels, impound contraband, and sanction the offending ships and shipping companies. Military exercises and preparation for a swift and overwhelming response to North Korean aggression are also critical to convincing Kim that the United States and its allies possess the capability and, if faced with a potential nuclear strike, the will to impose denuclearization militarily without his cooperation. The success of coercive diplomacy in the form of maximum pressure depends in part on Kim's belief that the United States and its allies are more motivated to achieve denuclearization than he is to hold on to nuclear weapons and missiles.[46]

Interlocutors with the Kim regime and its enablers should continue to emphasize that the removal of the Kim regime is not a goal of U.S. policy. But they might also communicate that the goal could change if Kim refuses to denuclearize and if leaders conclude that the risk of a nuclear-armed North Korea is greater than the risk of collapsing the regime. Of course, all must acknowledge that aggressive action against the North could precipitate an escalation to a costly war. Terms like *limited strike* are misleading because North Korea would have a say in what happened after such a strike. That is why it is worth the effort to test the thesis that the combination of maximum pressure, security guarantees, and the prospect of a prosperous North Korea can achieve denuclearization. Still, it is prudent for U.S. leaders to discuss with allies, especially South Korea and Japan, scenarios that might lead them to conclude that the only way to remove an unacceptable threat of nuclear blackmail or a catastrophic attack is to act militantly against the Kim regime and its forces.

Some will argue that even such a discussion would encourage Kim to keep his nuclear weapons at all cost for deterrence. Quite possibly, Kim views the example of Libya—in particular, Muammar Gaddafi's decision to dismantle his program only to be overthrown and brutally murdered just one month before Kim took over from his deceased father—as a reason to hold on to nuclear weapons. But Gaddafi was overthrown by an internal uprising that was enabled by a NATO air campaign. If his own people or those around him conclude that Kim's policies are failing, he should be much more worried about internal than external threats.

Due to the nature of the Kim regime and its developing nuclear capability, how can the United States, South Korea, Japan, or any of the DPRK's neighbors be sure that a North Korean TEL rolling out of a tunnel and carrying a missile is on its way to a test rather than to an actual attack? And how could any of those nations know if the warhead it is carrying is inert, high explosive, chemical, or nuclear? The United States, South Korea, and Japan should expand surveillance and missile defense as well as land-, sea-, and air-based strike capabilities to deter and, if necessary, preempt a DPRK attack.

Diplomats should focus as much on allies as on North Korea and its enablers. Because of the grave danger to their people, South Korean and Japanese leaders must be full-time partners in developing ways to overcome the North Korea challenge. The relationship between the two U.S. allies deteriorated in 2019 over the South Korean Supreme Court's verdict that the victims of forced labor during Japanese occupation deserved rights to reparations from Japanese firms. In response, Japan imposed trade restrictions on a number of Japan-made industrial products, such as chemicals and precision machine tools, that are deemed essential to South Korea's high-tech firms.[47] Seoul then nearly canceled the awkwardly named intelligence sharing agreement between the two countries, the General Security of Military Information Agreement (GSOMIA). Many South Koreans began calling for boycotts of Japan, a move that resulted in canceled travel plans and a refusal to patronize Japanese-owned businesses or purchase Japanese products. The rift was a gift to Beijing, allowing China to pose as a mediator. Xi hosted Moon and Abe in Beijing in December 2019. China used this and subsequent trilateral meetings between Moon, Abe, and Premier Li Keqiang to depict China as the most influential power broker in the region.

Mediating an entente and gradually strengthening the relationship between South Korea and Japan should be a top priority for the United States. Their relationship is important not only to ensuring a cohesive approach to North Korea, but also to convincing China and Russia to play a more positive role. Chung, Yachi, and I pledged that every North Korean provocation would be seen as pushing Seoul, Tokyo, and Washington closer together. Because the old San Francisco System of alliances is the opposite of what Beijing and Moscow desire, Xi and Putin might conclude that North Korea's nuclear and missile programs are no longer serving their interests. Expanded South Korean and Japanese defense capabilities, such as in missile defense and medium-range conventional ballistic missiles, might drive home the point that the United States and its allies are becoming stronger in response to the threat from North Korea.

As always, information may be a more powerful instrument than even the best military or cyber capabilities. North Korea's Ministry for the Protection of the State maintains a total blackout of all information other than state media and persecutes those suspected of ideological crimes. Policy debates over North Korea often revolve around whether to pursue opening up to DPRK or maximizing pressure. It is a false choice. South Korea in particular should build on previous efforts to reach the North Korean people through radio broadcasts, leaflets, CDs, and USB flash drives. We should also take advantage of new technology to penetrate the North's information blockade. In 2015, Silicon Valley entrepreneurs and scientists demonstrated such technological innovations, including compact satellite dishes designed by a pair of Korean American teenagers who partnered with a former Google engineer; smart balloons that can carry USBs to targeted areas; and a mesh network to distribute digital contraband through tiny, daisy-chained computers connected to peer-to-peer Wi-Fi.[48]

And if we breach the North's information defenses, what should our messages convey to the North Koreans? We might first counter the regime's narrative and the Juche ideology. Deprivation is not a sign of virtue, and it has not been inflicted on North Koreans from the outside. North Korea's neighbors and the United States are not hostile to the North Korean people; nor are they the cause of North Koreans' poverty and isolation. There is an alternative to living in their physical and psychological gulag. To the "masters of money" surrounding Kim and living in Pyongyang, we might describe an alternative future in which they and their families could be forgiven for past crimes and thrive. But most important, content delivered to North Koreans should expose them to alternative views so they might regain their ability to form opinions other than those approved by the regime.[49] President Moon prioritized the removal of guard posts and mines along the DMZ, but it may prove a much more difficult, yet worthy, endeavor to begin to erase the psychological and perceptual lines that divide the two very different systems.

South Korea should take every opportunity to draw a stark contrast between its free and open society and the North's failing, closed, authoritarian system. One way to do that is by publicizing the stories of North Korean escapees and ensuring that those escapees receive a warm welcome in the form of employment and educational opportunities. Those who flee North Korea may also form a cadre of experts whose knowledge would be vital to help North Korea transition after the collapse or transformation of the regime.[50]

Many North Korea scholars believe that the regime is unsustainable. It is not clear, however, that it would collapse or transform before Kim Jong-un presented an unacceptable danger to the world. What the United States and its allies and partners can do is prepare for a range of scenarios. The period following regime collapse could be violent and difficult. Estimates are that reunification would cost upward of $3 trillion. The two Koreas have grown apart not only economically, but also culturally and intellectually. Thus, what may prove most important to a post-Kim Korea are initiatives in education and efforts to manage the psychological trauma and humiliation among the North Korean people once they are faced with a relative lack of skills and social standing compared to South Koreans, who have built a successful society over multiple generations. However, as North Korea's nuclear and missile programs progress, there is no time to wait for the regime's collapse. In the near term, the strategy of maximum pressure should endeavor to convince Kim that he is safer, from both external and internal threats, without them.

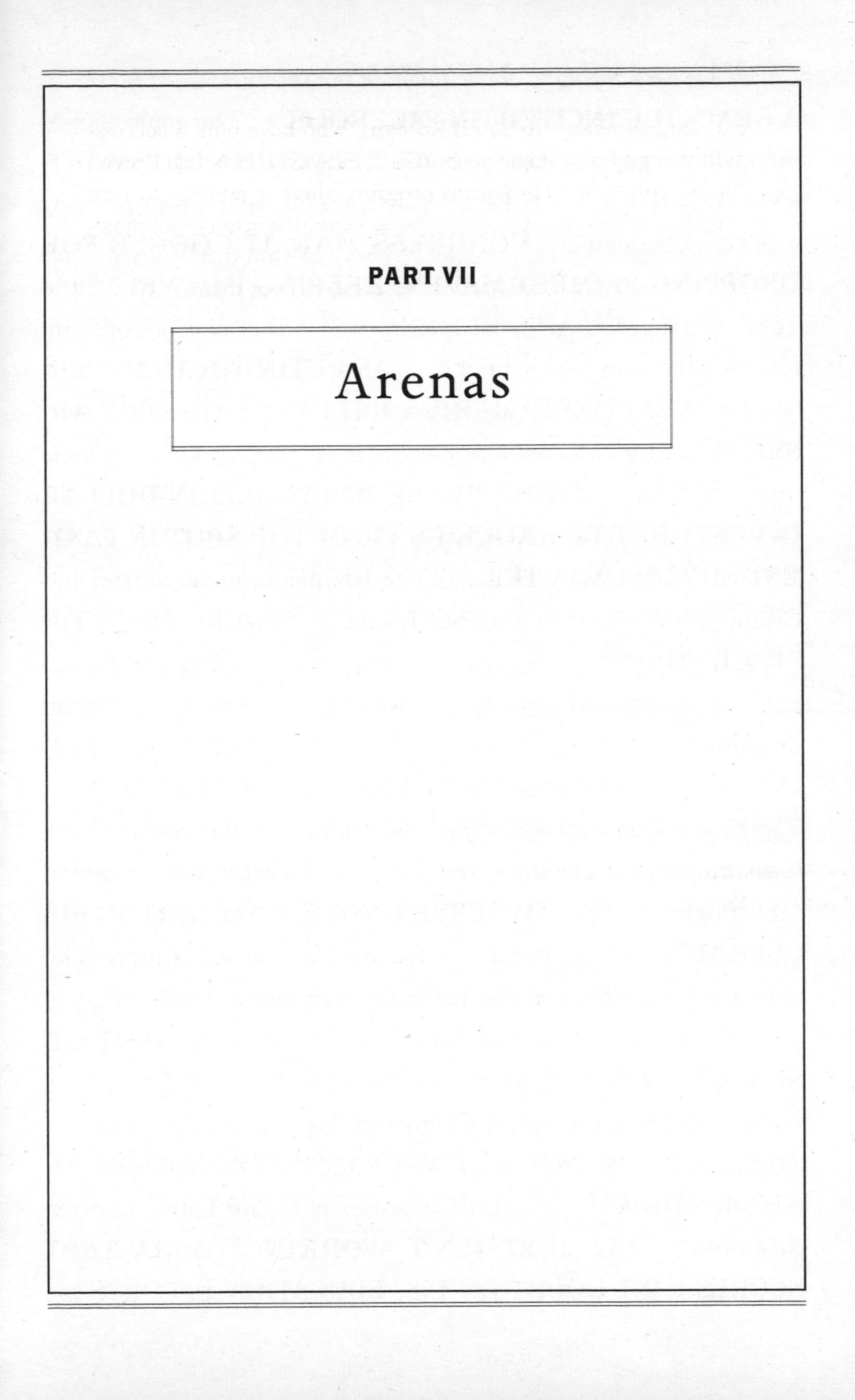

PART VII

Arenas

China has been trying to crack down on the Internet. Good luck! ... **A GENOCIDE INCITED ON FACEBOOK** ... The genie of freedom will not go back into the bottle ... **IS CHINA OUTSMARTING AMERICA IN A.I.?** ... *We have to wake up to the fiery urgency of the now* ... **CONGRESS MAD AT GOOGLE FOR DROPPING PROJECT MAVEN, KEEPING HUAWEI** ... the greatest danger of Artificial Intelligence is that people conclude too early that they understand it ... **AS PUTIN TOUTS HYPERSONIC WEAPONS, AMERICA PREPARES ITS OWN ARSENAL** ... *China's dominance of 5G networks puts U.S. economic future at stake* ... **CHINA PLANS MULTIBILLION-DOLLAR INVESTMENT TO KNOCK US FROM TOP SPOT IN FASTEST SUPERCOMPUTER** ... The Islamic Republic of Iran and China are standing in a united front ... **SPACE FORCE OR SPACE CORPS?** ... *Transform knowledge of ourselves, our planet, our solar system, our universe* ... We are almost as dependent on satellites as we are on the sun itself ... **TENCENT'S STARTUP INVESTMENT FRENZY NOW REACHES OUTER SPACE** ... *There's one issue that will define the contours of this century more dramatically than any other, and that is the urgent threat of a changing climate* ... **TRUMP SERVES NOTICE TO QUIT PARIS CLIMATE AGREEMENT** ... We are the first generation to be able to end poverty, and the last generation that can take steps to avoid the worst impacts of climate change ... **AT U.N. CLIMATE SUMMIT, FEW COMMITMENTS AND U.S. SILENCE** ... *India's air pollution rivals China's as the world's deadliest* ... *THE ARCTIC IS "NOT UP FOR GRABS," SAYS NORWEGIAN AMBASSADOR* ... The United States is losing Latin America to China ... **"IT JUST ISN'T WORKING": PISA TEST SCORES CAST DOUBT ON U.S. EDUCATION EFFORTS** ...

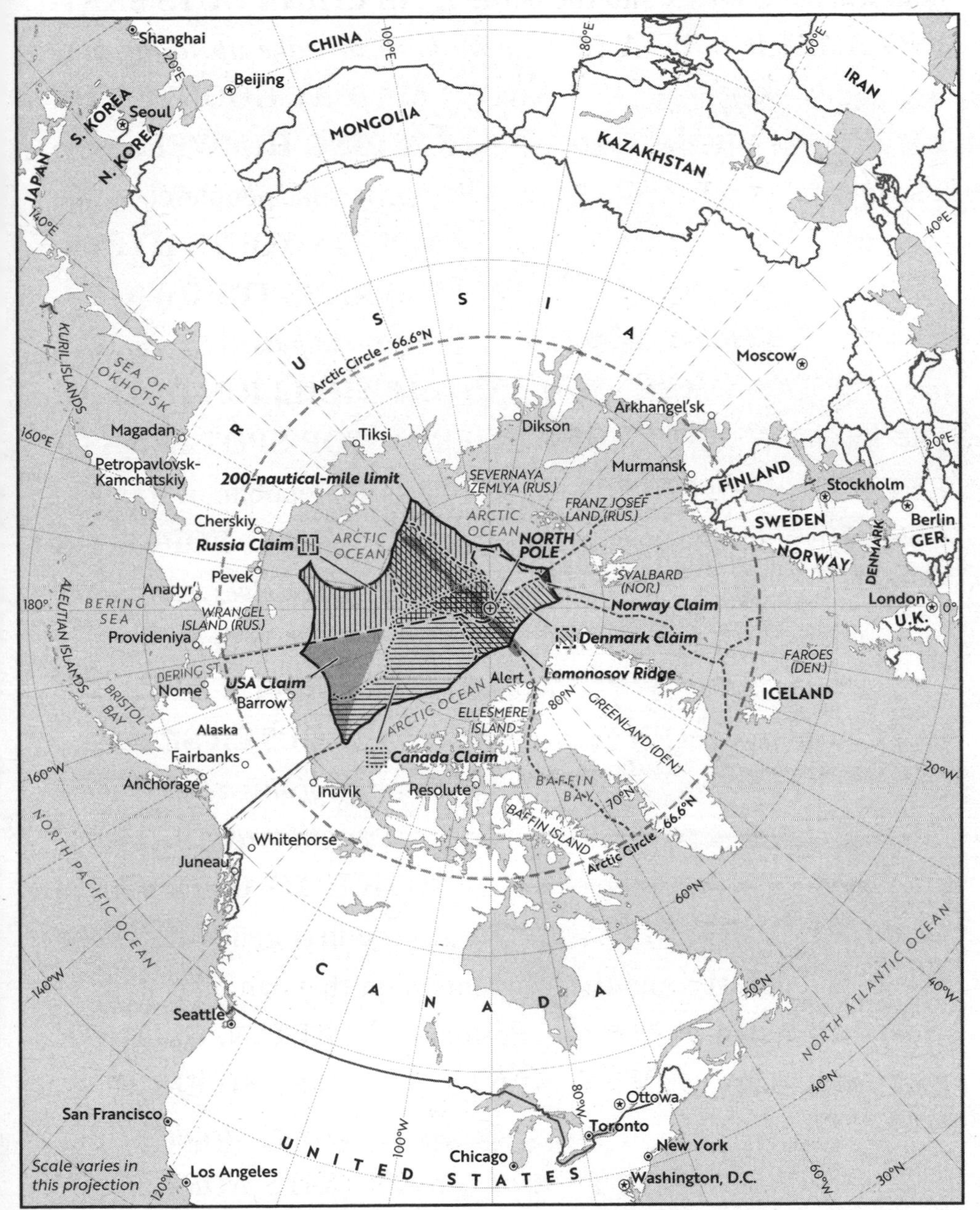

Arctic claims source: IBRU, Durham University

Map copyright © 2020 Springer Cartographics LLC

CHAPTER 13

Entering the Arena

> *What is required is a holistic approach that does not seek to isolate open systems from their environment, but apprehends their profound interconnectedness.*
>
> —**ANTOINE BOUSQUET, *THE SCIENTIFIC WAY OF WARFARE*, 2009**

ONLY SIX months after I visited the Foreign Policy Research Institute in Philadelphia to discuss the army's study of Russia new-generation warfare, the NSC staff and I became objects of a new facet of the Kremlin's sustained campaign of political subversion. Building on the vitriolic political discourse on social media surrounding Donald Trump and his new administration, Russian intelligence agents employed many of the same bots, trolls, and American accomplices it had used during the 2016 presidential election in an effort to undermine the effectiveness of the U.S. government. Members of the NSC staff, especially those who did not possess last names of northern European origin, were slandered and harassed on social media. The emphasis was on reinforcing the "deep state" narrative, which asserts that disloyal civil servants were actively undermining President Trump's agenda. The Kremlin apparently saw me and a well-functioning NSC staff as impediments to its foreign policy agenda, from Syria to Ukraine to Afghanistan, and to its efforts to secure relief of U.S. and European Union sanctions. The Russians accordingly took advantage of what the Atlantic Council's digital forensics laboratory described at the time as "the most well-organized campaign in the history of the alt-right" to remove me from the

White House and undermine the confidence of and confidence in the NSC staff. The alt-right, like the Russians, saw me as an obstacle to advancing its agenda, so it collaborated using social media under the #FireMcMaster campaign.

Consistent with the Russian "firehose of falsehood," the conspiracy theories and slanderous, bigoted content of the Fire McMaster campaign were often inconsistent. For example, one caricature on social media portrayed me as a puppet of billionaire George Soros and the Rothschild family (both of whom are frequent targets of anti-Semitic conspiracy theories), while articles in the pseudo-media charged me and others on the NSC staff with being "anti-Israel" and soft on Iran. Because I believed that calling terrorists "Islamic" masked their perversion of the religion and reinforced the terrorist narrative of fighting infidels to establish the caliphate, I was, despite having fought directly against terrorist organizations for many years, portrayed as soft on jihadist terrorism as well.[1]

Although I was concerned for the targeted members of the NSC staff, I paid scant attention to the attacks because there was work to do. But the experience amplified the urgency of entering new arenas of competition. The cyber-enabled effort to erode the effectiveness of U.S. foreign policy through disparagement and dissension would, with the advent of deep fakes and other new technologies, become only more prevalent and more dangerous. We would have to understand better these emerging technologies and how adversaries were likely to employ them against us.

In February 2017, the NSC staff worked with the president's cabinet to identify crucial challenges to national security. Some were geostrategic. Others centered on functional competitions, such as those in space or cyberspace. As we developed integrated strategies for those challenges, it became clear that the same elements of strategic narcissism that had disadvantaged the United States in competitions with revisionist powers, jihadist terrorists, and hostile states had also put us behind in technological and economic contests important to future security and prosperity.

Our team prioritized the competitive domains of cyberspace

and space, but we also worked with colleagues on the National Economic Council to determine how to promote American security and prosperity in areas such as energy and trade and across what we labeled the National Security Innovation Base (NSIB). We defined the NSIB—every initiative in government seems to require initials—as the network of knowledge, capabilities, and people, including academia, National Laboratories, and the private sector, that turns ideas into innovations, transforms discoveries into successful commercial products and companies, and protects and enhances the American way of life. We recognized that technologies (such as those associated with fifth-generation communications (5G), artificial intelligence, quantum computing, and biogenetics) would be vital to maintaining America's advantages in defense and in the global economy. But staying competitive did not mean foreclosing on cooperation, especially with the private sector and other free and open societies. We would even have to find new ways to cooperate with adversaries and rivals, especially on preventing the proliferation of the most destructive weapons on earth and tackling the interconnected problems of climate change, pollution, health, and food and water security. Competing effectively while fostering cooperation required a conscious effort to overcome narcissistic tendencies, in particular by rejecting optimism bias and wishful thinking.

* * *

THE ADVENT of the internet initially generated tremendous optimism. It transformed the global economy and accelerated communications and the transfer of data. But it also had unanticipated political implications. The internet was supposed to make autocracy untenable. At the turn of the twenty-first century, U.S. President Bill Clinton scoffed at the Chinese Communist Party's efforts to control the internet. "Good luck," he said in a speech to students at the Paul H. Nitze School of Advanced International Studies at Johns Hopkins University. "That's sort of like trying to nail Jell-O to the wall." He predicted that, in the new century, "liberty will spread by cell phone and cable modem."[2] Clinton was not alone in assuming that the internet would change China. It would have been difficult to imagine at the time the extent to which the CCP and other authoritarian regimes would change the internet. But like all technologies, the internet was neutral. What mattered most was *how* people would use it. In China and elsewhere, rather than foster political empowerment and freedom, the exponential increase in internet usage and smartphones actually gave authoritarian regimes new tools for repression and controlling their populations.[3]

In democratic societies, the free and universal internet was supposed to be liberating and empowering. And in many ways, it was, unleashing dramatic change through instant access to an endless supply of information and connecting people electronically in a way that had profound and positive effects on social interaction, productivity, and education. But it was not an unmitigated good. Social media companies lured citizens into forfeiting their privacy as those companies mined personal information to manipulate behavior for profit. Internet platforms proved ideal for amplifying hate speech, fomenting division, and even inciting violence, as in the case of the Rohingya genocide in Myanmar starting in 2016.[4] Moreover, as people became better connected than ever electronically, they disconnected from each other socially and emotionally. The internet was supposed to foster humaneness, but terrorist propaganda and other distressing content glorified the murder of innocents and desensitized people to violence. A significant number of young people abandoned playgrounds and athletic fields for their game controllers and smartphones.[5]

By 2020 it was clear that the cyber warfare threats to us and other democracies extended far beyond Vladimir Putin's Russia. For example, in 2017, Iran began using its state media to pose as independent news outlets to promote anti-Saudi, anti-Israeli, and pro-Palestinian sentiment in the United States.[6] North Korea launched disinformation campaigns to influence international negotiations, such as the inter-Korean dialogue and denuclearization talks with the United States.[7] Beginning in 2019, the CCP initiated campaigns to discredit Hong Kong protesters on the mainland as well as a failed operation to influence the outcome of the presidential election in Taiwan.[8]

But it is not only dictators who exploit personal data and violate privacy. So do internet companies who sell that data or use it for highly lucrative advertising. The United States and like-minded countries should demand standards and develop technologies that preserve privacy and prevent the misuse of personal data. Europe and the state of California have passed laws to protect personal data, but those should be refined and expanded to other governments that value privacy and due process of law. Improvements like protecting data through encryption and defaults to privacy are vital to protecting citizens from external as well as internal cyber-related threats to freedom. They are also vital for individual and societal health.

Viewing the internet and social media as an arena of competition rather than an unmitigated good is a mind-set we need in order to take advantage of the free exchange of information while protecting against dangers. In addition to the defensive measures identified in chapter 2, defense against cyber-enabled information warfare should also have an offensive component, to introduce information into closed systems, counter disinformation, and challenge government-approved narratives. Democracies should develop the means to bypass control mechanisms such as China's Great Firewall or Iran's internet restrictions. Technologies such as space-based broadband communication may make it harder for dictators to shut off access to information.

Perhaps most important, citizens should not wait for political leaders or the media to counter cyber-enabled information warfare. Individuals can decide to reject the toxicity and disinformation in the social media ecosystem and reintroduce civility into the discussions important to a thriving democracy. Engaging with those who think differently should be valued as part of a vibrant civic life, not only in cyberspace, but also in classrooms, cafés, town halls, basketball courts, and rugby pitches. As people argue about the issues that divide them, they might devote equal time to celebrate what they have in common. Citizens of free and open societies might cherish the freedoms and opportunities their forbears bequeathed to them while acknowledging that no democracy or free-market economy is perfect and that all are works in progress. And all might take an interest in self-education concerning the crucial challenges to their security, health, and prosperity as a means of inoculating themselves against disinformation.

* * *

THE CYBER warfare threat will only get worse due to advances in artificial intelligence. AI technologies allow systems to perform tasks usually reserved for humans. Machines will learn from data and use algorithms to make decisions free from human intervention. Combined with high-speed mobile communications networks such as 5G, supercomputers, and the "internet of things" (i.e., the internet of computing devices embedded in everyday objects), this could affect everything from power grids to public transportation to financial transactions to global logistics to driverless cars to home appliances.

AI technologies could make cyber attacks easier as more of the physical world becomes connected to cyberspace and the malicious actors who operate within it. In December 2019 alone, ransomware attacks (attacks that present the victim with a choice of either paying to regain access to their network and data or incurring millions of dollars in costs to restore them) crippled America's largest wire and cable manufacturer in Georgia, a health network in New Jersey, and the city governments of Riviera Beach, Florida, and New Orleans, Louisiana.[9] In 2019, the city of Baltimore chose not to pay a ransom of $75,000 and incurred an estimated cost of $18 million.

To deter attacks, the United States and its allies must be prepared to act against hostile cyber actors beyond the cyber domain. But sanctions or other threats of punitive actions are often inadequate. They require holding something of value to an adversary or an enemy at risk, and that proves difficult with elusive terrorists or criminals whose organizations hide their leadership and other important assets. And as hostile regimes like Iran and North Korea come under increased international and internal pressure, their leaders may conclude that they have little to lose. That is why deterrence by denial—that is, convincing adversaries that they cannot accomplish their objectives through a cyber attack—is essential.

Deterrence by denial requires a combination of offensive and defensive capabilities, improved resilience of systems, and a high degree of cooperation across government, businesses, and academia. Unfortunately, such cooperation is a challenge in our decentralized, democratic systems. According to Director of National Intelligence James Clapper, when North Korea hacked Sony studios in 2014, the response had to go "through some other country's infrastructure, the lawyers went nuts, so we didn't do anything on the cyber front." Instead, "We ended up sanctioning a bunch of North Korean generals."[10]

During the first year of the Trump administration, our NSC staff worked to remove these bureaucratic impediments. I was frustrated with the slow progress, but once appropriate authorities were granted, the United States became more responsive and competitive. Cyber defense of the 2018 midterm elections, directed under Gen. Paul Nakasone, the hypercompetent director of the NSA and commander of U.S. Cyber Command, was effective. As Nakasone reported to Congress in February 2019, "We created a persistent presence in cyberspace to monitor adversary actions and crafted tools and tactics to frustrate [the Russians'] efforts."[11]

A counterintuitive but key defensive action is ensuring that cyber networks and systems are designed for graceful degradation under the assumption that they will be attacked relentlessly. When Russia attacked Ukraine's power grid in December 2015, the antiquated nature of the system actually proved an asset, as it permitted the restoration of electrical services in less than six hours through an analog backup. Exquisite systems based on the latest technology may be prone to catastrophic failure. Resiliency must be a critical design parameter for communications, energy, transportation, and financial infrastructure. Resiliency requires keeping suspect hardware and software off networks and continuously identifying and, when appropriate, preempting enemy attacks. A first step is to recognize that allowing companies such as China's Huawei or ZTE into our communications networks is tantamount to opening Troy's gates to the mythical Trojan horse. Vigilance should be habitual and integrated into company and governmental operational culture. A best practice is to reward hackers who expose flaws. Microsoft, for example, changed its policy from threatening hackers with lawsuits to inviting them to security conferences and paying them "bug bounties" for uncovering security vulnerabilities.

There must also be a high degree of cooperation across the public and private sectors. As Jason Healey, an expert in cybersecurity at Columbia University, observed, "America's cyber power is not focused in Fort Meade with NSA and U.S. Cyber Command. The center of U.S. cyber power is instead in Silicon Valley; on Route 128 in Boston; in Redmond, Washington; and in all of our districts where Americans are creating and maintaining cyberspace."[12] U.S. government relations with the technology sector, however, are often contentious.

Competing effectively in the cyber domain requires common understanding. It is important for engineers at tech firms to know how adversaries use cyberspace and emerging technologies and to be aware that their firms are competing against not only other companies, but also hostile nations. Companies that reject opportunities to work with the United States and other democratic governments, while helping authoritarian regimes repress their own people, may not realize the dangers they promote. The decision by Google employees to protest the company's participation in a U.S. intelligence contract while Google was simultaneously helping the CCP empower its surveillance state must have been based in part on ignorance of what was at stake in the U.S. competition with the CCP.[13]

Private-sector companies that specialize in cybersecurity and countering cyber espionage hold promise for bridging the divide between the tech sector and government. One example is Strider, a cybersecurity company founded by Greg Levesque, who has experience in both government and industry. Strider uses proprietary data sets, machine learning, and human intelligence to combat intellectual property theft inside companies. More and more private-sector companies will likely conclude that they need to be active on adversary networks to detect and preempt attacks on their systems or intellectual property. And private-sector efforts that overlap with those of governments could lead to better civil-military coordination and cyber defense burden sharing. Because companies that go offensive in cyberspace risk incurring foreign government penalties, assuming liability for harm inflicted on innocent third parties, and sparking an escalation to armed conflict, public- and private-sector coordination is essential for integrating offense and defense in cyberspace.

⋆ ⋆ ⋆

IN THE last century, space was the new competitive domain. In 1957, when the Soviet Union launched *Sputnik* into orbit, U.S. leaders feared that they were losing that competition. Fear inspired a range of reforms, including a rejuvenation of science education, an intensified focus on missile development, and the creation of the National Aeronautics and Space Administration (NASA).[14] In the wake of the Cold War, however, U.S. commitment to leadership in space waned as some assumed that space would become a benign environment in which the world's powers would cooperate for mutual benefit.[15] Showing optimism bias similar to that about the free and open internet, the United States assumed that if it chose not to weaponize space assets, that others would follow its example. Predictably, this bout of strategic narcissism applied to space caused the United States to fall behind. After the retirement of the Space Shuttle in 2011, the United States became dependent on Russia for manned spaceflight. International cooperation in space did expand, but recognizing that space capabilities gave the United States significant economic and military advantages, Russia and China chose not only to develop their own programs, but also to build weapons to disable or destroy those of the United States and its allies.[16]

In 2007, China shot down one of its own satellites with a missile. In the ensuing years, Russia and China developed a range of disruptive counter-space capabilities, which ranged from anti-satellite laser weapons and missiles to orbiting weapons to electronic warfare jammers.[17] Countries friendly to the United States are developing their own capabilities to deter hostile actors. In March 2019, for example, India used an anti-satellite missile to blow up one of its own satellites and show the world that it, too, had offensive counter-space weapons.

In 2017, recognizing the need to compete more effectively in space across the government and commercial sectors, the Trump administration reestablished the National Space Council under Vice President Mike Pence. I asked the NSC staff to work with Vice President Pence's team to develop a strategy for reinvigorating our space program. Our team got to work under an extremely knowledgeable and effective air force officer, then-Maj. Gen. Bill Liquori. Bill understood that the stakes were high and that cooperation with allies and the private sector was necessary to combat the potential dangers stemming from the militarization of space and to take full advantage of opportunities associated with the commercialization of space. The Space Council established objectives to deter and, when necessary, defeat adversary space and counter-space threats; to ensure that American companies continued to lead in innovative space technologies; and to use space exploration to "transform knowledge of ourselves, our planet, our solar system, and the universe."[18]

As in cyberspace, deterring aggression depends on convincing an adversary that it cannot accomplish objectives through offensive action against U.S. space assets. The U.S. government and industry should protect technology that might assist China or Russia in developing advanced space capabilities that could be used against us. U.S. companies should be suspicious of foreign investors like the Chinese company Tencent, which has taken large stakes in U.S. space start-ups such as Moon Express, Planetary Resources, NanoRacks, and World View Enterprises.[19] Tencent, the company that owns WeChat and QQ, two of the largest social media applications, acts as an extension of the CCP by censoring, monitoring, and reporting private communications and personal data. It will continue to act as an extension of the party in space.

Despite these dangers, space competition provides real opportunities to improve security and prosperity and address some of earth's most pressing needs. Systems delivered into space will deliver persistent global access to the cloud as nongeostationary (NGSO) satellites (satellites that move in relation to Earth's surface) process data and communications. These satellites can provide real-time persistent remote sensing of the surface of Earth, which can contribute to environmental protection and rapid response to natural and man-made disasters. Planet, a company founded in 2010 by a team of ex-NASA scientists, aims to image the entire Earth every day and make changes visible, accessible, and actionable. The transparency its 150 earth observation satellites provide can identify diverse activities important to security, such as missile activity in North Korea, destruction of rainforests in Brazil, wildfires in Australia, pollution and damage to ecosystems in India and China, and protests in Iran. More opportunities to use space for solving problems on Earth are reaching technical feasibility and economic viability. One example is space-based solar power generation.

To take full advantage of opportunities and protect against dangers in space and cyberspace requires an understanding of how technologies interact with one another and humanity. Too often the application of a promising technology lags because it is viewed in isolation of others that, when combined, unleash tremendous potential. That is why collaboration among scientists and between scientists and policy makers is vital for innovation.

The need for collaboration on crucial challenges to national security is growing because technology-based innovation is shifting away from governments and toward the private sector.

* * *

PRIOR TO the end of the Cold War, the U.S. model of technological development was relatively closed, meaning that the government funded and controlled access to major initiatives such as nuclear weapons, jet fighters, and precision-guided munitions. These programs were protected by security classifications, patents, and copyrights. When the government decided to declassify technologies such as microchips, touch screens, and voice-activated systems, private-sector engineers and entrepreneurs combined and refined those technologies to kick-start new industries such as the smartphone.

In the twenty-first century, technological innovation truly opened up. Innovations increasingly derive from diffuse publicly financed research. Meanwhile, China has implemented its top-down military-civilian fusion strategy to steal technology and direct investments with the intention of surpassing the United States in strategic emerging industries (SEIs) and military capabilities. A new model for applying new technologies to national security challenges is overdue lest the United States and its allies find themselves at a significant disadvantage.

Much of academia, the private sector, and the government has been oblivious to how adversaries can steal and apply technologies developed in the United States to threaten security and human rights. I have discussed many of these in my earlier chapter on China, but I would emphasize here that U.S. capital is accelerating the CCP's efforts to surpass the United States in a range of critical emerging technologies, such as AI technologies and others, important to achieving military superiority. Seven hundred Chinese companies, the majority of which are state-owned or -controlled, are traded in the U.S. debt and equity markets. U.S. citizens fund companies that are building the next generation of the PLA's military aircraft, ships, submarines, unmanned systems, and airborne weapons. In 2018, U.S. venture capital investment in Chinese AI companies exceeded investment in U.S. companies. Many U.S. and allied executives and financiers go beyond the quotation attributed to Vladimir Lenin that "The capitalists will sell us the rope with which to hang them." They are actually financing the CCP's acquisition of the rope.

In 2017, it was clear to Matt Pottinger, Nadia Schadlow, Brig. Gen. Robert Spalding, and others on the NSC staff that it was past time for the United States to reenter the arena of technological competition. Any decisions involving technological and infrastructure development must consider how the proposed technology and infrastructure would interact with geopolitical competitions. One of the most important competitions is over control of data. Whoever controls 5G hardware will have access to data flows and influence over establishing data protocols that could not only impinge on privacy but also bestow unfair economic advantage. Control of data, when combined with AI technologies, can permit dominance of key sectors of the global economy.

But even an expansive view of Chinese designs on AI and military technologies may be too myopic, as the CCP's ambitious strategy is to control physical as well as digital infrastructure to achieve dominance of future global logistics and supply chains. The vanguard of this twenty-first-century conquest is China's state-owned and state-sponsored enterprises, including telecommunications, port management, and shipping companies. Democratic, free-market economies continue to furnish the CCP with "rope" as China has set about acquiring a global maritime infrastructure that complements its control of communications infrastructure. China has targeted EU countries and other U.S. allies such as Israel for control of ports. And many of these ports under Chinese control, such as Antwerp, Trieste, Marseille, and Haifa, are located near clusters of scientific and industrial research facilities. By 2020, according to China's Ministry of Transport, fifty-two ports in thirty-four countries were managed or constructed by Chinese companies, and that number was growing.[20]

The United States and other nations are at a disadvantage due to a failure to understand China's ambition holistically and the growing cultural, philosophical, and business process gap between their national security communities and innovation ecosystems such as Silicon Valley. That cannot continue. In the United States, tech executives and senior government officials are beginning to acknowledge that their lack of cooperation has helped shift power from free societies and free-market economic systems to closed, authoritarian systems. They have identified three obstacles to cooperation: the misalignment of government and business processes, a lack of understanding among scientists and engineers concerning the security implications of technological competitions, and the difficulty of moving people between public- and private-sector positions. Overcoming those obstacles requires action. Organizations like the Defense Innovation Unit, which gives the U.S. Department of Defense a presence in Silicon Valley, is an organizational best practice that could easily be replicated.

The United States, China, and the European Union are all taking different approaches to the degree to which the state, companies, or individuals control data. The United States should work with like-minded countries on a policy that ensures access to needed data while safeguarding privacy and maintaining consumer trust. Free and open societies should have common standards for how their governments interact with the private sector and with one another when it comes to how data is managed and how it is collected, processed, stored, and shared.[21] There is a growing rift between free and autocratic models of data governance. The United States and other free nations should agree on common standards consistent with their democratic principles.

★ ★ ★

SOME CHALLENGES require cooperation not only with allies and partners, but also with competitors and adversaries. Cooperation is necessary to prevent the spread or use of the most destructive weapons on earth. Halting the proliferation of nuclear weapons to hostile regimes like Iran and North Korea should be in all nations' interests. So should nuclear arms control agreements that put in place confidence-building measures critical for preventing a misunderstanding or miscalculation that could lead to the use of nuclear weapons. Arms control agreements and international conventions can limit nuclear stockpiles and arrest the development of, or eliminate, destabilizing classes of weapons such as chemical or biological weapons. The Chemical Weapons Convention of 1997 outlaws the production, stockpiling, and use of chemical weapons and their precursors. The New START treaty of 2010 between the United States and Russia reduced the number of strategic missile launchers each country had by half while also establishing a new inspection and verification regime. The Intermediate-Range Nuclear Forces (INF) Treaty of 1988 eliminated land-based intermediate-range nuclear weapons, but since 2014, Russia has violated the agreement, and China, which was not a signatory, has developed the prohibited missiles.

Obviously, an arms control agreement to which only one party adheres is an imaginary one. In 2017, the State Department let Russia know of the U.S. intention to withdraw from the INF in the hope of motivating the Kremlin to return to compliance. It did not work, and in February 2019 the Trump administration announced the U.S. withdrawal from the treaty, a decision I believed was right and long overdue. Concerns mounted that the New START, which expires in 2021, would suffer the same fate. By early 2020, it seemed that efforts to renew START and engage both Russia and China on a successor to the INF Treaty were strong possibilities. Arms control agreements that limit nuclear weapons and other weapons of mass destruction, if well monitored and enforced, besides helping to reduce the risk of the unimaginable, can make funds available for projects that benefit humanity rather than threaten us with Armageddon.

But as nuclear weapons and long-range missiles proliferate, missile defense becomes even more important for safeguarding the American homeland as well as allies and U.S. citizens overseas.

In 2019, the Trump administration completed a missile defense review that concluded that there was a need for significant investment in improving homeland and regional missile defense. It should not be controversial to support science and technology research programs to deliver cost-effective solutions to expanding missile threats. As the 2019 review directed, these solutions should integrate "offensive and defensive capabilities for deterrence" and ensure the ability to "intercept missiles in all phases of flight after launch."[22]

Other threats that place civilian populations and infrastructure at risk are proliferating. Autonomous aerial and subsurface vehicles pose a significant danger, for which defenses are immature. The 2019 swarm drone attack on the Saudi Arabian Aramco facility, which cut oil production by about half (approximately 5 percent of global oil production), should serve as a warning, as should the drone activity around London's Gatwick airport that shut down airport traffic in December 2018. Autonomous and swarm attacks threaten to force our citizenry to live in fear reminiscent of that which Londoners experienced during the Blitz in World War II from V1 and V2 rockets, first-generation drones launched by Nazi Germany.

★ ★ ★

A THEME in this book, strategic narcissism, and the corresponding tendency to artificially separate interconnected problem sets, encourages short-term, simplistic solutions to complex problems. Bias against the long-term approach, like other maladies affecting U.S. policy, stems from a lack of empathy. Jamil Zaki, professor of psychology at Stanford University and the director of the Stanford Social Neuroscience Lab, observes in his book, *The War for Kindness: Building Empathy in a Fractured World*, that both time and distance diminish empathy because humans' "caring instincts are short sighted." Our ability to feel empathy about future developments is limited because "we tend not to feel for our future selves. It goes against our instincts, therefore, to tackle problems that we have not yet been forced to confront. If the consequences of action or inaction are far off and afflict strangers yet to be born, we are less likely to sacrifice or invest today." That tendency is evident across the globe on the interrelated problems of climate change, pollution, energy security, and food and water security.

Like discussions with President Trump concerning the Iran Nuclear Deal, those concerning the Paris climate accord were animated. I was far from an expert on the subject, but it was clear to me that the issue of climate change tended to move people toward polar extremes. Climate activists endorsed impractical measures, while climate deniers and skeptics disregarded compelling evidence that global warming is happening, that it is caused by humans, and that, if unchecked, it will have disastrous consequences. Perhaps naïvely, I thought that if we simply focused on what Americans agreed on, we could develop options for a climate strategy, get beyond disagreements between those on the fringes of the issue, and make real progress.

I recommended to the president that the United States stay in the Paris Agreement, an environmental accord adopted by nearly every nation in 2015 to reduce global greenhouse gas emissions and limit the global temperature increase in the twenty-first century to 2 degrees Celsius above preindustrial levels. I believed that climate change was manmade and that we had to develop a sound, multinational solution. The agreement was nonbinding, so I did not see the downside of staying in it. Besides, I felt that withdrawing would result in a loss of American influence not only on the range of climate-related issues, but also on other challenges that required multinational efforts.

Those who argued for leaving, however, believed that if the United States failed to meet its targets for carbon reduction in the agreement, activists would initiate litigation against the government and industries. They also believed that meeting the targets would limit economic growth and impose costs on the American people even as the world's greatest polluters not only agreed to less ambitious goals, but also received payments from the United States and others as an incentive to convert to renewable energy sources. The agreement, like the Iran nuclear deal, was not ratified by Congress; many saw it as an infringement on sovereignty. In Trump's Rose Garden speech in June 2017 announcing his decision to initiate the withdrawal process, he said the deal "disadvantages the United States to the exclusive benefit of other countries, leaving American workers, who I love, and taxpayers to absorb the cost in terms of lost jobs, lower wages, shuttered factories, and vastly diminished economic production."[23] In short, the president believed that the accord would be an economic burden to the United States and, additionally, that it would result in a large-scale loss of jobs, putting the U.S. at an international disadvantage, all the while doing nothing to stop climate change in the long term.

I disagreed with the decision, but as I learned more about what action was needed to protect the global environment, reduce carbon emissions and methane, ensure access to energy, protect against health risks, and improve food and water security, I concluded that the cloud over America's reputation that followed the withdrawal from the agreement had a silver lining. Withdrawing might draw attention to the agreement's inadequacies and help persuade not only the United States, but also other nations to take a fundamentally different approach to climate change. The Paris Agreement, I came to believe, represented a danger because it fostered complacency. Although being a signatory allowed proponents to feel good about themselves, the nonbinding, unenforceable nature of the agreement did nothing to reduce what would be the greatest sources of human-induced global warming in coming decades: burgeoning carbon emissions from China and India and potential future emissions in developing economies in Africa. In the four years after the signing of the Paris Agreement, emissions increased by 1.5 percent per year.[24] What we needed were inexpensive and profitable solutions that could reduce greenhouse gas emissions in China, India, and across all developing economies.

In December 2019, a conference in Madrid, Spain, intended to finalize rules arising from the Paris Agreement, ended in disappointment. Like previous efforts to address global warming, discussions focused largely on nonsolutions that were either impractical or inadequate to address the complex problem set associated with climate change. Conference attendees voiced support for the Green New Deal, an unrealistic proposal that called for fulfilling all U.S. power demand through zero-emission energy sources; eliminating all greenhouse gas emissions from U.S. transportation, manufacturing, and agriculture; renovating all existing buildings and constructing new ones to improve energy efficiency; guaranteeing adequate housing, high-quality health care, and jobs with a family-sustaining wage; and mandating family and medical leave, paid vacations, and retirement security for everyone in the United States. Some proposed that Europe adopt a similarly unrealistic approach. A fundamentally different proposal was needed, but the trend seemed to be toward doubling down on non-solutions. Although advocates for environmental justice recommended vast capital transfers to developing economies to compensate for historical competitive disadvantages, energy solutions that are clean and profitable are the best way to right previous wrongs, create opportunities for economic growth, and address issues of environmental justice. The goal of reducing the ratio of carbon dioxide emissions to GNP would highlight the need for solutions that are applicable to the developing world rather than proposals to transfer capital from the taxpayers of developed economies to countries that are negatively affecting the earth and their own people.

Once again, strategic narcissism obscures solutions as some climate activists imagine a world consistent with what they want to achieve, but take no practical steps to seize opportunities to address the problem. Their conceit leads them to overlook political and economic realities that would shatter their dreams. Climate deniers evince a different form of strategic narcissism; theirs is based on willful ignorance. What the world needs is a comprehensive strategy based on the recognition that countries will not suppress their security and economic interests to join an international agreement. Proposals must have broad commercial and political appeal not only in prosperous nations, but also in developing economies.[25] And those solutions must avoid focusing on only one aspect of this complex problem set and thereby creating problems in other areas.

One common flaw with many climate proposals is that they pose single-country solutions to a global problem. Because pollution does not respect borders, solutions must apply globally. Climate scientists generally agree that global coal power generation needs to be reduced by 70 percent by 2030 and completely by 2050. Coal supplies 72 percent of India's power. And between 2006, when China surpassed the United States as the biggest source of carbon dioxide emissions, and 2020, China built the equivalent of fifty to seventy large coal-burning power plants every year. China is now the world's largest coal user, and in 2019 it had 121 gigawatts of coal plants under construction, more than the rest of the world combined. Each plant burns about a ton of coal every ten seconds.[26] Although Xi Jinping talks a good game on the environment, in 2019 China produced more carbon emissions than the United States and the European Union combined. Even worse, China is exporting more than 260 coal-fired power plants across Asia and Africa. One of those plants, in Kenya, fifteen miles from a UNESCO World Heritage Site, will be the largest source of pollution in the country.[27] The polluting effect on the African environment is as devastating as the increased carbon emissions are to the global environment. The trends in India are just as bad.

Other proposals are flawed because they avoid a systemic and holistic understanding and choose to focus on a single aspect of the problem. But challenges of the environment, climate, energy, health, food and water security, and even poverty and migration are interconnected. For example, agriculture affects climate, and climate affects agriculture. If cattle were a country, they would be the third-largest emitter of greenhouse gases. Deforestation is undertaken to make land available for cattle or to grow crops that are converted into biofuels. But deforestation removes trees that pull CO_2 from the air, and when burned, these trees release all their carbon and greenhouse gases back into the atmosphere.

Climate change induces food and water scarcity and spreads human and agricultural disease. The effects can be immediate as well as long term, especially given that human-induced stress factors, such as war and criminality, can both be triggered by food and water scarcity and aggravate it.[28] The ongoing tragedy in Yemen is a case in point. The effect of these interrelated problems on migration affects not only Africa, but also the Greater Middle East, Europe, and Central and North America. As a recent Hoover Institution study concluded, "[W]e are seeing a global movement of peoples, matching the transformative movement of goods and of capital in recent decades."[29]

Efforts to address only one challenge can exacerbate others and perpetuate rather than ameliorate threats to security and prosperity. In China, the anticipated explosion of electric vehicles will actually increase carbon emissions and worsen already deplorable air quality because an electric car that charges its batteries with electricity from a coal-burning plant produces more CO_2 per mile than a gasoline-powered car.[30] In India, President Narendra Modi promised drinking water to every household in the country, but potable water accounts for only 4 percent of the total water used in the country. Nearly 80 percent goes toward the irrigation of cropland. An estimated 200,000 people perish each year in India because of acute water scarcity, and approximately 600 million others endure severe water stress. By 2030, the demand for water is expected to be twice the available supply.[31] So, focusing just on water security without enacting reforms in agricultural techniques treats only symptoms while the underlying problem grows.

We need a dose of strategic empathy to develop solutions that address the interrelated nature of this problem set. And these solutions must generate incentives that lead to their broad adoption. A realistic approach that focuses on the following four objectives should minimize polarized political discussions that provide ammunition to climate deniers, obscure points of agreement, and delay initiatives that could make the choice between economic growth and survival of the planet a false dilemma.

- First, to ensure that the United States remains the global leader in reducing pollution while expanding our economy, any plan must cut across energy, agriculture, manufacturing, transportation, and construction. Renewable energy will remain important, but bolder, short-term action, such as encouraging a worldwide shift from coal to natural gas and an expansion of nuclear power, is essential.
- Second, the United States should lead the world in the development of integrated solutions, especially with the use of new technologies, including improved renewable energy sources, next-generation nuclear reactors, and better batteries and carbon-capture technologies. The costs of many of these technologies are plummeting.
- Third, the United States must work with other nations on the development of clean energy, new agricultural techniques and supply chains, water security, transportation, environmental stewardship, and health security. We might initiate a large scale, long-term partnership with India on those interconnected problems, with an emphasis on results.
- Fourth, all should prioritize the conservation of energy, food, and other resources, including making vehicles more energy-efficient.[32] A carbon credit system could encourage populations to emit less and conserve more. Conservation of food through limiting waste will be crucial. The water used to grow the food that we waste is greater than the water use of any single nation, and carbon emissions from that wasted food are greater than two times the emissions of all cars and trucks in the United States.[33]

There are many innovative solutions that advance toward these four objectives. One with which I am familiar is Zume, a company headquartered in San Francisco. I serve on its advisory board. Zume strives to integrate existing and mature technologies into new systems to revolutionize the entire food supply chain by better balancing supply and demand, saving on food and transportation waste, and integrating new agricultural technology to emit much less carbon. Integrated solutions such as those that Zume is pursuing could have a dramatic positive influence internationally across the nexus of environment, energy, climate change, health, and water and food security.

While international organizations and forums can help coordinate and inspire multinational solutions, the real progress must happen inside sovereign states. It is worth noting that during the global climate protests of September 2019, massive crowds totaling an estimated four million people worldwide called for action in New York, Seoul, Kabul, Istanbul, and elsewhere.[34] There were, however, no protests in China.[35] International organizations and nations whose citizens are victimized by irresponsible behavior must bring pressure to bear on nations that fail to be good stewards of the planet. Informing populations of potential solutions and of the consequences of inaction may help generate social pressure on autocratic regimes from within.

Discussions on these topics should begin with points of agreement. The vast majority of Americans can agree that climate change is a problem; that it is caused by humans and associated with carbon emissions; that there is no one exquisite solution; that solutions need to create economic incentives that lead to widespread adoption; and that the conversion of coal-burning plants to natural gas or another low-emission source should be an urgent near-term priority. And to generate the empathy necessary to impel action, we need to think about our children and grandchildren while recognizing that this isn't a future problem set. It is a right-now set of intertwined problems that include climate, energy, environment, health, food, and water security.

* * *

ENERGY IS the largest of the components associated with climate change, and may afford the greatest opportunity. The largest reduction in greenhouse gas emissions, for example, came not from a large government program or regulation, but instead from the fracking revolution in the United States. Fracking is the process of drilling into rock and then injecting highly pressurized liquid into the holes to create fissures, which allow gas and oil to escape and be captured. It was an unforeseen technological innovation that suddenly made cheap natural gas available in large quantities. This cheap supply provided an incentive to make the capital investments necessary to convert coal-fired plants to natural gas. Coal's share of U.S. electricity generation fell from 48 percent in 2008 to 22 percent in 2020. U.S. utilities do not build coal-fired power plants because natural gas and renewable power plants generate cheaper electricity.[36] The conversion of coal to natural gas worldwide presents the greatest near-term opportunity for carbon emissions reduction in the power and industrial sectors.

Advanced nuclear reactors provide another opportunity to achieve dramatic reductions in greenhouse gases without slowing economic growth. These Energy Multiplier Module (EM^2) reactors provide a grid-capable power source that takes just over three years to build. They are also more efficient and safer than standard reactors, which send spent fuel out for geologic storage. EM^2 reactors recycle used fuel after removal of some of the fission products. No liquid reprocessing is necessary, and no heavy metals are separated. Even better, these fission products require only about five hundred years of storage before decaying to background levels, compared to the ten thousand years or longer required for current designs. In addition, EM^2 has the potential to reduce our waste stockpile because it can be powered with this spent fuel.[37]

Advanced reactors would also provide an opportunity for the United States to regain its ability to compete with others, including China and Russia, in the global nuclear power market. Only two nuclear reactors are under construction in the United States, while China is rapidly building out its nuclear capacity. Russia dominates nuclear reactor exports, though China is catching up.[38] The last successful export and construction of an American reactor design was to China, which modified that design and now exports it. If we build reactors that are safer, cheaper, newer, and better, we can regain the lead in the industry and contribute to the reduction of greenhouse emissions.[39]

One can easily identify other competitions that are critical to defending the free world. There are geographically based competitions for influence and control of resources, such as those ongoing in the Arctic, Antarctica, and the Eastern Mediterranean. Ideological competitions between free-market capitalism and socialism are intensifying in the Western Hemisphere. Consider the autocratic regimes in Cuba and Nicaragua; the failing dictatorship in Venezuela; discontented populations trying to end statist economic practices in Ecuador and Bolivia; and those advocating for their return in Mexico, Argentina, and, to a lesser extent, Chile. Battlegrounds of criminality and organized crime in Mexico and Central America are perpetuating state weakness, inflicting human suffering, and driving large-scale migration. Many African states are battlegrounds between aspiring, young populations who are demanding a say in how they are governed and those who would expand autocratic governance to preserve power for the privileged few. The novel coronavirus pandemic in 2020 highlights the need for international cooperation on health. Preserving peace and prevailing in these and other competitions will require a rejection of strategic narcissism and an effort to foster understanding of complex challenges as the first step toward crafting solutions.

* * *

I CONCLUDE with what may be the most important competition critical to future security and prosperity: education. It is time for a new initiative similar to the National Defense Education Act, passed in 1958 in response to the Soviet Union's launching of *Sputnik*. That event motivated bipartisan efforts to prioritize not only science, but also history, political science, and language. Lawmakers back then recognized that education was a national security matter. Given China's advantages in AI and other emerging technologies, lawmakers should approach education reform with similar urgency today.

We might remember that education is not only for the young. As the historian of technology Elting Morrison observed in 1966, "the development of the instruments of industrial organization and our emotional and intellectual responses to them—cannot be learned once and for all in high school, college, graduate school, one Sloan Fellow year, or ten weeks in a senior executive development program. To live safely in our society, let alone manage it, will require a continuous education until a man dies."[40] Education is also critical to preserving our competitive advantages, because educated citizens are entrepreneurs who start new businesses and scientists who create medical breakthroughs and develop solutions to complex problems like climate change or the coronavirus. Educated citizens learn languages to connect with other societies, foster strategic empathy, and build a peaceful world. Educated citizens appreciate the great gifts of our free and open society as well as what we must do together to improve it. Educated citizens are best equipped to foil efforts to divide communities and pit them against each other. They are also best prepared to exercise their sovereignty in our democratic system by electing principled, thoughtful leaders and holding them accountable to strengthen our republic.

Conclusion

Many of our contemporaries are extraordinarily reluctant to acknowledge the reality of past time and prior events, and stubbornly resistant to all arguments for the possibility or utility of historical knowledge.

—DAVID HACKETT FISCHER, *HISTORIANS' FALLACIES: TOWARD A LOGIC OF HISTORICAL THOUGHT* (1970)

THE RAMP of the Marine Corps MV-22 Osprey aircraft dropped. I thanked the crew, yelling above the roar of the spinning tilt rotors, grabbed my two bags, and walked out to meet members of my new Secret Service detail, who, over the next thirteen months, would become like members of our family. The team took me directly to the White House to begin my first day as national security advisor. There had not been much time to prepare for the new job.

It had been a whirlwind twenty-four hours since President Trump announced my appointment in front of the press pool at Mar-a-Lago on President's Day. Now, I had accompanied him, the First Lady, and the presidential entourage on Air Force One back to Andrews Air Force Base. There, Ospreys positioned on the tarmac flew me back to Fort Eustis, Virginia. Maj. Kevin Kilbride, my aide-de-camp, met me and gave me a ride home. He, my executive officer, Col. Neal Corson, and my enlisted aide, Sfc. Juan Sanchez would help manage my very sudden departure from my job as deputy commander, futures, at the U.S. Army's Training and Doctrine Command. I would return to my home at Fort Eustis only once before I moved with Katie to Washington two months later.

At home, between a torrent of phone calls (mostly from people wishing me luck in the new job), I discussed with Katie and our daughter's fiancé, Lt. Lee Robinson, the transition to what was certain to be an interesting and challenging tour of duty. Lee, who was serving in the U.S. Army's Seventy-fifth Ranger Regiment, had been driving through Virginia on his way back to Fort Benning, Georgia, and had stopped to spend the night. The next day, after we got back from an early morning workout, he asked me why I was spending more time packing books than clothes.

I explained to him that I intended to draw on history to help frame contemporary challenges to national security. An important first step in developing policy and strategy, I believed, was to understand how the past produced the present. I also believed that the history of how previous presidents, their cabinets, and the National Security Council staff made decisions, developed policies, and crafted strategies held lessons for how to deliver the best advice and sound options to the president. For me, history was an avocation. I spent my spare time writing articles, reviewing books, and serving as a contributing editor to *Survival: Global Politics and Strategy*. As a general officer, I found that examining the history of a new position helped me ask the right questions and understand better the possibilities and difficulties associated with current challenges. For example, as commander of Fort Benning, I had based our educational reform effort on the changes that then-Lt. Col. George Marshall implemented there after World War I. And the army study I commissioned in 2015 on Russia's annexation of Crimea and invasion of Ukraine was modeled on the study that Gen. Donn Starry initiated on the 1973 Arab-Israeli War. For me, history came to life when it was applied to contemporary challenges and circumstances. As a regimental commander preparing our team for Iraq in 2004, I had consulted a wide range of literature on counterinsurgency to identify best practices. In the 1991 Gulf War, our cavalry troop used battle drills (rehearsed responses to a predictable set of circumstances in combat) based on my reading of U.S. major general Ernest

Harmon's and German field marshal Erwin Rommel's accounts of armored warfare in North Africa in World War II. For military leaders, reading and thinking about history is an integral part of our sacred duty to our nation and our fellow soldiers. Because the stakes in war involve life and death, combat leaders who choose to learn exclusively from personal experience are irresponsible. So, it was not a stretch for me to regard military and diplomatic history as foundational to improving U.S. strategic competence—that is, our ability to integrate elements of national power and the efforts of like-minded partners to advance and protect America's vital interests.

My soon-to-be son-in-law, Lee, got a longer answer than he had anticipated; it was the price he paid for asking a historian about the value of history. And readers who have stayed with me until this point know that the importance of history to understanding and coping with contemporary challenges is a major thrust of this book.

WALKING INTO the West Wing of the White House should be humbling for any person fortunate enough to serve there, but walking into the office occupied fifty-two years earlier by McGeorge Bundy, one of the principal characters in a book I had written twenty years earlier, was particularly so. In *Dereliction of Duty: Lyndon Johnson, Robert McNamara, the Joint Chiefs of Staff, and the Lies that Led to Vietnam*, I seek to explain how and why Vietnam became an American war. As I wrote about the national security decision-making process, I had no idea that I would one day become responsible for that process. Before I walked into that office and met members of the NSC staff, I had in mind four resolutions to improve U.S. strategic competence. All four are based on pitfalls that, during the Johnson administration, contributed to unwise decisions, a fundamentally flawed strategy, and, ultimately, a lost war that took the lives of 58,000 Americans and well over a million Vietnamese, consumed billions of American dollars, and inflicted on the United States one of the greatest political traumas since the Civil War.[1]

First, our NSC process would deliver options to advance and protect the interests of the American people and overcome national security challenges. The story in *Dereliction of Duty* is, in large measure, one of abdication of responsibility. President Lyndon Johnson made wartime decisions based primarily on his domestic political agenda: getting elected in his own right in 1964 and passing the Great Society legislation in 1965. Because he viewed Vietnam principally as a danger to those goals, he chose a middle course that he hoped would allow him to avoid difficult decisions. Johnson's path of least resistance proved unsustainable because it was built largely on lies aimed at the American people and their representatives in Congress. Once Americans realized that they had been misled about the scale and cost of the American military intervention in Southeast Asia, many lost faith in that effort. Now, in our work to develop integrated strategies for our most pressing national security challenges, we would provide the president with options differentiated by their level of risk to American interests and citizens, the resources required, and the prospect of their progressing toward national security and foreign policy goals and objectives. The president would then hear from his cabinet officials, who would recommend their favored course of action and their rationale for that recommendation. The NSC process would not consider the effect of policy decisions on partisan political concerns, the assumption being that successful policies serve all American people. Besides, the presentation of multiple options would give plenty of opportunity for political advisors to offer their assessments and recommendations.

Second, we would spend more time understanding and framing the nature of the problems and challenges we faced, viewing them through the lens of vital U.S. interests and crafting overarching goals and more specific objectives. During the period in which Vietnam became an American war, McGeorge Bundy argued that the objectives in Southeast Asia should be kept ambiguous, to give the president flexibility should the war effort fail.[2] The lack of clearly understood objectives, combined with the primacy of domestic political considerations, resulted in a strategy for Vietnam based on what its Washington, DC, purveyors preferred rather than on what the situation in Vietnam demanded. The next steps up the "ladder" of graduated pressure (such as the initiation of covert operations against North Vietnam in early 1964, the beginning of the Rolling Thunder bombing campaign in February 1965, or the deployment of large U.S. combat units to South Vietnam that summer) had all been taken without meaningful discussion of how those decisions fit into an overall strategy designed to achieve a clear and agreed policy goal. To ensure that we did not repeat the Vietnam War mistake of confusing activity with progress, our staff would institute "framing sessions," which I believed were necessary to foster understanding, before we developed options for the president. These sessions would result in succinct analyses of a particular challenge to national security; the "so what," or a description of the effect of that challenge on American security, prosperity, and influence; and the recommended goal and objectives. Consistent with Ancient Greek philosopher Aristotle's observation that it is only worth discussing what is in our power, the framing would include an assumption of the degree of influence the United States and others had over that challenge. Only after key members of the president's cabinet had discussed, modified, and approved the framing would they share ideas and give guidance concerning how to integrate all elements of national power, and the efforts of like-minded partners, toward the agreed objectives.

Third, we would insist on the presentation of multiple options to the president, as a means of providing best advice from across all departments and agencies of government. When it became clear that President Johnson wanted a strategy that allowed him to avoid difficult decisions on Vietnam, McGeorge Bundy and Robert McNamara delivered "graduated pressure" to placate both those who advocated for resolute military intervention in the war and those opposed to intervention, whom Johnson called "the sob sisters and peace societies."[3] Yet implicit and flawed assumptions that underpinned the strategy went unchallenged.

It is important to provide any president with multiple options because, unlike members of the cabinet or NSC staff, the president is elected and should be the person to set the course for U.S. foreign policy and national security strategy. Presenting a single option designed either to tell a president what he or she wants to hear or to present the consensus position of the cabinet is doing him or her a disservice.

Fourth, we would not assume linear progress toward our objectives and would instead acknowledge the degree of agency that others (whether the authoritarian powers Russia and China, transnational threats such as jihadist terrorist organizations, hostile states such as Iran or North Korea, or multiple actors in emerging arenas of competition in cyberspace or space) had over the future course of events. Two war games in 1964 exposed as false the principal assumption underpinning graduated pressure: "By applying limited, graduated military actions, reinforced by political and economic pressures against a nation providing support for an insurgency, we could cause that nation to decide to reduce greatly, or eliminate altogether, its support for the insurgency. The objective of the attacks and pressures is not to destroy the nation's ability to provide support but rather to affect its calculation of interests."[4] This vintage example of strategic narcissism is striking for its utter disconnection from the ideology and aspirations that drove the North Vietnamese and Vietnamese Communist leaders and for the implicit assumption that the principal cause of insecurity in South Vietnam was external support from the North. The last turn of the war game imagined the situation three years later, in 1968. The United States had more than five hundred thousand troops in Vietnam and no hope for success. Popular opposition to the war was growing. The war games and their eerily prophetic results, however, were ignored. Bundy thought the findings were too harsh. Therefore, to ensure that the president received the best assessments, we would include measures of effectiveness for every approved strategy. Assessments would go to the president periodically or when an event occurred that presented a new hazard or an opportunity. And we would scrutinize the assumptions on which the strategies were based and be prepared to reframe challenges if assumptions were invalidated.

PRACTICAL EXPERIENCE as well as the study of history shaped my approach to my new responsibilities. Experience in Afghanistan and Iraq convinced me that our inconsistent and flawed strategies in those wars did not satisfy the simple definition of strategy taught in the U.S. military's professional education system: the intelligent identification, use, and coordination of resources (or ways and means) for the successful attainment of a specific objective, or end. But strategy in war extends beyond logic and reason because it has a moral component. I believed that the strategies in ongoing wars had become morally untenable because they did not explain to the American people how the exertions of their sons and daughters would achieve outcomes worthy of the cost in blood and treasure.[5] As in Vietnam, the wars of 9/11 suffered initially from a form of strategic narcissism based on the conceit that American military technological prowess obviated the need to think deeply about the nature of the enemy or the political and human complexities of those wars. That conceit was made possible through the neglect of history and, in particular, of continuities in the very nature of war. It is easy to ignore continuities and assume that future wars or future competitions short of war will be fundamentally different from those of the past. I would therefore do my best to encourage the development of options for the wars in South Asia and the Middle East consistent with four fundamental continuities in the nature of war.

First, war is political. In Afghanistan and Iraq and, later, in Syria, war strategies violated the dictum articulated by eighteenth-century philosopher of war Carl von Clausewitz that "war should never be thought of as something autonomous, but always as an instrument of policy."[6] Just as there was no simple or purely military solution to the problem of Vietnam, there was no military-only solution for the contemporary wars in Afghanistan and the Middle East. Instead of learning from the failure to put the strength and legitimacy of the South Vietnamese government and armed forces at the center of that effort, later American leaders viewed Vietnam as a mistake to be avoided.[7] They assumed that the consolidation of military gains politically in Afghanistan and Iraq were not an integral part of war. Yet, successful military operations against ISIS in Syria and Iraq are not ends in and of themselves; they are the results of only one instrument of power that must be coordinated with others to achieve and sustain political goals. The lesson to learn from the American experience in Vietnam, Afghanistan, and Iraq is to be skeptical of concepts that divorce war or competitions short of war from their enduring political nature, particularly concepts that promise fast, cheap victory through technology.

Second, war is human. People fight today for the same fundamental reasons the Greek historian Thucydides identified nearly 2,500 years ago: fear, honor, and interest. In Vietnam, as predicted, covert raids and tit-for-tat bombing did not convince Ho Chi Minh and the leaders of North Vietnam to desist from supporting the Vietnamese Communist insurgency in the South. Vietnamese Communist leaders were committed to winning even at an extraordinarily high price; they had demonstrated that commitment not only in the Johnson administration's war games, but also during the First Indochina War against the French. In Afghanistan, Iraq, and the broader war against jihadist terrorists, strategies that simply target enemy leaders or forces do not address the human as well as the political drivers of violence. That is why breaking the cycle of violence, restoring hope, reforming education, and isolating vulnerable populations from jihadist ideology are essential to defeating those who foment hatred to justify violence against innocents. It is also why strategic empathy and, in particular, the effort to understand how emotion and ideology drive and constrain the other is fundamental to improving strategic competence.

Third, war is uncertain. War is uncertain because it is political and human and because it is interactive. Neither the future course of events nor our enemies will conform to announcements of our linear plans, such as declaring a time line for withdrawal of forces years in advance. As Professor Hew Strachan observed, "One sort of war can turn into another."[8] As the wars in Afghanistan and Iraq evolved across nearly two decades, the United States was slow to adapt, in part due to a fundamental misunderstanding of war—a failure to grasp that the future course of events depends not only on what one decides to do next, but also on enemy reactions and initiatives that are difficult to predict. The North Vietnamese did not impose limits on themselves and conform to the strategy of graduated pressure, but instead intensified the war effort and exploited American restrictions.[9] Under the concept of graduated pressure, the United States selected military actions based on its readily available military capabilities rather than on the effects the application of military force might achieve. In Afghanistan and Iraq, the "light footprint" approach to those wars allowed determined enemies to regain strength and wage sophisticated insurgencies. The lesson is that in war and in competitions short of war, America does not control the future course of events, and strategies must not only be sustained over time, but also adapt continuously to retain the initiative.

Fourth, war is a contest of wills. As Gen. George Marshall observed in his address to the annual meeting of the American Historical Association in 1939, "In our democracy where the government is truly an agent of the popular will," foreign policy and military policy are "dependent on public opinion," and our policies and strategies "will be as good or bad as the public is well informed or poorly informed regarding the factors that bear on the subject."[10] With Vietnam, as Americans watched their first televised war, they realized not only that they had been misled, but that their government had not developed a strategy to achieve a desired outcome at an acceptable cost. With Afghanistan and Iraq, the determination to avoid another Vietnam encouraged not only the light footprint approach, but also a short-term mentality and early declarations of "mission accomplished" in both wars. The unexpected length and difficulty of those wars sapped American will, as did inconsistent strategies based on flawed assumptions that ran counter to war's political, human, and interactive nature. Moreover, U.S. leaders did not devote sufficient effort to explaining what was at stake in those wars, or how the sacrifices of Americans' fellow citizens were contributing to a worthy outcome. The lack of wartime leadership encouraged narcissism among the public, who understood neither their enemies nor the experiences of their sons and daughters engaged with those enemies. War reporting focused on casualties or troop levels while portraying soldiers as victims who had no authorship over their fate. It is thus that the post-9/11 "endless wars" became conflated with the trauma of Vietnam and began to drain America's will.

WHAT SOME have called the "Vietnam syndrome" (a belief that the United States should simply avoid military intervention abroad) was the most prominent and immediate manifestation of the widely held interpretation that that war was unjustified and unwinnable. The mantra of "no more Vietnams" often muted discussion of what might be learned from that experience. The analogy to Vietnam was applied indiscriminately as well as superficially. Across the three decades following the 1973 Paris Peace Accords that ended American involvement in the war, assertions that any use of force abroad would lead to "another Vietnam" appeared in connection with military operations in Latin America, the Horn of Africa, the Balkans, Southwest Asia, and Central Asia. President George H. W. Bush declared after the First Gulf War that America had "kicked the Vietnam syndrome once and for all."[11] But under the guise of ending endless wars, the Vietnam analogy became conflated with Afghanistan and Iraq analogies to produce something like the Vietnam syndrome on steroids.

Simplistic interpretations of the American experiences in Afghanistan and Iraq obscure the differences in the character of those conflicts. Some interpretations point to American pursuit of "armed domination" or an effort to remake the world in America's image. These interpretations overlook the fact that the United States and its allies invaded Afghanistan after the most devastating terrorist attack in history. And while the majority of Americans might now argue that the invasion of Iraq was unwise—or, that at least it was unwise to think regime change in Baghdad would be easy—the arguments for retrenchment do not acknowledge the consequences of America's precipitate disengagement from Iraq in 2011 as giving rise to ISIS, or the U.S. halting withdrawal from Syria in 2019 as setting conditions for an intensification of that multiparty conflict and complicating efforts to bring about ISIS's enduring defeat. We should be aware that simplistic interpretations of the American experience in Afghanistan and Iraq cloud understanding and can be used to justify flawed policies and bad decisions. Just as the memory of America's divisive military intervention in Vietnam, and the strong emotions that tainted many early interpretations of that war, clouded understanding and left plenty of room for manipulating the historical record, America's understanding of more recent experiences in Afghanistan and Iraq has become more symbolic than historical; as with the Vietnam syndrome, the wars of 9/11 are used to evoke emotion rather than promote understanding.

Many who are deeply skeptical of U.S. military engagement abroad self-identify as part of a realist school of international relations. But *realist* is the wrong word. They get the world wrong because they start from an ideologically driven approach to U.S. engagement with the world. They are against any form of military intervention abroad and for the withdrawal of U.S. forces not only from the wars in Iraq, Syria, and Afghanistan, but also from the preponderance of other military commitments overseas. Rather than viewing the Vietnam syndrome and the overconfidence in American military technology of the 1990s as setting the United States up for the difficulties experienced in Afghanistan and Iraq, many who adhere to this school of thought argue that America's conceit is to pursue "liberal hegemony," an effort to turn as many countries as possible into liberal democracies. One of the school's proponents, Professor John Mearsheimer, alleges that America's "crusader mentality" drives a misguided, costly, and self-defeating foreign policy designed to "remake the world in its own image."[12] The realist school has found common cause with those who adhere to the New Left interpretation of history, which became more influential in academia during and after the Vietnam War. The realists and the New Left have been bolstered by a large influx of cash from billionaires George Soros and Charles Koch, who share little common ground politically except their advocacy for American retrenchment. The two pumped millions of dollars into new think tanks such as the Quincy Institute for Responsible Statecraft and funded programs within existing think tanks such as the Atlantic Council and RAND.[13] The cash and appeal to emotion gained traction despite what Professor Paul Miller described as an effort to set up and then knock down a straw man of liberal hegemony with "historically myopic, morally stunted, and strategically incoherent" arguments.[14]

Because adherents to realism and the New Left both believe that the United States is the principal cause of the world's problems, they argue that if the United States withdrew from competitions overseas, we would be safer. Their bywords are *restraint* and *offshore balancing*, which are meant to communicate a reduced emphasis on U.S. alliances and a diminished military posture overseas. But their views make them paragons of strategic narcissism due to their tendency to disregard the agency that the "other" has over the future course of events. In their view, the United States causes others to act; our presence abroad creates enemies; our absence abroad would restore harmony. Other states, according to this orthodoxy, only react to the United States and have no aspirations or objectives of their own. The United States, therefore, is to blame for antagonizing Russia and China, the former through the expansion of NATO and the latter through an excessive U.S. military presence in the Indo-Pacific. America, they believe, is to blame for jihadist terrorism because the offense of Americans' presence in Muslim holy lands generated a natural backlash against us infidels. The United States is the cause of nuclear proliferation, they feel, because states like Iran and North

Korea need those weapons to defend against an overly aggressive United States; a U.S. policy of conciliation with both countries would transform those states into responsible actors and even convince their leaders that they no longer need to brutally repress their own people.[15]

These twenty-first-century realists and fellow travelers of the New Left believe that American retrenchment would not only make the world safer, but also save money that could be applied to domestic needs. But as the history of the challenges in this book makes clear, American behavior did not cause Russian and Chinese aggression, jihadist terrorism, or the hostility of Iran and North Korea. Nor would disengagement make any of those challenges easier to overcome. America would have paid a much cheaper price for maintaining a military presence on the Korean Peninsula in 1950 than the cost of the Korean War, just as sustained engagement in Iraq beyond 2011 would have cost far less than the post-2014 campaign to liberate Iraqi and Syrian territory from ISIS. It is also much cheaper to deter Russia in Europe through U.S. presence today than to restore security after aggression tomorrow. And it is easier to ensure freedom of navigation and overflight in the South China Sea and elsewhere now than to fight to restore them later.

The "realist" argument for retrenchment appeals to those deeply skeptical about efforts to promote democracy—skepticism due in part to excessive hope. The promise of the 1990s was that the world was progressing inexorably toward liberal-democratic governments and a globally harmonious system of states. Globalization would lead to the convergence of states as democratization advanced. That optimistic worldview was unsustainable. When it failed, it gave way to retrenchment and resignation. Much of the unrealistic optimism about the arc of history stemmed from the assumptions some made after the collapse of Communist authoritarian governments in 1989 that the regime changes in eastern Europe were replicable in the Middle East, Africa, and Asia. Yet, this thinking did not give due consideration to local context or, in particular, to political, social, cultural, and religious dynamics that complicate majority rule, the protection of minority rights, and the rule of law. It is clear that the United States can influence, but cannot determine, the evolution of the world order in favor of free and open societies. As nineteenth-century philosopher John Stuart Mill observed, "[T]he virtues needed for maintaining freedom must be cultivated by the people themselves."[16] It is also true, however—as protests in 2019 and 2020 in Hong Kong, Moscow, Tehran, Baghdad, Khartoum, Caracas, and Beirut attest—that people want a say in how they are governed.

The existence of free and open societies abroad benefits security because such societies are natural defenses against hostile, aggressive, authoritarian powers. As argued in this book, support for democracy and the rule of law is the best means of promoting peace and competing with those who promote authoritarian, closed systems. The United States and other nations should also continue to promote basic and unalienable rights as captured in the Universal Declaration of Human Rights approved by the UN General Assembly in Paris on December 10, 1948, while recognizing that America and its allies cannot be the guarantor of those rights. And those who self-identify as realists are right to be skeptical about the ability of international organizations to promote peace, justice, and prosperity across the globe. Because authoritarian and hostile regimes do their best to co-opt organizations like the United Nations, strong nations governed under the principle of popular sovereignty are the best advocates for the oppressed. As stated in the 2017 National Security Strategy of the United States, it is possible to recognize that a "world that supports American interests and reflects our values makes America more secure and prosperous" and to affirm "America's commitment to liberty, democracy, and the rule of law" while also acknowledging "that the American way of life cannot be imposed upon others, nor is it the inevitable culmination of progress."[17]

IT IS my hope that this book contributes to improving U.S. strategic competence through enabling a better understanding both of the history of how crucial challenges to national security developed and of the ideology, emotions, and aspirations that drive the other. But to preserve our competitive advantages, Americans need to focus inwardly as well as outwardly. For example, the best means to counter Putin's playbook is to strengthen our democratic institutions and processes and restore faith in our democratic principles and free-market economies. To improve our strategic competence, we need to develop leaders who can think in time and who understand what it takes to implement ideas and strategies on the ground. It is vitally important to understand the local realities and know the actual people. Competitions are ultimately about human behavior. And although countering the Chinese Communist Party's campaign of co-option, coercion, and concealment is important, an exclusively defensive stance would not only consign America to second place in critical competitions, but also play into the CCP's narrative that the United States is trying to keep China down. Although investments in research and development, military capabilities, and infrastructure are vital, improvement in education may be the most important initiative to ensuring that future generations are able to innovate and create opportunities for their children and grandchildren.

Prevailing on today's battlegrounds requires an unprecedented degree of cooperation among government, academia, and the private sector. To attempt to direct that cooperation would cut against America's democratic and decentralized nature. Understanding today's competitions, however, and especially what is at stake in them, can serve as a basis for joint action to counter hostile behavior such as cyber-enabled information warfare and to work together to maintain competitive advantages in technology and in the emerging data-driven economy. Encouraging public service and creating easier ways to move into and out of public service should be a top priority. Because new arenas of competition transcend the limits of geography and reach into society and industry, it is important for every citizen to understand the nature of those competitions. Cooperative efforts are essential both to defend freedom and to work on interconnected problem sets, such as those associated with climate, the environment, energy, and food and water security. The threats from Russia's cyber-enabled information warfare, China's industrial espionage and influence campaigns, jihadist terrorist efforts to direct or inspire attacks on our homeland, Iran's offensive cyber capability, and North Korea's missiles reach across borders and attempt to exploit vulnerabilities across all segments of society. That is why government-only efforts to defend the free world are passé.

To compete more effectively in war and in competitions short of war, the United States and other free nations should invest in strategic competence. Educating the public about the battlegrounds of today and tomorrow is an especially important task. A reinvigoration of history in higher-level education is particularly important, as many courses in diplomatic and military history have been displaced by theory-based international relations courses, which tend to mask the complex causality of events and obscure the cultural, psychological, social, and economic elements that distinguish cases from one another. Some theories risk sapping students of strategic empathy and encouraging them to reduce complex problem sets into frameworks that create only the illusion of understanding. A growing interest in applied history in some universities is a promising development.

But many universities do not teach military and diplomatic history, or they teach it only in relation to social history. After the Vietnam War, many gave in to the antiwar movement's tendency to confuse the study of war with militarism. Thinking clearly about problems of diplomacy, national security, and defense, however, is both a necessity and the best way to prevent war. The analogy drawn by the late historian Dennis Showalter is apt: no one would ever accuse an oncologist of being an advocate for the disease he or she studies.

The Foreign Policy Research Institute defines geopolitics as "an approach to contemporary international affairs that is anchored in the study of history, geography and culture." With the new arenas of competition discussed in chapter 13, technology might be added to that definition—to educate future leaders about how to maximize the potential and minimize the danger of emerging technologies, in the context of geopolitical competitors.[18]

Contemporary challenges to security demand a concerted effort to deter conflict. There are two fundamental ways to deter conflict. First, by the threat of punitive action that would inflict pain that exceeds the attacker's anticipated gains. This form of deterrence requires leaders to convince the potential enemy that the target of his contemplated aggression possesses both the will and the capability to retaliate. It also requires the ability to hold at risk something of value to the potential aggressor. Iran was able to escalate its forty-year proxy war with the United States on its own terms because its leaders were conditioned to assume that Washington would not retaliate directly against Iran in response to proxy attacks on U.S. people and facilities in the Middle East. That is why both the strike that killed Qasem Soleimani and Abu Mahdi al-Muhandis and the economic sanctions imposed in response to continued Iranian aggression were aimed, in part, at restoring deterrence of Iran. But deterrence by the threat of punitive action later is often unsuccessful.

For ideological regimes or adversaries that do not possess valuable assets that the United States can hold at risk, a second form of deterrence is more appropriate. "Deterrence by denial" is based on the ability to convince adversaries that they cannot accomplish their objectives through the use of force or other forms of aggression. Applied to Putin's playbook, deterrence by denial would entail closing fissures in our societies and restoring confidence in democratic principles, institutions, and processes. Deterrence by denial in cyberspace entails requiring resilient systems and an effective, layered, active defense of critical networks and infrastructure. Deterring potential military offensives, such as a Russian invasion of Baltic states or a Chinese invasion of Taiwan, requires convincing the would-be aggressors that defenses are strong enough to prevent their success.[19]

America's allies on the Eurasian landmass and across the islands of the Indo-Pacific region are invaluable not only to deterring aggression, but also to engaging in critical competitions short of conflict.[20] Strengthening ally defenses and augmenting them with American defense capabilities reduces burdens on the United States while bolstering allies' security. But our allies provide us with competitive advantages that extend beyond deterring potential aggression. From a geopolitical perspective, allies on the "rimland" of the Eurasian landmass pose dilemmas for Russia and China and have the greatest potential to prevent what would be devastating wars. Alliances are also tools for accumulating greater moral as well as military capability than the United States has on its own. Allies magnify America's voice and make it more difficult for countries to infringe on the sovereignty of their neighbors or their own citizens. And when facing forms of economic aggression, such as Russia's use of energy for coercion or China's effort to coopt nations and companies, allies can reach out to like-minded partners and help convince them not to compromise principles or their security for the lure of short-term profits or a great discount on 5G communications infrastructure. The Trump administration was right to demand a higher degree of burden sharing by allies and reform within NATO to cope with twenty-first-century threats, but expressions of doubt about the value of allies when Russia and China are doing their best to break alliances apart are counterproductive. Allies, along with knowledge of history and technology and collaboration across the military, government, industry, and academia, bolster our strategic competence.

STILL, THE ability to compete effectively requires confidence as well as competence. A general understanding of America's role in the world is important because war and competitions short of war (such as defeating RNGW or the CCP's sophisticated strategy of co-option, coercion, and concealment) are, fundamentally, contests of will. Over-optimism and retrenchment both stem from strategic narcissism and, in particular, the failure to acknowledge the degree of agency and control that the "other" has over the course of events. I had hoped, as we developed a

National Security Strategy and integrated strategies for our most pressing challenges, to help the president develop a pragmatic approach based on neither over-optimism nor resignation. And because American influence depends, in large measure, on the confidence others have in our ability to execute an effective, long-term foreign policy, our team worked to develop options that were sound and sustainable such that they could garner support from the American people.

Partisan vitriol among America's political leadership gives friends and foes alike the impression that the United States is incapable of competing effectively based on a bipartisan foreign policy. As the late professor and philosopher Richard Rorty observed, "National pride is to countries what self-respect is to individuals: a necessary condition for self-improvement." If we lack national pride, how can we possess the confidence necessary to fight effectively in war or implement a competitive foreign policy? In the United States, civics education might try to reverse the shift toward micro-identities and the focus on victimhood to foster what political scientist Francis Fukuyama describes as "broader and more integrative identities."[21] Every time Americans talk or tweet about issues that divide them, they might devote at least equal time to what unites them—especially our commitment to the fundamental individual liberties contained in our Declaration of Independence, our Constitution, and our Bill of Rights. The academy's role in restoring our confidence and national pride may include a review of history, literature, and philosophy curricula to ensure that they contain not only self-criticism and a broad range of cultural perspectives, but also an acknowledgment of the nobility and accomplishments of our great unfinished American experiment in democracy and liberty. There is important work to do in primary and secondary education to rekindle in our youth an understanding of our history, including not only the contradictions and imperfections in our experiment, but also the virtues and great promise of America. Our teachers should not overlook blights on our history—the profound failures in the forcible subjugation of Native Americans, in slavery, in the internment of Japanese Americans during World War II, in institutionalized racism, in inequality for women, or in the mistreatment of other minorities—but they might place those stories in context. They might offer a progressive narrative that illuminates the advantages and resiliency of an American Constitution that placed sovereignty in the hands of the people and extended equal rights to previously excluded communities consistent with the nation's founding principles.[22] Besides education, public service is another means of strengthening strategic confidence and national pride. Working to eliminate opportunity inequality through educational and economic reforms is integral to the fight to defend the free world, as confidence in our democracies and free-market economies is essential to maintaining our will to compete.[23]

Immigrants have been and remain one of America's greatest competitive advantages. Oppressed peoples who come to the United States, a self-selecting group, have the intrepidity to start a new life and are appreciative of the freedom and opportunity in America. A way to help overcome fractures in our society would be to talk less about who we do not want to come to America and more about whom America needs. Those who believe in our Constitution, the rule of law, and the opportunity to work hard to create a better life should be welcomed into our liberal-democratic culture.

An effort to restore confidence must extend to other free and open societies. Gains made in representative government and economic reform in the Western Hemisphere should not be taken for granted. And because Europe is particularly important to U.S. security, supporting its effort to overcome its struggle with identity politics and with Russia's attempt to sow dissention within and among European nations should remain a top priority for U.S. diplomats.

I WROTE this book on the eleventh floor of the Hoover Tower at the center of the Stanford University campus. Herbert Hoover founded the institution that bears his name a century ago after witnessing the horrors of the Great War. Hoover, an orphan who graduated from Stanford University's inaugural class and who would later become America's thirty-first president, led a massive relief effort at the end of World War I in Belgium that was credited with saving more than ten million people from starvation.[24] It was there that he bore witness to the horrors of war and resolved to do all he could to help prevent another one. The experience of World War I, a conflict that took the lives of more than sixteen million people, highlighted the need to understand the political and historical basis for violent conflict as critical to preserving peace and ending wars. Hoover founded the Hoover War Collection, later named the Hoover Institution on War, Revolution, and Peace, as a place where scholars might study past wars to prevent future conflicts. As we know, however, the "war that was to end all wars" was instead the first of two world wars that marked the bloodiest century in modern world history. The tower that contains the vast Hoover Library and Archives, a collection meant to provide scholars with materials that might help explain the origins of wars and uncover prospects for peace, was completed in 1941, the year the United States entered World War II. It is in the spirit of that archive that I wrote this book, in an effort to use the study of the past to illuminate the present as the best way of influencing the future.

As historian Zachary Shore observed, "the greatest source of national strength is an educated populace."[25] It is my hope that this book will make a small contribution to the strength of our nation and other nations of the free world. Writing it was a continuation of my own education. I will judge it to have been worthwhile if it inspires vibrant, thoughtful, and respectful discussion of how we can best defend the free world and preserve a future of peace and opportunity for generations to come.

ACKNOWLEDGMENTS

WHEN I joined the Hoover Institution at Stanford University, it was my first career change since I entered West Point at age seventeen. When I retired from the U.S. Army after thirty-four years, I knew I would miss serving alongside dedicated and talented young men and women. I am grateful now to work with graduate and undergraduate students at Stanford who also possess a strong desire to contribute to our nation and all humanity. I could not have completed a project of this scope and ambition without the assistance of a tremendous team of student research assistants who filled out innumerable evidence sheets, cleaned up the text and references, organized reviewer criticisms, and deciphered my cursive handwriting, which, for a generation who grew up with computers, must have been the equivalent of transcribing cuneiform. Our seminar-like discussions of the challenges to national and international security and the connections between them were enriching and helped me impose order on complex problems. I am particularly indebted to Chelsea Burris Berkey, Kate Yeager, Jeffery Chen, Sri Muppidi, Sylvie Ashford, Edouard Asmar, Lee Bagan, Megan Chang Haines, Rand Duarte, Eddy Rosales Chavez, Aron Ramirez, Nolan Matcovich, Griffin Bovée, Taek Lee, Kyle Duchynski, Sophia Boyer, Jonathan Deemer, David Jaffe, Samantha Thompson, Emma Bates, Lisa Einstein, William Howlett, Leah Matchett, Isaac Kipust, Cyrus Reza, Carter Clelland, Katherine Du, Hiroto Saito, Theo Velaise, and James Kanoff. Getting to know these talented young people bolstered my confidence in our ability to overcome the challenges that are the subject of this book.

There is no better place to research and write than at the Hoover Institution at Stanford University. My colleagues at Hoover, the Freeman Spogli Institute, and the Stanford Graduate School of

Business provided sage advice and contributed through their scholarship and research. I am grateful for assistance from Nadia Schadlow, Jakub Grygiel, Michael Auslin, Victor Davis Hanson, Toomas Hendrik Ilves, Russell Berman, Michael McFaul, Abbas Milani, Kathryn Stoner, Daniel Sneider, Kevin Warsh, Niall Ferguson, Amy Zegart, James Timbie, Condoleezza Rice, David Mulford, Scott Sagan, George Shultz, Larry Diamond, Stephen Kotkin, Timothy Garton Ash, Michael Bernstam, Arye Zvi Carmon, Glenn Tiffert, Tim Kane, Peter Robinson, Karl Eikenberry, Herb Lin, Alex Stamos, Raj Shah, David Berkey, Saumitra Jha, and Charles O'Reilly. Erik Jensen from the Stanford School of Law also provided valuable advice. My thanks to the entire Hoover staff who, under director Tom Gilligan, provided encouragement and the ideal environment to write. Special thanks to Eryn Witcher Tillman, Denise Elson, Jeff Jones, Mandy MacCalla, Juanita Rodriguez, Erika Monroe, Silvia Sandoval, the late Celeste Szeto, Laurie Garcia, Dan Wilhelmi, Shana Farley, Megan Ring, James Shinbashi, and Rick Jara. Writing this book made me appreciate even more the privilege of holding the Fouad and Michelle Ajami Senior Fellowship, a position named for a man I admired and made possible by the generosity of Michelle Ajami and Patrick Byrne.

Friends, former colleagues, and subject matter experts provided early guidance, read portions of the manuscript, and made helpful corrections and suggestions. I learned much from Zachary Shore, Fiona Hill, Marin Strmecki, Clare Lockhart, Michael Brown, Janan Mosazai, Hussain Haqqani, Seth Center, John "Mick" Nicholson, Norine MacDonald, Melissa Skorka, Gretchen Peters, Rob Kee, Larry Goodson, Fernando Lujan, Lisa Curtis, Michael Bell, Matthew Pottinger, Toby Dodge, Tristan Abbey, Thomas Lafleur, Joe Wang, Omar Hossino, Kenan Rahmani, Regis Matlak, Emma Sky, Dana Eyre, Kenneth Pollack, David Pearce, Ryan Crocker, Alton Buland, Diana Sterne, Kirsten Fontenrose, Ali Ansari, Yll Bajraktari, Ylber Bajraktari, Matt Turpin, Jeremie Waterman, Chas Freeman, Charles Eveslage, Jimmy Goodrich, Orville Schell, Alexander Bernard, Donald Sparks, Joseph Byerly, Christopher Starling, Scott Moore, Saad Mohseni, and Jordan Grimshaw.

I am indebted to Mark Dubowitz, Juan Zarate, Samantha Ravich, Bradley Bowman, Bill Roggio, Thomas Joscelyn, Cliff Rogers, and the team at the Foundation for Defense of Democracies for the opportunity to learn from their efforts to improve our strategic competence. This project also benefited tremendously from the opportunity to work with Ken Weinstein, Patrick Cronin, Ben Gillman, Taro Hayashi, and Masashi Murano at the Hudson Institute's Japan Chair.

Other academic institutions hosted discussions that helped improve particular chapters. My thanks to Graham Allison and Fred Logevall of the Applied History Working Group at Harvard University for convening a group of talented scholars, including Josh Goldstein, Justin Winokur, Calder Walton, Carl Forsberg, Philip Balson, Paul Behringer, Anne Karalekas, Eugene Kogan, Charles Maier, Chris Miller, Nathaniel Moir, Laurie Slap, Peter Slezkine, and Emily Whalen.

I benefited tremendously from discussions that Michael Horowitz convened at the University of Pennsylvania's Perry World House. Christian Ruhl, LaShawn Renee Jefferson, Avery Goldstein, Alexander R. Weisger, Amy E. Gadsden, Scott Michael Moore, Mitchell Orenstein, Shira Eini Pindyck, Christopher William Blair, Casey William Mahoney, Joshua A. Schwartz, Michael Noonan, and Duncan Hollis provided constructive feedback and advice on early drafts. The errors and flaws that remain are mine alone.

I could not have asked for a better team of professionals to help bring this project to fruition. I am grateful for the assistance of Rafe Sagalyn and Amanda Urban at ICM Partners. And it was a pleasure to work with my editor, Jonathan Jao, Sarah Haugen, Jenna Dolan, Tina Andreadis, and the tremendous team at HarperCollins. The late Marion (Buz) Wyeth, with whom I was privileged to work two decades ago, would have been proud. My thanks, too, to Anne Withers, Mike Smith, and Keegan Barber at the White House.

I could not have completed this project without the love and support of my family. Thank you, Katie, for your encouragement, understanding, and for providing the foundation for our growing family. And thanks to my daughters, Katharine, Colleen, and Caragh, sons-in-law Alex and Lee, and my sister, Letitia, who all read portions of the manuscript and made helpful suggestions. And finally, thanks to my grandsons, Henry and Jack Robinson, who brought our family such joy and reminded me of our generation's duty to provide future generations with an inheritance of peace and freedom.

NOTES

Introduction

1 Alan R. Millet and Williamson Murray, *Military Innovation in the Interwar Period* (Cambridge, UK: Cambridge University Press, 2007); Eugenia C. Kiesling, *Arming Against Hitler: France and the Limits of the Military Planning* (Lawrence: University Press of Kansas, 1996); Robert A. Doughty, *The Breaking Point: Sedan and the Fall of France, 1940* (Mechanicsburg, PA: Stackpole Books, 2014).

2 Trudy Rubin, *Willful Blindness: The Bush Administration and Iraq* (Philadelphia, PA: Philadelphia Inquirer, 2004).

3 On the Iron Curtain, see Phil McKenna, "Life in the Death Zone," *Nova*, PBS, February 18, 2015, https://www.pbs.org/wgbh/nova/article/european-green-belt/; John Pike, "2nd Stryker Cavalry Regiment," Globalsecurity.org, January 13, 2012, https://www.globalsecurity.org/military/agency/army/2acr.htm.

4 For more on the Battle of 73 Easting, see H. R. McMaster, "Eagle Troop at the Battle of 73 Easting," The Strategy Bridge, February 26, 2016, https://thestrategybridge.org/the-bridge/2016/2/26/eagle-troop-at-the-battle-of-73-easting.

5 Hans Morgenthau and Ethel Person, "The Roots of Narcissism," *Partisan Review* 45, no. 3 (Summer 1978): 337–47, Howard Gotlieb Archival Research Center, http://archives.bu.edu/.

6 Francis Fukuyama, *The End of History and the Last Man* (New York: Free Press, 2006).

7 Statement of Richard N. Haass, President, Council on Foreign Relations Before the Committee on Foreign Relations, United States Senate, On U.S.-China Relations in the Era of Globalization, May 15, 2008, U.S. Senate: Committee on Foreign Relations, March 15, 2008, https://www.foreign.senate.gov/imo/media/doc/HaassTestimony080515p.pdf.

8 Frederick W. Kagan, *Finding the Target: The Transformation of American Military Policy* (New York: Encounter Books, 2007); Herbert R. McMaster, "Crack in the Foundation Defense Transformation and the Underlying Assumption of Dominant Knowledge in Future War," July 2003, https://doi.org/10.21236/ada416172.

9 Linda D. Kozaryn, "U.S. Aircrew Detained in China Heads Home," U.S. Department of Defense, April 12, 2001, https://archive.defense.gov/news/newsarticle.aspx?id=44964.

10 Lawrence Wright, *The Looming Tower: Al-Qaeda and the Road to 9/11* (New York: Vintage Books, 2007).

11 On this point, see David Kilcullen, *The Dragons and the Snakes: How the Rest Learned to Fight the West* (New York: Oxford University Press, 2020);

Douglas Jehl, "C.I.A. Nominee Wary of Budget Cuts," *New York Times*, Feb. 3, 1993.

12 David Kilcullen, *The Dragons and the Snakes: How the Rest Learned to Fight the West* (New York: Oxford University Press, 2020).

13 The White House, "A National Security Strategy for a Global Age," December 2000, https://history.defense.gov/Historical-Sources/National-Security-Strategy/.

14 Institute for the Analysis of Global Security, "How Much Did the September 11 Terrorist Attack Cost America?," Institute for the Analysis of Global Security, http://www.iags.org/costof911.html.

15 The bailout was part of the Emergency Economic Stabilization Act of 2008.

16 "Obama on Afghan War Drawdown: 'The Tide of War Is Receding,'" *PBS NewsHour*, PBS, June 23, 2011, https://www.pbs.org/newshour/show/obama-on-afghan-troop-drawdown-the-tide-of-war-is-receding#transcript.

17 Obama quotes from Jeffrey Goldberg, "The Obama Doctrine," *The Atlantic*, June 25, 2018, https://www.theatlantic.com/magazine/archive/2016/04/the-obama-doctrine/471525/.

18 Zachary Shore, *A Sense of the Enemy: The High-Stakes History of Reading Your Rival's Mind* (New York: Oxford University Press, 2014), 258.

19 Nadia Schadlow, "Competitive Engagement: Upgrading America's Influence," *Orbis*, September 13, 2013, https://www.sciencedirect.com/science/article/pii/S0030438713000446.

20 Niall Ferguson, *Kissinger: 1923–1968: The Idealist* (New York: Penguin Press, 2015).

21 Sun Tzu and Thomas F. Cleary, *The Art of War* (Boston, MA: Shambhala, 2005).

22 Winston S. Churchill, "Painting as a Pastime," in *Amid These Storms* (New York: Charles Scribner's Sons, 1932).

Chapter 1: Fear, Honor, and Ambition: Mr. Putin's Campaign to Kill the West's Cow

1 On Russia's alleged interference, see "Assessing Russian Activities and Intentions in Recent US Elections," Intelligence Community Assessment, Office of the Director of National Intelligence, January 6, 2017, https://www.dni.gov/files/documents/ICA_2017_01.pdf; Becky Branford, "Information Warfare: Is Russia Really Interfering in European States?" BBC, March 31, 2017, https://www.bbc.co.uk/news/world-europe-39401637.

2 Sarah Marsh, "US joins UK in blaming Russia for NotPetya cyber-attack," *The Guardian*, February 15, 2018, https://www.theguardian.com/technology/2018/feb/15/uk-blames-russia-notpetya-cyber-attack-ukraine.

3 Andy Greenberg, "The Untold Story of NotPetya, the Most Devastating Cyberattack in History," *Wired*, December 7, 2018, https://www.wired.com/story/notpetya-cyberattack-ukraine-russia-code-crashed-the-world/.

4 Andrew Kramer, "Russian General Pitches 'Information' Operations as a Form of War," *New York Times*, March 2, 2019, https://www.nytimes.com/2019/03/02/world/europe/russia-hybrid-war-gerasimov.html.

5 "Statement from Pentagon Spokesman Capt. Jeff Davis on U.S. Strike in Syria," Department of Defense, April 6, 2017, https://www.defense.gov/Newsroom/Releases/Release/Article/1144598/statement-from-pentagon-spokesman-capt-jeff-davis-on-us-strike-in-syria/.

6 On Gerasimov and the Syrian Civil War, see Kramer, "Russian General Pitches 'Information' Operations as a Form of War." On Syrian use of nerve agents, see Daryll Kimball and Kelsey Davenport, "Timeline of Syrian Chemical Weapons Activity, 2012–2019," Fact Sheets & Briefs, Arms Control Association, March 2019, https://www.armscontrol.org/factsheets/Timeline-of-Syrian-Chemical-Weapons-Activity. On Putin's intervention in Syria, see Maksymilian Czuperski et al., "Distract Deceive Destroy: Putin at War in Syria," *The Atlantic*, April 2016, https://publications.atlanticcouncil.org/distract-deceive-destroy/assets/download/ddd-report.pdf; and Thomas Gibbons-Neff, "How a 4-Hour Battle Between Russian Mercenaries and U.S. Commandos Unfolded in Syria," *New York Times*, May 24, 2018, https://www.nytimes.com/2018/05/24/world/middleeast/american-commandos-russian-mercenaries-syria.html.

7 Neil MacFarquhar, "Yevgeny Prigozhin, Russian Oligarch Indicted by U.S., Is Known as 'Putin's Cook,'" *New York Times*, February 16, 2018, https://www.nytimes.com/2018/02/16/world/europe/prigozhin-russia-indictment-mueller.html.

8 Mariam Tsvetkova, "Russian Toll in Syria Battle Was 300 Killed and Wounded: Sources," Reuters, February 15, 2018, https://www.reuters.com/article/us-mideast-crisis-syria-russia-casualties/russian-toll-in-syria-battle-was-300-killed-and-wounded-sources-idUSKCN1FZ2DZ.

9 Lavrov was the mouthpiece of the regime abroad, a legendary diplomat who spoke flawless English and used his diplomatic cunning in foreign capitals to cover up the Kremlin's abuses. Yet it was not clear if Lavrov's Ministry of Foreign Affairs had the bureaucratic weight in the Russian system, which was dominated by the security services, to put back together the pieces of a bilateral relationship that the security services had broken with Russia's 2016 election interference. Patrick Jackson, "Europe | Profile: Putin's Foreign Minister Lavrov," BBC News, June 29, 2007, http://news.bbc.co.uk/2/hi/europe/6242774.stm. On Patrushev's rise to power, see Andrew Monaghan, "Power in Modern Russia" (Manchester, UK: Manchester University Press, 2017), 21.

10 Quotations in the previous paragraph are from Fiona Hill and Clifford G. Gaddy, *Mr. Putin: Operative in the Kremlin* (Washington, DC: Brookings Institution Press, 2015), 185–89, 388. On Putin and Patrushev's climb, see Hill and Gaddy, 41, 185–89.

11 This was in violation of the 1997 NATO-Russia Founding Act, an agreement that demonstrates a commitment to respecting the sovereignty of each state and its right to choose the best means to ensure its security

while promising a peaceful settlement of disputes. These three concerns are also consistent with the analysis of Russian policy assumptions in Monaghan, "Power in Modern Russia," 26–27; North Atlantic Treaty Organization, "Summary: Founding Act on Mutual Relations, Cooperation and Security Between NATO and the Russian Federation," May 27, 1997, https://www.nato.int/cps/en/natohq/official_texts_25470.htm?selectedLocale=en. & NATO; "Founding Act on Mutual Relations, Cooperation and Security Between NATO and the Russian Federation Signed in Paris, France," https://www.nato.int/cps/en/natohq/official_texts_25468.htm.

12 Jakub J. Grygiel and A. Wess Mitchell, *The Unquiet Frontier: Rising Rivals, Vulnerable Allies, and the Crisis of American Power* (Princeton, NJ: Princeton University Press, 2017), 49.

13 On Kerry, see John Kerry, "Face the Nation Transcripts March 2, 2014: Kerry, Hagel," *Face the Nation*, March 2, 2014, https://www.cbsnews.com/news/face-the-nation-transcripts-march-2-2014-kerry-hagel/. On *maskirovka*, see "Maskirovka: From Russia, with Deception," October 30, 2016, https://www.realcleardefense.com/articles/2016/10/31/maskirovka_from_russia_with_deception_110282.html. For more on World War I, see Niall Ferguson, *The Pity of War* (New York: Basic Books, 1999).

14 For more on sanctions, see "President Donald J. Trump Is Standing Up to Russia's Malign Activities," Fact Sheets, The White House, April 6, 2018, https://www.whitehouse.gov/briefings-statements/president-donald-j-trump-standing-russias-malign-activities/; Office of the Spokesperson, "Sanctions Announcement on Russia," U.S. Department of State, December 19, 2018, https://www.state.gov/sanctions-announcement-on-russia/.

15 On the attempted coup, see Andrew E. Kramer and Joseph Orovic, "Two Suspected Russian Agents Among 14 Convicted in Montenegro Coup Plot," *New York Times*, May 9, 2019, https://www.nytimes.com/2019/05/09/world/europe/montenegro-coup-plot-gru.html; Adam Casey and Lucan Ahmad, "Russian Electoral Interventions, 1991–2017," Scholars Portal Dataverse, December 15, 2017, https://dataverse.scholarsportal.info/dataset.xhtml?persistentId=doi:10.5683/SP/BYRQQS. On sanctions, see Jeremy Herb, "Senate Sends Russia Sanctions to Trump's Desk," CNN, July 27, 2017, https://www.cnn.com/2017/07/27/politics/russian-sanctions-passes-senate/index.html.

16 Tak Kumakura, "North Koreans May Have Died in Israel Attack on Syria, NHK Says," Bloomberg, April 27, 2008, https://web.archive.org/web/20121103011551/http://www.bloomberg.com/apps/news?pid=newsarchive&sid=aErPTWRFZpJI&refer=japan; IDF, "The Secret Operation Revealed a Decade Later," IDF Press Center, March 21, 2018, https://www.idf.il/en/articles/operations/the-secret-operation-revealed-a-decade-later/.

17 Vladimir Putin, "Russia at the Turn of the Millennium." https://pages.uoregon.edu/kimball/Putin.htm.

18 Donald Kagan, *Thucydides: The Reinvention of History* (New York: Penguin Books, 2010), 1, 9, 14–16.

19 Vladimir Putin, "Annual Address to the Federal Assembly of the Russian Federation," President of Russia, April 25, 2005, http://en.kremlin.ru/events/president/transcripts/22931.

20 On Russia in the late 1990s, see Michael McFaul, "Russia's Unfinished Revolution: Political Change from Gorbachev to Putin" (Ithaca, NY: Cornell University Press, 2001). See also Chrystia Freeland, *Sale of the Century: The Inside Story of the Second Russian Revolution* (London: Little, Brown and Company, 2000).

21 On the use of assistance, see U.S. General Accounting Office, "Foreign Assistance: International Efforts to Aid Russia's Transitions Have Had Mixed Reviews." GAO, November 2000, 33, https://www.gao.gov/products/GAO-01-8. On Russia under the Obama administration, see Carter Ash, *Inside the Five-Sided Box: Lessons from a Lifetime of Leadership in the Pentagon* (New York: Dutton, 2019), 272–77.

On Putin's speech, see *Washington Post* Editorial Board, "After the Fall of the Soviet Union, the U.S Tried to Help Russians," *Washington Post*, May 4, 2015, https://www.washingtonpost.com/opinions/after-the-fall-of-the-soviet-union-the-us-tried-to-help-russians/2015/05/04/cc4f7c20-f043-11e4-8666-a1d756d0218e_story.html.

22 On Putin quote, see Russian media source: PNA HOBOCTN, "Путин: 'цветные революции' в ряде стран—это урок для России," PNA HOBOCTN, November 20, 2014, https://ria.ru/20141120/1034329699.html. For estimates on civilian deaths and torture, see Simon Shuster. "Putin's Secret Agents," *Time*, https://time.com/putin-secret-agents/; Eli Lake, "Georgia's Democracy Recedes into Russia's Shadow," Bloomberg Opinion. September 14, 2018. https://www.bloomberg.com/opinion/articles/2018-09-14/georgia-s-rose-revolution-recedes-into-russia-s-shadow. On the color revolutions, see Lake, "Georgia's Democracy Recedes Into Russia's Shadow"; also, William Schneider, "Ukraine's 'Orange Revolution,'" *The Atlantic*, December 2004, https://www.theatlantic.com/magazine/archive/2004/12/ukraines-orange-revolution/305157/; Anthony H. Cordesman, "Russia and 'The Color Revolution': A Russian Military View of a World Destabilized by the U.S. and the West," Center for Strategic and International Studies, May 28, 2014, https://www.csis.org/analysis/russia-and-%E2%80%9Ccolor-revolution%E2%80%9D. On Putin's reaction to the color revolutions, see Leonid Bershidsky, "Why 'Color Revolutions' Can't Be Exported," Bloomberg Opinion, February 14, 2018, https://www.bloomberg.com/opinion/articles/2018-02-15/saakashvili-and-why-color-revolutions-can-t-be-exported. On Putin's election, see "Putin Declared President-elect," RT, March 5, 2012, https://www.rt.com/news/putin-win-presidential-election-813/.

23 On Putin's view of the color revolutions and their motivations, see Dmitri Simes, "Senior Kremlin Official Accuses NATO of Plotting 'Color Revolutions' in Russia's Neighborhood," CNSNews.com, July 5, 2019,

https://www.cnsnews.com/news/article/dimitri-simes/senior-kremlin-official-accuses-nato-plotting-color-revolutions-russias; Julia Gurganus and Eugene Rumer, "Russia's Global Ambitions in Perspective," Carnegie Endowment for International Peace, February 20, 2019, https://carnegieendowment.org/2019/02/20/russia-s-global-ambitions-in-perspective-pub-78067.

24 Statistics from the World Bank, "GDP per Capita—United States" and "GDP per Capita—Russian Federation," World Bank, https://data.worldbank.org/indicator/NY.GDP.PCAP.CD. U.S. GDP per capita in 2017 was $59,927.93. Statistics gathered from World Bank, "Corruption Perceptions Index 2017," Transparency International, February 21, 2018, https://www.transparency.org/news/feature/corruption_perceptions_index_2017. On Russia's corruption, see Transparency International, "Corruption Perceptions Index 2017."

25 On projected population decline in Russia, see David Holloway, "Russia and the Solecism of Power," Governance in an Emerging World, Fall Series, Issue 118, October 3, 2018. https://www.hoover.org/research/russia-and-solecism-power. For statistics on health, see Rachel Nuwer, "Why Russian Men Don't Live as Long," *New York Times*, February 17, 2014, https://www.nytimes.com/2014/02/18/science/why-russian-men-dont-live-as-long.html. Life expectancy data accessed from the World Bank, "Life Expectancy at Birth, Total (Years): Russian Federation," World Bank, https://data.worldbank.org/indicator/SP.DYN.LE00.IN?locations=RU&name_desc=false.

26 I am indebted to Dr. Kathyrn Stoner of Stanford University for this and many other insights into Russian strategy. For more on Putin's preference for destroying rather than rebuilding order, see Jakub Grygiel, "The Geopolitical Nihilist," *American Interest*, December 10, 2014, https://www.the-american-interest.com/2014/12/10/the-geopolitical-nihilist/.

27 Valery Gerasimov, "The Value of Science Is in the Foresight: New Challenges Demand Rethinking the Forms and Methods of Carrying Out Combat Operations," trans. Robert Coalson, *Military Review*, January–February 2016 (originally published in *Military-Industrial Kurier*, February 27, 2013), https://www.armyupress.army.mil/Portals/7/military-review/Archives/English/MilitaryReview_20160228_art008.pdf.

28 Christopher Paul and Miriam Matthews, "The Russian 'Firehose of Falsehood' Propaganda Model: Why It Might Work and Options to Counter It," RAND Corporation, 2016, https://www.rand.org/pubs/perspectives/PE198.html; Margaret L. Taylor, "Combating Disinformation and Foreign Interference in Democracies: Lessons from Europe," Brookings Institution, July 31, 2019, https://www.brookings.edu/blog/techtank/2019/07/31/combating-disinformation-and-foreign-interference-in-democracies-lessons-from-europe/.

29 Amy Zegart, "The Dark Arts of Deception: What's Old? What's New? What's Next?" Global Populisms Conference, March 1–2, 2019, Stanford University, https://fsi-live.s3.us-west-1.amazonaws.com/s3fs-public/zegart_populisms_memo_2.21.2019_1.pdf.

30 See Stanford Internet Observatory Paper, "Evidence of Russia-Linked Influence Operations in Africa," https://cyber.fsi.stanford.edu/io/news/prigozhin-africa.

31 On the Ukrainian presidential elections, see Office for Democratic Institutions and Human Rights, "Ukraine Presidential Election 31 October, 21 November and 26 December 2004: OSCE/ODIHR Election Observation Mission Final Report," May 11, 2005, https://www.osce.org/odihr/elections/ukraine/14674?download=true; and Steven Pifer, *The Eagle and the Trident: U.S.-Ukraine Relations in Turbulent Times* (Washington, DC: Brookings Institution Press, 2017), 274.

32 On Russian interference in Moldova, see Andrey Makarychev, "Russia's Moldova Policy: Soft Power at the Service of Realpolitik," ponarseurasia.org, March 2010, http://www.ponarseurasia.org/sites/default/files/policy-memos-pdf/pepm_094.pdf.

33 Gabe Joselow, "Election Cyberattacks: Pro-Russia Hackers Have Been Accused in Past," NBC, November 3, 2016, https://www.nbcnews.com/mach/technology/election-cyberattacks-pro-russia-hackers-have-been-accused-past-n673246.

34 On Russian subversion in each of these countries, see Jamie Doward, "Malta Accuses Russia of Cyber-attacks in Run-up to Election," *The Guardian*, May 27, 2017, https://www.theguardian.com/world/2017/may/27/russia-behind-cyber-attacks-says-malta-jseph-muscat; Oren Dorell, "Alleged Russian Political Meddling Documented in 27 Countries Since 2004," *USA Today*, September 7, 2017, https://www.usatoday.com/story/news/world/2017/09/07/alleged-russian-political-meddling-documented-27-countries-since-2004/619056001/; Ann M. Simmons, "Russia's Meddling in Other Nations' Elections Is Nothing New. Just Ask the Europeans," *Los Angeles Times*, March 30, 2017, https://www.latimes.com/world/europe/la-fg-russia-election-meddling-20170330-story.html; and Larry Diamond, "Russia and the Threat to Liberal Democracy," *The Atlantic*, December 9, 2016, https://www.theatlantic.com/international/archive/2016/12/russia-liberal-democracy/510011/.

35 On the Montenegrin election, see David Shimer, "Smaller Democracies Grapple with the Threat of Russian Interference," *The New Yorker*, December 8, 2018, https://www.newyorker.com/news/news-desk/smaller-democracies-grapple-with-the-threat-of-russian-interference.

36 Milivoje Pantovic, "Vucic: Serbia Arrests People Involved in 'Illegal Acts' in Montenegro," *Balkan Insight*, October 25, 2016, https://balkaninsight.com/2016/10/25/serbian-pm-failed-to-explain-coup-in-montenegro-10-24-2016/.

37 Kramer and Orovic, "Two Suspected Russian Agents."

38 On Americans' economic outlook, see Eduardo Porter, "Where Were Trump's Votes? Where the Jobs Weren't." *New York Times*, December 13, 2016, https://www.nytimes.com/2016/12/13/business/economy/jobs-economy-voters.html. Michelle Ver Ploeg, "Access to Affordable, Nutritious Food Is Limited in 'Food Deserts,'" U.S. Department of Agriculture Economic Research Service, March 1, 2010, https://www.ers.usda

.gov/amber-waves/2010/march/access-to-affordable-nutritious-food-is-limited-in-food-deserts/.

39 Aaron Blake, "More Young People Voted for Bernie Sanders than Trump and Clinton Combined—By a Lot," *Washington Post*, June 20, 2016, https://www.washingtonpost.com/news/the-fix/wp/2016/06/20/more-young-people-voted-for-bernie-sanders-than-trump-and-clinton-combined-by-a-lot/.

40 For quote on Trump and historical perspective on Trump's election, see Victor Davis Hanson, *The Case for Trump* (New York: Basic Books, 2020); Colleen Kelley, *A Rhetoric of Divisive Partisanship: The 2016 American Presidential Campaign Discourse of Bernie Sanders and Donald Trump* (Lanham, MD: Lexington Books, 2018), 15.

41 These and the statistics on the IRA in the following paragraphs from Renee DiResta, et al., "The Tactics and Tropes of the Internet Research Agency," New Knowledge, 2018, https://disinformationreport.blob.core.windows.net/disinformation-report/NewKnowledge-Disinformation-Report-Whitepaper.pdf.

42 The New Knowledge report on "The Tactics and Tropes of the Internet Research Agency" concluded that the purpose was "to undermine citizens' trust in government, exploit societal fractures, create distrust in the information environment, blur the lines between reality and fiction, undermine trust among communities, and erode confidence in the democratic process." Renee DiResta et al., "The Tactics and Tropes of the Internet Research Agency."

43 Niall Ferguson, "Silicon Valley and the Threat to Democracy," The Daily Beast, January 21, 2018, https://www.thedailybeast.com/social-media-shreds-the-social-fabric-one-click-at-a-time; Report of the Select Committee on Intelligence, "U.S. Senate: Russian Active Measures Campaigns and Interferences in the 2016 U.S. Election, Volume 2: Russia's Use of Social Media with Additional Views," U.S. Senate, n.d., https://www.intelligence.senate.gov/sites/default/files/documents/Report_Volume2.pdf.

44 Select Committee on Intelligence, "Russian Active Measures Campaigns and Interferences in the 2016 U.S. Election, Volume 2: Russia's Use of Social Media with Additional Views."

45 Harvey Klehr and William Tompson, "Self-determination in the Black Belt: Origins of a Communist Party," *Labor History* 30, no. 3 (1989): 355, https://doi.org/10.1080/00236568900890231.

46 On the IRA's use of false identities, see Robert S. Mueller, United States of America v. Viktor Borisovich Netyksho, Defendants: Case 1:18-cr-00215-ABJ, U.S. Justice Department, https://www.justice.gov/file/1080281/download. On Guccifer 2.0, see Robert S. Mueller, "Report on the Investigation into Russian Interference in the 2016 Presidential Election," U.S. Department of Justice, Washington, DC, March 2019, https://www.justice.gov/storage/report.pdf (hereafter cited as "Mueller Report"); on memes and their repurposing, see New Knowledge report, "The Tactics and Tropes of the Internet Research Agency"; and U.S. Jus-

tice Department, "United States of America v. Internet Research Agency," July, 2018, .pdf at https://www.justice.gov/file/1035477/download.

47 See David Folkenflik, "Behind Fox News' Baseless Seth Rich Story: The Untold Tale," NPR, August 1, 2017, https://www.npr.org/2017/08/01/540783715/lawsuit-alleges-fox-news-and-trump-supporter-created-fake-news-story; and U.S. Senate Select Committee on Intelligence, "New Reports Shed Light on Internet Research Agency's Social Media Tactics," Washington, DC, December 2018, https://cryptome.org/2018/12/ssci-ru-sm-aid-trump.pdf; Janine Zacharia, "Facebook, Others Must Do More to Protect National Security," *San Francisco Chronicle*, October 17, 2017, http://janinezacharia.net/reporting/facebook-and-others-must-protect-national-security/.

48 See Mueller, United States of America v. Viktor Borisovich Netyksho.

49 Tom LoBianco, "Trump Falsely Claims 'Millions of People Who Voted Illegally' Cost Him Popular Vote," CNN, November 28, 2016, https://www.cnn.com/2016/11/27/politics/donald-trump-voter-fraud-popular-vote/index.html.

50 Adam Goldman, Jo Becker, and Matt Apuzzo, "Russian Dirt on Clinton? 'I Love It,' Donald Trump Jr. Said," *New York Times*, July 11, 2017, https://www.nytimes.com/2017/07/11/us/politics/trump-russia-email-clinton.html. Also see Mueller Report.

51 On Cambridge Analytica, see Nicholas Confessore, Matthew Rosenberg, and Sheera Frenkel, "Facebook Data Collected by Quiz App Included Private Messages," *New York Times*, April 10, 2018, https://www.nytimes.com/2018/04/10/technology/facebook-cambridge-analytica-private-messages.html.

52 Shimer, "Smaller Democracies"; Amy Zegart and Michael Morrell, "Why U.S. Intelligence Agencies Must Adapt or Fail," *Foreign Affairs*, May/June 2019, https://www.foreignaffairs.com/articles/2019-04-16/spies-lies-and-algorithms.

53 Philip Rucker, Anton Troianovski, and Seung Min Kim, "Trump Hands Putin a Diplomatic Triumph by Casting Doubt on U.S. Intelligence Agencies," *Washington Post*, July 16, 2018, https://www.washingtonpost.com/politics/ahead-of-putin-summit-trump-faults-us-stupidity-for-poor-relations-with-russia/2018/07/16/297f671c-88c0-11e8-a345-a1bf7847b375_story.html.

54 On the DNC's reaction to the hacks, see Donna Brazile, *Hacks: The Inside Story of the Break-ins and Breakdowns that Put Donald Trump in the White House* (New York: Hachette Books, 2017), 95–103. On Obama, see Philip Bump, "What Obama Did, Didn't Do, and Couldn't Do in Response to Russian Interference," *Washington Post*, February 21, 2018, https://www.washingtonpost.com/news/politics/wp/2018/02/21/what-obama-did-didnt-do-and-couldnt-do-in-response-to-russian-interference/. On Trump's quote, see Paul Waldman, "Trump Sucks Up to Putin, Embarrassing Us Yet Again," *Washington Post*, June 28, 2019, https://www.washingtonpost.com/opinions/2019/06/28/trump-sucks-up-putin-embarrassing-us-yet-again/.

55 Anton Troianovski and Joby Warrick, "Agents of Doubt: How a Powerful Russian Propaganda Machine Chips Away at Western Notions of Truth," *Washington Post*, December 10, 2018, https://www.washingtonpost.com/graphics/2018/world/national-security/russian-propaganda-skripal-salisbury/.

56 On the Skripal attack, see Andrew Roth and Vikram Dodd, "Salisbury Novichok Suspects Say They Were Only Visiting Cathedral," *The Guardian*, September 13, 2018, https://www.theguardian.com/uk-news/2018/sep/13/russian-television-channel-rt-says-it-is-to-air-interview-with-skripal-salisbury-attack-suspects; Troianovski and Warrick, "Agents of Doubt."

57 "Statement from the Press Secretary on the Expulsion of Russian Intelligence Officers," Statements and Releases, The White House, March 26, 2018, https://www.whitehouse.gov/briefings-statements/statement-press-secretary-expulsion-russian-intelligence-officers/.

58 Carl Gershman, "Remembering a Journalist Who Was Killed for Standing Up to Putin," *Washington Post*, October 6, 2016, https://www.washingtonpost.com/opinions/remembering-a-journalist-who-was-killed-for-standing-up-to-putin/2016/10/06/d3a9e176-8bf7-11e6-bff0-d53f592f176e_story.html; David Filipov, "Here Are 10 Critics of Vladimir Putin Who Died Violently or in Suspicious Ways," *Washington Post*, March 23, 2017, https://www.washingtonpost.com/news/worldviews/wp/2017/03/23/here-are-ten-critics-of-vladimir-putin-who-died-violently-or-in-suspicious-ways/.

59 BBC News, "Syria War: What We Know About Douma 'Chemical Attack,'" BBC, July 10, 2018, https://www.bbc.com/news/world-middle-east-43697084; Sheena McKenzie, "Suspected Syria Chemical Attack Might Have Affected 500 People, WHO Says," CNN, April 11, 2018, https://www.cnn.com/2018/04/11/middleeast/syria-chemical-attack-500-affected-who-intl/index.html.

60 On the Wagner Group, see Mike Giglio, "How a Group of Russian Guns for Hire Are Operating in the Shadows," BuzzFeed News, April 19, 2019, https://www.buzzfeednews.com/article/mikegiglio/inside-wagner-mercenaries-russia-ukraine-syria-prighozhin. On the Malaysia Airlines flight, "Russia's Role in Shooting Down an Airliner Becomes Official," *The Economist*, May 30, 2018, https://www.economist.com/europe/2018/05/30/russias-role-in-shooting-down-an-airliner-becomes-official.

61 Peter Ford, "Russia's Retreat Ends Chechnya War but Leaves a Long-Term Impact," *The Christian Science Monitor*, January 6, 1997, https://www.csmonitor.com/1997/0106/010697.intl.intl.2.html.

62 "Alert (TA18-074A) Russian Government Cyber Activity Targeting Energy and Other Critical Infrastructure Sectors," CISA, March 16, 2018, https://www.us-cert.gov/ncas/alerts/TA18-074A.

63 Nicole Perlroth and David E. Sanger, "Cyberattacks Put Russian Fingers on the Switch at Power Plants, U.S. Says," *New York Times*, March 15, 2018, https://www.nytimes.com/2018/03/15/us/politics/russia-cyberattacks.html.

64 Alexander Cooley, "Whose Rules, Whose Sphere? Russian Governance and Influence in Post-Soviet States," Carnegie Endowment for International Peace, June 30, 2017, https://carnegieendowment.org/2017/06/30/whose-rules-whose-sphere-russian-governance-and-influence-in-post-soviet-states-pub-71403.

65 "Nord Stream: The Gas Pipeline Directly Connecting Russia and Europe," Gazprom, http://www.gazprom.com/projects/nord-stream/; Editorial Board, "The Right (and Wrong) Way to Deal with Nord Stream 2," Bloomberg Opinion, November 27, 2018, https://www.bloomberg.com/opinion/articles/2018-11-27/nord-stream-2-the-right-and-wrong-response-for-america.

66 George Frost Kennan, *Russia and the West Under Lenin and Stalin* (Boston, MA: Little, Brown and Company, 1961), 13.

67 Philipp Ther, *Europe Since 1989: A History*, trans. Charlotte Hughes-Kreutzmüller (Princeton, NJ: Princeton University Press. 2016), 302–3.

68 On the role of the Kremlin in each of these crises, see Michael Stott, "Russia Blames U.S. for Global Financial Crisis," Reuters, June 7, 2008, https://www.reuters.com/article/us-russia-forum-medvedev/russia-blames-u-s-for-global-financial-crisis-idUSL0749277620080607; Timothy Heritage, "Russia Waits in Wings as Greek Debt Crisis Deepens," Reuters, July 3, 2015, https://www.reuters.com/article/us-eurozone-greece-russia/russia-waits-in-wings-as-greek-debt-crisis-deepens-idUSKCN0PD0YH20150703; Peter Walker, "Russia 'Spreading Fake News about Refugees to Sow Discord in Europe' Says Ex-Spy," *The Independent*, March 22, 2017, https://www.independent.co.uk/news/world/europe/russia-europe-threat-refugee-crisis-europe-aggravate-propaganda-kremlin-farenc-katrei-hungarian-spy-a7642711.html; David D. Kirkpatrick, "Signs of Russian Meddling in Brexit Referendum," *New York Times*, November 15, 2017, https://www.nytimes.com/2017/11/15/world/europe/russia-brexit-twitter-facebook.html; Andrew Roth and Angelique Chrisafis, "Gilets Jaunes: Grassroots Heroes or Tools of the Kremlin?" *The Guardian*, December 17, 2018, https://www.theguardian.com/world/2018/dec/17/gilets-jaunes-grassroots-heroes-or-kremlin-tools; Matt Bradley, "Europe's Far-Right Enjoys Backing from Russia's Putin," NBC News, February 12, 2017, https://www.nbcnews.com/news/world/europe-s-far-right-enjoys-backing-russia-s-putin-n718926.

69 Tony Judt, *Postwar: A History of Europe Since 1945* (New York: Penguin Books, 2005), 737.

70 For European perspectives on the American pivot, see Bjonar Sverdrup-Thygeson, Marc Lanteigne, and Ulf Sverdrup, "'For Every Action . . .': The American Pivot to Asia and Fragmented European Responses," Project on International Order and Strategy, Brookings and the Norwegian Institute of International Affairs, January 27, 2016, https://www.brookings.edu/wp-content/uploads/2016/07/The-American-pivot-to-Asia-and-fragmented-European-responses-2.pdf. On the Macron quote, see Alexandra Ma, "French President Macron Dunked on Trump for Pulling out of Syria Without Telling His NATO Allies," Business Insider,

n.d., https://www.businessinsider.sg/macron-trump-withdraw-syria-without-telling-nato-economist-2019-11/; Maegan Vazquez and Allie Malloy, "Macron Refuses to Back Down After Trump Attack," CNN, December 4, 2019. https://www.cnn.com/2019/12/03/politics/donald-trump-nato/index.html.

71 Andrew Rawnsley, "Interview: Madeleine Albright: 'The Things That Are Happening Are Genuinely, Seriously Bad,'" *The Guardian*, July 8, 2018, https://www.theguardian.com/books/2018/jul/08/madeleine-albright-fascism-is-not-an-ideology-its-a-method-interview-fascism-a-warning.

Chapter 2: Parrying Putin's Playbook

1 House Permanent Select Committee on Intelligence, House Committee on Oversight and Reform, House Committee on Foreign Affairs, "Excerpts from Joint Deposition: Dr. Fiona Hill Former Deputy Assistant to the President and Senior Director for Europe and Russia, National Security Council," Washington, DC, October 14, 2019, https://intelligence.house.gov/uploadedfiles/20191108_-_hill_transcript_excerpts_-_137591.pdf.

2 Bill Keller, "Major Soviet Paper Says 20 Million Died as Victims of Stalin," *New York Times*, February 4, 1989, https://www.nytimes.com/1989/02/04/world/major-soviet-paper-says-20-million-died-as-victims-of-stalin.html.

3 Keir Giles, *Moscow Rules: What Drives Russia to Confront the West* (Washington, DC: Brookings Institution Press), 38.

4 According to the U.S. Department of Commerce's Bureau of Economic Analysis, Texas's GDP was ~ $1.8 trillion at the end of 2018. Heading into 2019, Russia's nominal GDP was ~ $1.65 trillion while Italy's GDP was ~ $2 trillion. Figures gathered from the World Bank, "Gross Domestic Product by State, First Quarter 2019," July 25, 2019.

5 Nan Tian et al., "Trends in World Military Expenditure, 2018," April 2019, https://www.sipri.org/sites/default/files/2019-04/fs_1904_milex_2018.pdf.

6 Timothy Garton Ash, "Europe's Crises Conceal Opportunities to Forge Another Path," *Financial Times*, November 21, 2018, https://www.ft.com/content/160d11b6-ec25-11e8-89c8-d36339d835c0.

7 Jill Dougherty, "U.S. Seeks to 'Reset' Relations with Russia" CNN, March 7, 2009, http://edition.cnn.com/2009/WORLD/europe/03/07/us.russia/index.html; Sue Pleming, "U.S. and Russia Pledge Fresh Start to Relations," Reuters, March 6, 2009, https://www.reuters.com/article/us-russia-usa/u-s-and-russia-pledge-fresh-start-in-relations-idUSTRE52522420090307.

8 On Clinton's quotes, see Glenn Kessler, "Clinton 'Resets' Russian Ties—and Language," *Washington Post*, March 7, 2009, http://www.washingtonpost.com/wp-dyn/content/article/2009/03/06/AR2009030600428.html. On Obama, see J. David Goodman, "Microphone Catches a Can-

did Obama," *New York Times*, March 27, 2012, https://www.nytimes.com/2012/03/27/us/politics/obama-caught-on-microphone-telling-medvedev-of-flexibility.html.

9 Jillian Rayfield, "Obama: The '80s Called, They Want Their Foreign Policy Back," *Salon*, October 23, 2012, https://www.salon.com/2012/10/23/obama_the_80s_called_they_want_their_foreign_policy_back/.

10 "Press Conference by President Bush and Russian Federation President Putin," The White House, President George W. Bush, National Archives and Records Administration, June 16, 2001, https://georgewbush-whitehouse.archives.gov/news/releases/2001/06/20010618.html.

11 Peter Baker, "The Seduction of George W. Bush," *Foreign Policy*, November 6, 2013, https://foreignpolicy.com/2013/11/06/the-seduction-of-george-w-bush/.

12 Tyler Pager, "Putin Repeats Praise of Trump: He's a 'Bright' Person," Politico, June 17, 2016, https://www.politico.com/story/2016/06/putin-praises-trump-224485.

13 Sophie Tatum, "Trump Defends Putin: 'You Think Our Country's So Innocent?'" CNN, February 6, 2017, https://www.cnn.com/2017/02/04/politics/donald-trump-vladimir-putin/index.html.

14 "Remarks by President Trump in Press Gaggle Aboard Air Force One en Route Hanoi, Vietnam," The White House (U.S. Government), https://www.whitehouse.gov/briefings-statements/remarks-president-trump-press-gaggle-aboard-air-force-one-en-route-hanoi-vietnam/; and Jeremy Diamond, "Trump Sides with Putin over U.S. Intelligence," CNN, July 16, 2018, https://www.cnn.com/2018/07/16/politics/donald-trump-putin-helsinki-summit/index.html.

15 On the relationship between the Bush administration and Putin, see Condoleezza Rice, *Democracy: Stories from the Long Road to Freedom* (New York: Twelve, 2018).

16 FP Staff, "Here's What Trump and Putin Actually Said in Helsinki," *Foreign Policy*, July 18, 2018, https://foreignpolicy.com/2018/07/18/heres-what-trump-and-putin-actually-said-in-helsinki/.

17 Scott Shane, "Stephen Bannon in 2014: We Are at War with Radical Islam," *New York Times*, February 2, 2017, https://www.nytimes.com/interactive/2017/02/01/us/stephen-bannon-war-with-radical-islam.html.

18 See, for example, Donald J. Trump, "Trump: I'm Not Pro-Russia, I Just Want Our Country Safe," interview by Tucker Carlson, *Tucker Carlson Tonight*, Fox News, July 17, 2018, https://www.youtube.com/watch?v=MB8etvUSag0; see also Anne Applebaum, "The False Romance of Russia," *The Atlantic*, December 12, 2019, https://www.theatlantic.com/ideas/archive/2019/12/false-romance-russia/603433/.

19 George Kennan, *Russia and the West Under Lenin and Stalin* (Boston, MA: Little, Brown and Company, 1961), 349–69.

20 Jason Schwartz, "Senate Approves Supplemental Lend-Lease Act, Oct. 23, 1941," Politico, October 23, 2017, https://www.politico.com/story/2017/10/23/senate-approves-supplemental-lend-lease-act-oct-23-1941-243990.

21 On neo-Nazis and Russia, see Vegas Tenold, "My Six Years Covering Neo-Nazis: 'They're All Vying for the Affections of Russia,'" interview by Lois Beckett, *The Guardian U.S.*, February 17, 2018, https://www.theguardian.com/books/2018/feb/17/vegas-tenold-everything-you-love-will-burn-q-and-a-nazis; on Hungary, see Rick Lyman and Alison Smale, "Defying Soviets, Then Pulling Hungary to Putin," *New York Times*, November 7, 2014, https://www.nytimes.com/2014/11/08/world/europe/viktor-orban-steers-hungary-toward-russia-25-years-after-fall-of-the-berlin-wall.html; on the Orthodox Church, see Ishaan Tharoor, "The Christian Zeal Behind Russia's War in Syria," *Washington Post*, October 1, 2015, https://www.washingtonpost.com/news/worldviews/wp/2015/10/01/the-christian-zeal-behind-russias-war-in-syria/; on Russian Orthodox Church ties to Russian intelligence, see Paul A. Goble, "FSB, SVR Divide Control of Moscow Patriarchate Church at Home and Abroad, Ukrainian Intelligence Official Says," Euromaidan Press, January 29, 2019, http://euromaidanpress.com/2019/01/29/fsb-svr-divide-control-of-moscow-patriarchate-church-at-home-and-abroad-ukrainian-intelligence-official-says/.

22 Mueller Report.

23 "Specially Designated Nationals List Update," news release, March 15, 2018, Resource Center, Office of Foreign Assets Control, U.S. Department of the Treasury, https://www.treasury.gov/resource-center/sanctions/OFAC-Enforcement/Pages/20180315.aspx.

24 Mueller Report.

25 Adam Goldman, Julian E. Barnes, Maggie Haberman, and Nicolas Fandos, "Lawmakers Are Warned That Russia Is Meddling to Re-elect Trump," *New York Times*, February 20, 2020.

26 U.S. Cyber Command, "Achieve and Maintain Cyberspace Superiority, Command Vision for U.S. Cyber Command," April, 2018, https://www.cybercom.mil/Portals/56/Documents/USCYBERCOM%20Vision%20April%202018.pdf?ver=2018-06-14-152556-010.

27 U.S. Cyber Command, "Achieve and Maintain Cyberspace Superiority."

28 Ellen Nakashima, "U.S. Cyber Command Operation Disrupted Internet Access of Russian Troll Factory on Day of 2018 Midterms," *Washington Post*, February 27, 2019, https://www.washingtonpost.com/world/national-security/us-cyber-command-operation-disrupted-internet-access-of-russian-troll-factory-on-day-of-2018-midterms/2019/02/26/1827fc9e-36d6-11e9-af5b-b51b7ff322e9_story.html.

29 On the false reports, see Damien McGuinness, "How a Cyber Attack Transformed Estonia," BBC, April 27, 2017, https://www.bbc.com/news/39655415. On Estonia's response to Russia's sustained disinformation campaign, see Emily Tamkin, "10 Years After the Landmark Attack on Estonia Is the World Better Prepared for Cyber Threats?" *Foreign Policy*, April 27, 2017, https://foreignpolicy.com/2017/04/27/10-years-after-the-landmark-attack-on-estonia-is-the-world-better-prepared-for-cyber-threats/.

30 Quotes from former president Toomas Henrik Ilves, interview by author, May 31, 2019.

31 The Security Committee, "Finland's Cyber Security Strategy," October 4, 2018, https://turvallisuuskomitea.fi/en/finlands-cyber-security-strategy/ and https://www.kyberturvallisuuskeskus.fi/en/.
32 "Empowering Users to Discover What Matters," Soap Public Media, https://www.getsoap.org/the-impact/.html.
33 "How Soap Works to Deliver Clarity," Soap Public Media, https://www.getsoap.org/how-does-soap-work/.
34 "Our Mission, Clean and Simple," Soap Public Media, https://www.getsoap.org/mission/.
35 Lionel Barber, Henry Foy, and Alex Barker, "Vladimir Putin Says Liberalism Has 'Become Obsolete,'" *Financial Times*, June 27, 2019, https://www.ft.com/content/670039ec-98f3-11e9-9573-ee5cbb98ed36.
36 Andrew Radin et al., "The Outlook for Russia's Growing Military Power," RAND Corporation, June 18, 2019, https://www.rand.org/pubs/research_briefs/RB10038.html.
37 Senate Judiciary Committee, "Testimony of William Browder to the Senate Judiciary Committee on FARA Violations Connected to the anti-Magnitsky Campaign by Russian Government Interests," July 26, 2017, https://www.judiciary.senate.gov/imo/media/doc/07-26-17%20Browder%20Testimony.pdf.
38 Robert Coalson, "Analysis: After 'Significant' Regional Elections, Russia's Opposition Looks to the Future," Radio Free Europe/Radio Liberty, September 17, 2019, https://www.rferl.org/a/russia-analysis-opposition-future-regional-election-putin-navalny-protests/30169846.html.
39 Matthew Luxmoore, "'With Smart Voting Strategy,' Russian Opposition Takes Aim at Putin's 'Party of Crooks and Thieves,'" Radio Free Europe/Radio Liberty, September 8, 2019, https://www.rferl.org/a/russia-smart-voting/30153235.html.
40 Ellen Barry, "Putin Contends Clinton Incited Unrest Over Vote," *New York Times*, December 8, 2011, https://www.nytimes.com/2011/12/09/world/europe/putin-accuses-clinton-of-instigating-russian-protests.html.
41 Rice, *Democracy*, 74.
42 On the United States' role in a post-Putin Russia, see Herman Pirchner, *Post Putin: Succession, Stability, and Russia's Future*, American Foreign Policy Council (London, UK: Rowman & Littlefield, 2019). Also see James M. Goldgeier and Michael McFaul, *Power and Purpose: U.S. Policy Toward Russia After the Cold War* (Washington, DC: Brookings Institution Press, 2003), 346, 351.
43 Rice, *Democracy*, 73–74.
44 Holly Ellyatt, "China's Xi Calls Putin His 'Best Friend' Against a Backdrop of Souring U.S. Relations," CNBC, June 5, 2019, https://www.cnbc.com/2019/06/05/putin-and-xi-meet-to-strengthen-ties-as-us-relations-sour.html.
45 Grygiel and Mitchell, *The Unquiet Frontier*, 61–74.
46 On the hundreds of violating flights, see the Heritage Foundation, "Russia: Assessing Threats to U.S. Vital Interests," October 30, 2019, https://

www.heritage.org/military-strength/assessing-threats-us-vital-interests/russia; Emma Chanlett Avery, Caitlin Campbell, and Joshua A. Williams, "The U.S.-Japan Alliance," Congressional Research Service, June 13, 2019, https://fas.org/sgp/crs/row/RL33740.pdf. On the July 2019 incident, see Mike Yeo, "Russian-Chinese Air Patrol Was an Attempt to Divide Allies, Says Top U.S. Air Force Official in Pacific," *DefenseNews*, August 23, 2019, https://www.defensenews.com/global/asia-pacific/2019/08/23/russian-chinese-air-patrol-was-an-attempt-to-divide-allies-says-top-us-air-force-official-in-pacific/.

47 James Dobbins, Howard Shatz, and Ali Wyne, "A Warming Trend in China-Russia Relations." RAND Corporation, April 18, 2019, https://www.rand.org/blog/2019/04/a-warming-trend-in-china-russia-relations.html.

Chapter 3: An Obsession with Control: The Chinese Communist Party's Threat to Freedom and Security

1 Gerald F. Seib, Jay Solomon, and Carol E. Lee, "Barack Obama Warns Donald Trump on North Korea Threat," *Wall Street Journal*, Dow Jones and Company, November 22, 2016, https://www.wsj.com/articles/trump-faces-north-korean-challenge-1479855286.

2 The Obama administration's China policy is summarized here: Cheng Li, "Assessing U.S.-China Relations Under the Obama Administration," Brookings Institution, September 5, 2016. https://www.brookings.edu/opinions/assessing-u-s-china-relations-under-the-obama-administration/.

3 John Fairbank, *The United States and China* (Cambridge: Harvard University Press, 1948), 9.

4 "Xi Jinping: 'Time for China to Take Centre Stage," BBC, October 18, 2017, https://www.bbc.com/news/world-asia-china-41647872.

5 On China's rise, see Gideon Rachman, *Easternisation: War and Peace in the Asian Century* (London: The Bodley Head, 2016).

6 On MacCartney's journey to visit Emperor Qianlong, see Howard French, *Everything Under the Heavens: How the Past Helps Shape China's Push for Global Power* (New York: Alfred A. Knopf, 2017), 5–8. As Wang Jisi, dean of the School of International Studies at Peking University, observed in 2015, "Ever since the founding of 'New China' in 1949… Foreign relations as well as trade and economic policy had to match the narrative of national greatness so the ruler, whether an eighteenth-century Emperor or a modern-day autocrat, could confirm his claim to rule." Quoted in French, *Everything Under the Heavens*, 7–8.

7 For China's sense of insecurity under Xi Jinping, see Michael D. Swaine and Ashley J. Tellis, *Interpreting China's Grand Strategy: Past, Present, and Future* (Santa Monica, CA: RAND Corporation, 2000), 12–13; Sulmaan Khan, *Haunted by Chaos: China's Grand Strategy from Mao Zedong to Xi Jinping* (Cambridge, MA: Harvard University Press, 2018), 7–8 and 209–35.

8 For example, the future premier, Zhou Enlai, who is best known as the urbane interlocutor of Henry Kissinger in the 1970s, oversaw the murder of his political rival's family in the 1930s. Benjamin Elman, *Civil Ex-*

aminations and Meritocracy in Late Imperial China (Cambridge, MA: Harvard University Press, 2013), 30.

9 The "Century of Humiliation" is remembered in China especially for the following grievances: defeat in the First and Second Opium Wars (1839–42 and 1856–60) by Great Britain; unequal treaties in the mid- to late nineteenth century; the Taiping Rebellion (1850–64); defeat in the Sino-French War (1884–85); defeat in the First and Second Sino-Japanese Wars (1894–95 and 1937–45); the Eight-Nation Alliance suppressing the Boxer uprising (1899–1901); the British invasion of Tibet (1903–4); the Twenty-one Demands by Japan (1915); and the Japanese invasion of Manchuria (1931–32). In many cases, China was forced to pay large amounts of reparations, open up ports for trade, lease or cede territories, and make various other concessions of sovereignty to foreign "spheres of influence" following military defeats.

10 For a history of the Cultural Revolution, see Khan, *Haunted by Chaos*, 111–26. On Deng's policies and reforms, see Ezra F. Vogel, *Deng Xiaoping and the Transformation of China* (Cambridge, MA: Belknap Press/Harvard University Press, 2013). The 1981 declaration is found in "Resolution on Certain Questions in the History of Our Party Since the Founding of the People's Republic of China," The Sixth Plenary Session of the Eleventh Central Committee of the Communist Party of China, June 27, 1981.

11 On Xi's upbringing of forced labor, see Chris Buckley and Didi Kirsten Tatlow, "Cultural Revolution Shaped Xi Jinping, From Schoolboy to Survivor," *New York Times*, September 24, 2015, https://www.nytimes.com/2015/09/25/world/asia/xi-jinping-china-cultural-revolution.html; and Evan Osnos, "Born Red," *The New Yorker*, March 30, 2015, https://www.newyorker.com/magazine/2015/04/06/born-red. On how this affects Xi's policies, see Patricia Thornton, *Disciplining the State: Virtue, Violence, and State-making in Modern China*. (Cambridge, MA: Harvard University Asia Center, 2007), 168–69.

12 Orville Schell and John Delury, *Wealth and Power: China's Long March to the Twenty-first Century* (New York: Random House, 2013), 386.

13 "Xi, unlike Mao, never grew into the Party, but always belonged to it. He has no existence separate from the culture of the Party, and no autonomy from it." Kerry Brown, *CEO, China: The Rise of Xi Jinping* (New York: I. B. Tauris, 2017), 230.

14 Timothy Beardson, *Stumbling Giant: The Threats to China's Future* (New Haven, CT: Yale University Press, 2013), 435.

15 On China's growth, see Wei Chen, Xilu Chen, Chang-Tai Hsieh, and Zheng Song, "A Forensic Examination of China's National Accounts," Brookings Papers on Economic Activity, March 7, 2019, https://www.brookings.edu/wp-content/uploads/2019/03/bpea_2019_conference-1.pdf. On SOEs, see Greg Levesque, "China's Evolving Economic Statecraft," *The Diplomat*, April 12, 2017, https://www.nytimes.com/2016/10/14/world/asia/china-soe-state-owned-enterprises.html.

16 Lily Kuo and Kate Lyons, "China's Most Popular App Brings Xi Jinping to Your Pocket," *The Guardian*, February 15, 2019, https://www.the

guardian.com/world/2019/feb/15/chinas-most-popular-app-brings-xi-jinping-to-your-pocket.

17 Austin Ramzy and Chris Buckley, "'Absolutely No Mercy': Leaked Files Expose How China Organized Mass Detentions of Muslims," *New York Times*, November 16, 2019, https://www.nytimes.com/interactive/2019/11/16/world/asia/china-xinjiang-documents.html.

18 "Mass Rally Thanks U.S. for 'Supporting Hong Kong,'" Radio Television Hong Kong, November 28, 2019, https://news.rthk.hk/rthk/en/component/k2/1494997-20191128.htm.

19 Benjamin Lim and Ben Blanchard, "Xi Jinping Hopes Traditional Faiths Can Fill Moral Void in China: Sources," Reuters, September 29, 2013, https://www.reuters.com/article/us-china-politics-vacuum/xi-jinping-hopes-traditional-faiths-can-fill-moral-void-in-china-sources-idUSBRE98S0GS20130929.

20 Christian Shepherd, "Disappearing Textbook Highlights Debate in China over Academic Freedom," Reuters, February 1, 2019, https://www.reuters.com/article/us-china-law/disappearing-textbook-highlights-debate-in-china-over-academic-freedom-idUSKCN1PQ45T.

21 On China's tributary system, see Christopher Ford, *The Mind of Empire: China's History and Modern Foreign Relations* (Lexington: University Press of Kentucky, 2010), 92–96; and French, *Everything Under the Heavens*, 10–12, 244. See also David Kang, *East Asia Before the West: Five Centuries of Trade and Tribute* (New York: Columbia University Press, 2010), 107.

22 U.S. Energy Information Administration, "World Oil Transit Chokepoints," July 25, 2017, https://www.eia.gov/beta/international/analysis_includes/special_topics/World_Oil_Transit_Chokepoints/wotc.pdf.

23 Jun Ding and Hongjin Cheng, "China's Proposition to Build a Community of Shared Future for Mankind and the Middle East Governance," *Asian Journal of Middle Eastern and Islamic Studies* 11, no. 4 (2017): 3.

24 On One Belt One Road, see Audrye Wong, "China's Economic Statecraft under Xi Jinping," Brookings Institution, January 22, 2019, https://www.brookings.edu/articles/chinas-economic-statecraft-under-xi-jinping/#footref-1; and Dylan Gerstel, "It's a (Debt) Trap! Managing China-IMF Cooperation Across the Belt and Road," Center for Strategic and International Studies, October 17, 2018, https://www.csis.org/npfp/its-debt-trap-managing-china-imf-cooperation-across-belt-and-road.

25 On China's efforts to influence the Maldives election, see Brahma Chellaney, "Beijing Loses a Battle in the Maldives—but the Fight for Influence Goes On," *Nikkei Asian Review*, September 25, 2018, https://asia.nikkei.com/Opinion/Beijing-loses-a-battle-in-the-Maldives-but-the-fight-for-influence-goes-on; Oki Nagai and Yuji Kuronuma, "Maldives Election Marks Setback for China's Belt and Road," *Nikkei Asian Review*, September 25, 2018, https://asia.nikkei.com/Spotlight/Belt-and-Road/Maldives-election-marks-setback-for-China-s-Belt-and-Road2.

26 Tom Wright and Bradley Hope, "WSJ Investigation: China Offered to

Bail Out Troubled Malaysian Fund in Return for Deals," *Wall Street Journal*, January 7, 2019.

27 David Ndii, "China's Debt Imperialism: The Art of War by Other Means?" Elephant, August 18, 2018, https://www.theelephant.info/op-eds/2018/08/18/chinas-debt-imperialism-the-art-of-war-by-other-means/.

28 Nicholas Casey and Clifford Krauss, "It Doesn't Matter if Ecuador Can Afford This Dam. China Still Gets Paid," *New York Times*, December 24, 2018, https://www.nytimes.com/2018/12/24/world/americas/ecuador-china-dam.html.

29 On the Maldives, see Simon Mundy and Kathrin Hille, "The Maldives Counts the Cost of Its Debts to China," *Financial Times*, February 10, 2019, https://www.ft.com/content/c8da1c8a-2a19-11e9-88a4-c32129756dd8. On Malaysia, see Tom Wright and Bradley Hope, "WSJ Investigation: China Offered to Bail Out Troubled Malaysian Fund in Return for Deals," *Wall Street Journal*, January 7, 2019, https://www.wsj.com/articles/how-china-flexes-its-political-muscle-to-expand-power-overseas-11546890449.

On Ecuador, see Nicholas Casey and Clifford Krauss, "It Doesn't Matter If Ecuador Can Afford This Dam. China Still Gets Paid," *New York Times*, December 24, 2018.

On Venezuela, see "China to Lend Venezuela $5 Billion as Maduro Visits Beijing," Bloomberg, September 13, 2018, https://www.bloomberg.com/news/articles/2018-09-13/china-to-give-venezuela-5-billion-loan-as-maduro-visits-beijing.

30 Erik Sherman, "One in Five U.S. Companies Say China Has Stolen Their Intellectual Property," *Fortune*, March 1, 2019, https://fortune.com/2019/03/01/china-ip-theft/.

31 William Hannas, James Mulvenon, and Anna Puglisi, *Chinese Industrial Espionage: Technology Acquisition and Military Modernisation* (Abingdon: Routledge, 2013), 165–71, 216–25, 230.

32 On nontraditional intelligence collection, see Hearing on China's Non-Traditional Espionage Against the United States: The Threat and Potential Policy Responses, Before the Senate Comm. on the Judiciary, 115th Congress (2018) (statement of John Demers, Assistant Attorney General, National Security Division, U.S. Department of Justice), https://www.judiciary.senate.gov/meetings/chinas-non-traditional-espionage-against-the-united-states-the-threat-and-potential-policy-responses. For more on CRI, see Greg Levesque, "Testimony Before the U.S.-China Economic and Security Review Commission Hearing on What Keeps Xi Up at Night: Beijing's Internal and External Challenges," U.S.-China Economic and Security Review Commission, February 2019, https://www.uscc.gov/sites/default/files/Levesque_USCC%20Testimony_Final_0.pdf.

33 "PRC Acquisition of U.S. Technology," U.S. National Security and the People's Republic of China, GovInfo, https://www.govinfo.gov/content/pkg/GPO-CRPT-105hrpt851/html/ch1bod.html; Office of the Under Secretary of Defense, "Defense Budget Overview: United States Department

of Defense Fiscal Year 2019 Budget Request," February 2018, https://dod.defense.gov/Portals/1/Documents/pubs/FY2019-Budget-Request-Overview-Book.pdf.

34 On the Department of Energy, see Department of Justice, "Former Sandia Corporation Scientist Sentenced for Taking Government Property to China," United States Attorney's Office, District of New Mexico, November 24, 2014, https://www.justice.gov/usao-nm/pr/former-sandia-corporation-scientist-sentenced-taking-government-property-china. On forced technology transfer, see Michael Brown and Pavneet Singh, "China's Technology Transfer Strategy," Defense Innovation Unit Experimental, January 2018, 19, https://admin.govexec.com/media/diux_chinatechnologytransferstudy_jan_2018_(1).pdf. On the Kuang-Chi Group, see Greg Levesque, "Testimony."

35 On the African Union, see John Aglionby, Emily Feng, and Yuan Yang, "African Union Accuses China of Hacking Headquarters," *Financial Times*, January 29, 2018, https://www.ft.com/content/c26a9214-04f2-11e8-9650-9c0ad2d7c5b5. For General Alexander quote, see Claudette Roulo, "Cybercom Chief: Culture, Commerce Changing Through Technology," U.S. Department of Defense, October 12, 2012, https://archive.defense.gov/news/newsarticle.aspx?id=118201. The study was conducted by the Council of Economic Advisors, "The Cost of Malicious Cyber Activity to the U.S. Economy," February 2018, 36, https://www.whitehouse.gov/wp-content/uploads/2018/03/The-Cost-of-Malicious-Cyber-Activity-to-the-U.S.-Economy.pdf .

36 United States of America v. Zhu Hua and Zhang Shilong, 2018 S.D.N.Y. (2018), https://www.justice.gov/opa/press-release/file/1121706/download.

37 A partial list of confirmed transfers and attempts includes: radiation-hardened microchips and semiconductor devices, military technical data for navigation and precision strike capabilities, technical specifications for the B-2 Stealth Bomber and cruise missiles, electronics used in military radar, and military encryption technology. For more details, see "Summary of Major U.S. Export Enforcement, Economic Espionage, Trade Secret and Embargo-Related Criminal Cases," Department of Justice, February, 2015, https://www.justice.gov/file/347376/download.

38 China Power Team, "Is China at the Forefront of Drone Technology?" China Power, May 29, 2018, https://chinapower.csis.org/china-drones-unmanned-technology/.

39 On China recruiting spies, see Mike Giglio, "China's Spies Are on the Offensive," *The Atlantic*, August 26, 2019, https://www.theatlantic.com/politics/archive/2019/08/inside-us-china-espionage-war/595747. On Hong Kong protests, see Steven Myers and Paul Mozur, "China Is Waging a Disinformation War Against Hong Kong Protesters," *New York Times*, August 13, 2019, https://www.nytimes.com/2019/08/13/world/asia/hong-kong-protests-china.html; and Tom Mitchell, Nicolle Liu, and Alice Woodhouse, "Cathay Pacific Crisis Ushers in Nervous New Era for Hong Kong Inc.," *Financial Times*, August 28, 2019, https://www.ft.com/content/cb6f5038-c7ac-11e9-a1f4-3669401ba76f.

40 On China's influence campaigns, see Tara Francis Chan, "A Secret Government Report Uncovered China's Attempts to Influence All Levels of Politics in Australia," Business Insider, May 28, 2018, https://www.businessinsider.com/secret-australian-government-report-uncovered-china-influence-campaign-2018-5; David Shullman, "Protect the Party: China's Growing Influence in the Developing World," Brookings, October 4, 2019, https://www.brookings.edu/articles/protect-the-party-chinas-growing-influence-in-the-developing-world/.

41 More information on all quotes and claims in this section is taken from Larry Diamond and Orville Schell, eds., *China's Influence and American Interests: Promoting Constructive Vigilance* (Stanford, CA: Hoover Institution Press, 2018), 20, 60 63–68, 146–51, and 169–73. For more on Chinese influence, see John Garnaut, "How China Interferes in Australia," *Foreign Affairs*, March 9, 2018, https://www.foreignaffairs.com/articles/china/2018-03-09/how-china-interferes-australia.

42 Hardina Ohlendorf, "The Taiwan Dilemma in Chinese Nationalism," *Asian Survey* 54 no. 3 (2014): 471–91.

43 On Taiwan's exports, see Da-Nien Liu, "The Trading Relationship Between Taiwan and the United States: Current Trends and the Outlook for the Future," Brookings Institution, November 2016, https://www.brookings.edu/opinions/the-trading-relationship-between-taiwan-and-the-united-states-current-trends-and-the-outlook-for-the-future/; and "TW's Top 10 Export Destinations," Bureau of Foreign Trade, Taiwan Ministry of Economic Affairs, https://www.trade.gov.tw/english/Pages/Detail.aspx?nodeID=94&pid=651991&dl_DateRange=all&txt_SD=&txt_ED=&txt_Keyword=&Pageid=0.

44 Jason Li, "China's Surreptitious Economic Influence on Taiwan's Elections," *The Diplomat*, April 12, 2019, https://thediplomat.com/2019/04/chinas-surreptitious-economic-influence-on-taiwans-elections/.

45 Chris Horton, "China, an Eye on Elections, Suspends Some Travel Permits to Taiwan," *New York Times*, July 31, 2019, https://www.nytimes.com/2019/07/31/world/asia/taiwan-china-tourist-visas.html.

46 For Wang Yi quote, see Thomas Wright, "Taiwan Stands Up to Xi," *The Atlantic*, January 15, 2020, https://www.theatlantic.com/ideas/archive/2020/01/taiwans-new-president-is-no-friend-of-beijing/605020/.

47 For Xi quote, see "Xi Jinping Says Taiwan 'Must and Will Be' Reunited with China," BBC, January 2, 2019, https://www.bbc.com/news/world-asia-china-46733174. On China's military preparations, see the Office of the Secretary of Defense, "Annual Report to Congress: Military and Security Developments Involving the People's Republic of China 2019," Office of the Secretary of Defense, 15, https://media.defense.gov/2019/May/02/2002127082/-1/-1/1/2019_CHINA_MILITARY_POWER_REPORT.pdf.

48 Andreo Calonzo, "Duterte Will Ignore South China Sea Ruling for China Oil Deal," Bloomberg, September 11, 2019, https://www.bloomberg.com/news/articles/2019-09-11/duterte-will-ignore-south-china-sea-ruling-for-china-oil-deal; Cliff Venzon, "Duterte Struggles to Sell

His China Pivot at Home," *Nikkei Asian Review*, October 9, 2019, https://asia.nikkei.com/Spotlight/Cover-Story/Duterte-struggles-to-sell-his-China-pivot-at-home.

49 Patrick M. Cronin and Ryan Neuhard, "Total Competition: China's Challenge in the South China Sea," Center for a New American Security, January 8, 2020, https://www.cnas.org/publications/reports/total-competition.

Chapter 4: Turning Weakness into Strength

1 Donovan Chau and Thomas Kane, *China and International Security: History, Strategy, and 21st-Century Policy* (Westport, CT: Praeger, 2014), 64.

2 Hillary Rodham Clinton, *Hard Choices* (New York: Simon and Schuster, 2014).

3 On this point, see Michael H. Hunt, *The Making of a Special Relationship: The United States and China to 1914* (New York: Columbia University Press, 1983), and John Pomfret, *The Beautiful Country and the Middle Kingdom: America and China, 1776 to the Present* (New York: Henry Holt, 2016), 570–71.

4 Department of State, Office of the Historian, "Document 12: Memorandum of Conversation," *Foreign Relations of the United States, 1969–1976*, Volume XVIII, *China, 1973–1976*, https://history.state.gov/historicaldocuments/frus1969-76v18/d12.

5 As Joseph Riley observed in his study of post–Cold War United States–China relations, "the vast majority of U.S. policy makers from the George H. W. Bush administration through the Obama administration have believed that broad economic, political, and cultural exchange with China would encourage Beijing to liberalize its mercantilist economic policies and authoritarian political structure." Joseph Riley, *The Great Gamble: Washington's Ill-Fated Attempt to Reform Beijing* (manuscript).

6 See "CRACKDOWN IN BEIJING: Excerpts from Bush's News Session," *New York Times*, June 6, 1989, https://www.nytimes.com/1989/06/06/world/crackdown-in-beijing-excerpts-from-bush-s-news-session.html.

7 "Clinton's Words on China: Trade Is the Smart Thing," *New York Times*, March 9, 2000, https://www.nytimes.com/2000/03/09/world/clinton-s-words-on-china-trade-is-the-smart-thing.html.

8 Yuka Koshino, "How Did Obama Embolden China? Comparative Analysis of 'Engagement' and 'Containment' in Post–Cold War Sino-American Relations," U.S.-Japan Research Institute, 2015, 14, http://www.us-jpri.org/wp/wp-content/uploads/2016/07/CSPC_Koshino_2015.pdf.

9 Susan Rice, "Remarks as Prepared for Delivery by National Security Advisor Susan E. Rice," Office of the Press Secretary, The White House, November 21, 2013, https://obamawhitehouse.archives.gov/the-press-office/2013/11/21/remarks-prepared-delivery-national-security-advisor-susan-e-rice.

10 "China Already Violating U.S. Cyber Agreement, Group Says," CBS News, October 19, 2015, https://www.cbsnews.com/news/crowdstrike-china-violating-cyberagreement-us-cyberespionage-intellectual-property/.

11 Del Quentin Wilber, "China 'Has Taken the Gloves Off' in Its Thefts of U.S. Technology Secrets," *Los Angeles Times*, November 16, 2018.

12 Jeffrey Goldberg, "The Obama Doctrine," *The Atlantic*, April 2016, https://www.theatlantic.com/magazine/archive/2016/04/the-obama-doctrine/471525/.

13 On satellite imagery, see the database at CSIS's Asia Maritime Transparency Initiative, including Asia Maritime Transparency Initiative, "A Look at China's SAM Shelters in the Spratlys," Center for Strategic and International Studies, February 23, 2017, https://amti.csis.org/chinas-sam-shelters-spratlys/. On the further militarization of the South China Sea, see "How Much Trade Transits the South China Sea?" Center for Strategic and International Studies, https://chinapower.csis.org/much-trade-transits-south-china-sea/#easy-footnote-bottom-1-3073; Jeremy Page, Carol E. Lee, and Gordon Lubold, "China's President Pledges No Militarization in Disputed Islands," *Wall Street Journal*, September 25, 2015, https://www.wsj.com/articles/china-completes-runway-on-artificial-island-in-south-china-sea-1443184818.

14 See Joseph Riley, *The Great Gamble* (manuscript).

15 Michael Pence, "Remarks by Vice President Michael Pence on the Administration's Policy Toward China," Remarks, The White House, October 4, 2018, https://www.whitehouse.gov/briefings-statements/remarks-vice-president-pence-administrations-policy-toward-china/.

16 United States Senate Committee on Homeland Security and Governmental Affairs, "Threats to the U.S. Research Enterprise: China's Talent Recruitment Plans," November 18, 2019, 31–32, https://www.hsgac.senate.gov/imo/media/doc/2019-11-18%20PSI%20Staff%20Report%20-%20China's%20Talent%20Recruitment%20Plans.pdf.

17 Edward Wong, "Competing Against Chinese Loans, U.S. Companies Face Long Odds in Africa," *New York Times*, January 13, 2019.

18 Pomfret, *The Beautiful Country*.

19 For example, the Committee on Foreign Investment in the United States (CFIUS) has helped safeguard sensitive technologies. In 2018, the Foreign Investment Risk Review Modernization Act expanded CFIUS jurisdiction and blocked loopholes that the CCP had exploited.

20 Sui-Lee Wee, "China Uses DNA to Track Its People, with the Help of American Expertise," *New York Times*, February 21, 2019, https://www.nytimes.com/2019/02/21/business/china-xinjiang-uighur-dna-thermo-fisher.html.

21 Roger Robinson Jr., "Why and How the U.S. Should Stop Financing China's Bad Actors." *Imprimis* 48, no. 10 (2019), https://imprimis.hillsdale.edu/roger-w-robinson-stop-financing-china/.

22 Michael Brown and Pavneet Singh, "China's Technology Transfer Strategy: How Chinese Investments in Emerging Technology Enable a Strategic Competitor to Access the Crown Jewels of U.S. Innovation," Defense Innovation Unit-Experimental, January 15, 2018,https://admin.govexec.com/media/diux_chinatechnologytransferstudy_jan_2018_(1).pdf.

23 In 2018, the CCP dedicated a session of the Eighteenth Party Congress to the rule of law (*fazhi*), but its version of the concept is based on the absolute leadership of the party, not on the more universal understanding that the state itself is accountable to laws that are promulgated publicly, enforced equally, and adjudicated independently. Ronald Alcala, Eugene Gregory, and Shane Reeves, "China and the Rule of Law: A Cautionary Tale for the International Community," *Just Security*, June 28, 2018, https://www.justsecurity.org/58544/china-rule-law-cautionary-tale-international-community/.

24 The U.S. Department of Justice China Initiative, launched in 2018, raised awareness of threats from trade secret theft, such as nontraditional collectors of intelligence in labs, universities, and the defense industry as well as risks to supply chains in telecommunications and other sectors. Katharina Buchholz, "Which Countries Have Banned Huawei?" *Statista*, August 19, 2019, https://www.statista.com/chart/17528/countries-which-have-banned-huawei-products/.

25 Department of Justice, "Chinese Telecommunications Conglomerate Huawei and Subsidiaries Charged in Racketeering Conspiracy and Conspiracy to Steal Trade Secrets," February 13, 2020.

26 For country-specific analyses, see Larry Diamond and Orville Schell, eds., *China's Influence and American Interests: Promoting Constructive Vigilance* (Stanford, CA: Hoover Institution Press, 2018), 163–209.

27 U.S. Department of Justice, "Two Chinese Hackers Associated with the Ministry of State Security Charged with Global Computer Intrusion Campaigns," U.S. Department of Justice Press Office, December 20, 2018, https://www.justice.gov/opa/pr/two-chinese-hackers-associated-ministry-state-security-charged-global-computer-intrusion.

28 Even the Cambodian dictator Han Sen could not avoid increased public scrutiny of wasteful and failed projects, such as the construction of a city-size casino resort with empty hotels and an unfinished casino that displaced thousands of people and caused severe environmental damage. And from 2018 to 2019, Australian lawmakers passed new laws to counter CCP influence by blocking foreign campaign contributions and restricting foreign investment in sensitive sectors of the economy. Yinka Adegoke, "Chinese Debt Doesn't Have to Be a Problem for African Countries," *Quartz*, May 13, 2018, https://qz.com/africa/1276710/china-in-africa-chinese-debt-news-better-management-by-african-leaders/.

29 Regarding subsidies, see Ellen Nakashima, "U.S. Pushes Hard for a Ban on Huawei in Europe, but the Firm's 5G Prices Are Nearly Irresistible," *Washington Post*, May 29, 2019, https://www.washingtonpost.com/world/national-security/for-huawei-the-5g-play-is-in-europe--and-the-us-is-pushing-hard-for-a-ban-there/2019/05/28/582a8ff6-78d4-11e9-b7ae-390de4259661_story.html. See also Huawei, "Huawei Investment & Holding Co., Ltd. 2018 Annual Report," Huawei.com, https://www.huawei.com/en/press-events/annual-report/2018.

On Huawei's expansion and role of the CCP, see Chuin-Wei Yap, "State Support Helped Fuel Huawei's Global Rise," *Wall Street Journal*,

December 25, 2019, https://www.wsj.com/articles/state-support-helped-fuel-huaweis-global-rise-11577280736. Huawei has repeatedly denied this. See Karl Song, "No, Huawei Isn't Built on Chinese State Funding," Huawei.com, February 25, 2020, https://www.huawei.com/ke/facts/voices-of-huawei/no-huawei-isnt-built-on-chinese-state-funding.

On the campaign of cyber espionage, see the U.S. Department of Justice's indictment, Department of Justice, "Chinese Telecommunications Conglomerate Huawei and Subsidiaries Charged in Racketeering Conspiracy and Conspiracy to Steal Trade Secrets," February 13, 2020, https://www.justice.gov/opa/pr/chinese-telecommunications-conglomerate-huawei-and-subsidiaries-charged-racketeering. See also Andrew Grotto, "The Huawei Problem: A Risk Assessment," *Global Asia* 14, no. 3 (2019): 13–15, http://www.globalasia.org/v14no3/cover/the-huawei-problem-a-risk-assessment_andrew-grotto; Klint Finley, "The U.S. Hits Huawei with New Charges of Trade Secret Theft," *Wired*, February 13, 2020, https://www.wired.com/story/us-hits-huawei-new-charges-trade-secret-theft/.

30 On the strategic benefits the CCP could gain from Huawei, see David E. Sanger, Julian E. Barnes, Raymond Zhong, and Marc Santora, "In 5G Race with China, U.S. Pushes Allies to Fight Huawei," *New York Times*, January 26, 2019, https://www.nytimes.com/2019/01/26/us/politics/huawei-china-us-5g-technology.html.

On the inseparability of Huawei from CCP influence, see Christopher Balding and Donald C. Clarke, "Who Owns Huawei?" *Social Science Research Network*, April 17, 2019, https://papers.ssrn.com/sol3/papers.cfm?abstract_id=3372669; Raymond Zhong, "Who Owns Huawei? The Company Tried to Explain. It Got Complicated," *New York Times*, April 25, 2019, https://www.nytimes.com/2019/04/25/technology/who-owns-huawei.html.

Huawei has denied these allegations. See Associated Press, "Huawei Denies U.S. Violations, 'Disappointed' by Criminal Charges," Associated Press, January 28, 2019, https://www.marketwatch.com/story/huawei-denies-us-violations-disappointed-by-criminal-charges-2019-01-28.

31 On the charges on circumvention of sanctions on Iran and North Korea, see Department of Justice, "Chinese Telecommunications Conglomerate Huawei and Subsidiaries Charged in Racketeering Conspiracy and Conspiracy to Steal Trade Secrets," U.S. Department of Justice, February 13, 2020, https://www.justice.gov/opa/pr/chinese-telecommunications-conglomerate-huawei-and-subsidiaries-charged-racketeering.

32 Chris Demchak and Yuval Shavitt, "China's Maxim—Leave No Access Point Unexploited: The Hidden Story of China Telecom's BGP Hijacking," *Military Cyber Affairs* 3, no. 1 (2018): 5–7.

33 On the simultaneous employment of Huawei employees at China's MOIS and the PLA, see Robert Mendick, "Huawei Staff CVs Reveal Alleged Links to Chinese Intelligence Agencies," *Telegraph*, July 5, 2019, https://www.telegraph.co.uk/news/2019/07/05/huawei-staff-cvs-reveal-alleged-links-chinese-intelligence-agencies/. Huawei has repeatedly

denied similar allegations. See Isobel Asher Hamilton, "Huawei's Security Boss Says the Company Would Sooner 'Shut Down' than Spy for China," *Business Insider*, March 6, 2019, https://www.businessinsider.com/huawei-would-sooner-shut-down-than-spy-for-china-2019-3?r s=US&IR=T.

For the report on how Huawei helped African autocrats use technology to spy on their political opponents, see Joe Parkinson, Nicholas Bariyo, and Josh Chin, "Huawei Technicians Helped African Governments Spy on Political Opponents," *Wall Street Journal*, August 15, 2019, https://www.wsj.com/articles/huawei-technicians-helped-african-govern ments-spy-on-political-opponents-11565793017?mod=breakingnews. For Huawei's denial of this report, see Huawei, "A Legal Demand Letter to The Wall Street Journal," Huawei.com, August 16, 2019, https://www.huawei.com/ke/facts/voices-of-huawei/a_legal_demand_letter_to_the_wall_street_journal.

34 Sources on Huawei: Kate O'Keeffe and Dustin Volz, "Huawei Telecom Gear Much More Vulnerable to Hackers Than Rivals' Equipment, Report Says," *Wall Street Journal*, June 25, 2019, https://www.wsj.com/articles/huawei-telecom-gear-much-more-vulnerable-to-hackers-than-rivals-equipment-report-says-11561501573; Arjun Kharpal, "Huawei Staff Share Deep Links with Chinese Military, New Study Finds," CNBC, July 8, 2019, https://www.cnbc.com/2019/07/08/huawei-staff-and-chinese-military-have-deep-links-study-claims.html. Joe Parkinson and Nicholas Bariyo, "Huawei Technicians Helped African Governments Spy on Political Opponents," *Wall Street Journal*, August 15, 2019, https://www.wsj.com/articles/huawei-technicians-helped-african-governments-spy-on-political-opponents-11565793017; Akito Tanaka, "China in Pole Position for 5G Era with a Third of Key Patents," *Nikkei Asian Review*, May 3, 2019, https://asia.nikkei.com/Spotlight/5G-networks/China-in-pole-position-for-5G-era-with-a-third-of-key-patents; Jeffrey Johnson, "Testimony Before the U.S.-China Economic and Security Review Commission Hearing on 'Chinese Investment in the United States: Impacts and Issues for Policy Makers,'" U.S.-China Economic and Security Review Commission, January 26, 2017, https://www.uscc.gov/sites/default/files/Johnson_USCC%20Hearing%20Testimony012617.pdf.

35 Graham Allison, *Destined for War: Can America and China Escape Thucydides's Trap?* (Melbourne, Australia:: Scribe Publications, 2019).

36 John Lee, "China's Economic Slowdown: Root Causes, Beijing's Response and Strategic Implications for the US and Allies," Hoover Institute, December 16, 2019.

37 James Legge, *Confucian Analects: The Great Learning and the Doctrine of the Mean* (Mineola, NY: Dover Publications, 1971), 263–64; Keegan Elmer, "U.S. Tells China: We Want Competition…but Also Cooperation." *South China Morning Post*, October 1, 2018, https://www.scmp.com/news/china/diplomacy/article/2166476/us-tells-china-we-want-competition-not-cooperation.

Chapter 5: A One-Year War Twenty Times Over: America's South Asian Fantasy

1 Original Source: Abdullah Azzam, "Al-Qa'idah al-Sulbah," *Al-Jihad* 41 (April 1988): 46. English source: Rohan Gunaratna, "Al Qaeda's Ideology," Hudson Institute, May 19, 2005, https://www.hudson.org/research/9777-al-qaeda-s-ideology. A note on the translation: "The original text in Arabic was translated into English by Reuven Paz, Academic Director, International Policy Institute for Counter Terrorism, Israel."

2 General Nicholson said that twenty groups were concentrated in Afghanistan-Pakistan in 2017. Brian Dodwell and Don Rassler, "A View from the CT Foxhole: General John W. Nicholson, Commander, Resolute Support and U.S. Forces—Afghanistan," *CTC Sentinel* 10, no. 2 (February 2017): 12–15, https://ctc.usmaedu/a-view-from-the-ct-foxhole-general-john-w-nicholson-commander-resolute-support-and-u-s-forces-afghanistan/.

3 St. Thomas Aquinas, *Summa Theologica* (AD 1265–1274), n.p.

4 Sun Tzu, *The Art of War* (Leicester, UK: Allandale Online Publishing, 2000), https://sites.ualberta.ca/~enoch/Readings/The_Art_Of_War.pdf.

5 Kevin Sullivan, "Embassy in Kabul Reopened by U.S." *Washington Post*, December 18, 2001, https://www.washingtonpost.com/archive/politics/2001/12/18/embassy-in-kabul-reopened-by-us/f89df7ec-a332-4156-98bc-81df3c951cfd/.

6 Sean Naylor, *Not a Good Day to Die: The Untold Story of Operation Anaconda* (New York: Berkley Caliber Books, 2006), 8.

7 Steve Coll, *Ghost Wars: The Secret History of the C.I.A., Afghanistan, and Bin Laden, from the Soviet Invasion to September 10, 2001* (New York: Penguin Press, 2004), 582; Steve Coll, *Directorate S: The C.I.A. and America's Secret Wars in Afghanistan and Pakistan* (New York: Penguin Press, 2018), 20–21.

8 For a specific time line of Kabul, the CIA arrived on September 26, the Taliban fled Kabul on November 12, and Northern Alliance leaders alongside the CIA, entered Kabul on November 14. Coll, *Directorate S*, 80, 93. The quote originally attributed to Sun Tzu is "Thus the good fighter is able to secure himself against defeat, but cannot make certain of defeating the enemy," Sun Tzu, *The Art of War*.

9 Estimate includes ISI personnel who fled. Seymour M. Hersh, *Chain of Command: The Road from 9/11 to Abu Ghraib* (New York: HarperCollins, 2005), 132; On those who escaped from Tora Bora, see Naylor, *Not a Good Day to Die*, 20–21.

10 Nadia Schadlow, *War and the Art of Governance: Consolidating Combat Success into Political Victory* (Washington, DC: Georgetown University Press, 2017), 220–26.

11 CNN, "Rumsfeld: Major Combat Over in Afghanistan," CNN, May 1, 2003, http://www.cnn.com/2003/WORLD/asiapcf/central/05/01/afghan.combat/.

12 For more, see Thomas J. Barfield, *Afghanistan: A Cultural and Political History* (Princeton, NJ: Princeton University Press, 2012).

13 Patrick Porter, *Military Orientalism: Eastern War Through Western Eyes* (New York: Oxford University Press, 2013).

14 Schadlow, *War and the Art of Governance*, 223.

15 Barfield, *Afghanistan: A Cultural and Political History*, 25, 50–51, 284–93.

16 Wright, *The Looming Tower*, 133.

17 For the statements themselves, see Thomas Joscelyn, "Al Qaeda Leader Argues Taliban's 'Blessed Emirate' a Core Part of New Caliphate," FDD's *Long War Journal*, August, 24, 2018, https://www.longwarjournal.org/archives/2018/08/al-qaeda-leader-argues-talibans-blessed-emirate-a-core-part-of-new-caliphate.php; Thomas Joscelyn, "Ayman al Zawahiri Pledges Allegiance to the Taliban's New Emir," FDD's *Long War Journal*, August 13, 2015, https://www.longwarjournal.org/archives/2015/08/ayman-al-zawahiri-pledges-allegiance-to-the-talibans-new-emir.php.

18 Coll, *Directorate S*, 311.

19 On NATO operations at the time, see NATO OTAN, "Resolute Support Mission (RSM): Key Facts and Figures," NATO, February, 2017, https://www.nato.int/nato_static_fl2014/assets/pdf/pdf_2017_02/20170209_2017-02-RSM-Placemat.pdf. Quote attributed to Winston Churchill at his residence at Chequers, UK, on April 1, 1945.

20 On the rise of ISIS, see Joby Warrick, *Black Flags: The Rise of ISIS* (New York: Doubleday, 2015), 303.

21 Shaun Gregory, "The ISI and the War on Terrorism," *Studies in Conflict and Terrorism Journal* 30, no. 12 (March 2007): 1013–31, DOI: 10.1080/10576100701670862.

22 On the Peshawar school attack, see Declan Walsh, "Taliban Besiege Pakistan School, Leaving 145 Dead," *New York Times*, December 16, 2014, https://www.nytimes.com/2014/12/17/world/asia/taliban-attack-pakistani-school.html. For more on the Pakistani Taliban, see Philip J. Crowley, "Designations of Tehrik-e Taliban Pakistan and Two Senior Leaders," U.S. State Department press release, September 1, 2010, http://www.state.gov/r/pa/prs/ps/2010/09/146545.htm. On Al-Qaeda operations along the Afghanistan-Pakistan border, see UN Security Council, "Tenth Report of the Analytical Support and Sanctions Monitoring Team Submitted Pursuant to Resolution 2255 (2015) Concerning the Taliban and Other Associated Individuals and Entities Constituting a Threat to the Peace, Stability and Security of Afghanistan," June 13, 2019, 22, https://www.undocs.org/S/2019/481.

23 Anahad O'Connor, "Weak Times Sq. Car Bomb Is Called Intentional," *New York Times*, July 21, 2010, https://www.nytimes.com/2010/07/21/nyregion/21bomb.html. On Shahzad: Coll, *Directorate S*, 450–52; and on drones in northern Waziristan, Coll, *Directorate S*, 438.

24 Jibran Ahmed and Yeganeh Torbati, "U.S. Drone Kills Islamic State Leader for Afghanistan, Pakistan: Officials," Reuters, August 12, 2016, https://www.reuters.com/article/us-afghanistan-islamicstate-idUSKCN10N21L.

25 Mujib Mashal and Fahim Abed, "After Deadly Attack on Kabul Hospital, 'Everywhere Was Full of Blood,'" *New York Times*, March 8, 2017, https://

www.nytimes.com/2017/03/08/world/asia/kabul-military-hospital-in-afghanistan-comes-under-attack.html.

26 Ian S. Livingston and Michael O'Hanlon, "Afghanistan Index," Brookings Institution, September 29, 2017, 4, https://www.brookings.edu/afghanistan-index/.

27 "Clinton Extends Hand to the Taliban," ABC News, July 15, 2009, https://www.abc.net.au/news/2009-07-16/clinton-extends-hand-to-taliban/1355022.

28 On Obama's framing of the Taliban, see President Obama, "Statement by the President on Afghanistan," The White House, October 15, 2015, https://obamawhitehouse.archives.gov/the-press-office/2015/10/15/statement-president-afghanistan. On specific limitations of U.S. military actions, see Rowan Scarborough, "Rules of Engagement Limit the Actions of U.S. Troops and Drones in Afghanistan," *Washington Times*, November 16, 2013, https://www.washingtontimes.com/news/2013/nov/26/rules-of-engagement-bind-us-troops-actions-in-afgh/. For a detailed account of the connections between the Taliban and ISI, see Coll, *Directorate S*.

29 On the number of Afghan soldiers wounded, see Special Inspector General for Afghanistan Reconstruction, "January 30, 2017, Quarterly Report to the United States Congress," January 30, 2017, 98, https://www.sigar.mil/pdf/quarterlyreports/2017-01-30qr.pdf; Special Inspector General for Afghanistan Reconstruction, "April 30, 2016 Quarterly Report to the United States Congress," April 30, 2016, 94, https://www.sigar.mil/pdf/quarterlyreports/2016-04-30qr.pdf. On civilian casualties, the United Nations Assistance Mission in Afghanistan (UNAMA) reported 2,315 civilian deaths attributed to antigovernment elements in 2015 and 2,131 in 2016, for a total of 4,446. UNAMA, "Protection of Civilians in Armed Conflict Annual Report 2015," United Nations Human Rights Office of the High Commissioner, February 2016, 33, https://unama.unmissions.org/protection-of-civilians-reports; UNAMA, "Protection of Civilians in Armed Conflict Annual Report 2016," United Nations Human Rights Office of the High Commissioner, February 2017, 50, https://unama.unmissions.org/protection-of-civilians-reports.

30 Coll, *Directorate S*, 371; "ARG (Presidential Palace)," Islamic Republic of Afghanistan, Office of the President, https://president.gov.af/en/history-of-arg-presidential-palace/nggallery/image/bg1o8456-1/.

31 Mark Mazzetti and Jane Perlez, "C.I.A. and Pakistan Work Together, Warily," *New York Times*, February 24, 2010, https://www.nytimes.com/2010/02/25/world/asia/25intel.html.

32 George W. Bush, *Decision Points* (New York: Crown Publishers, 2010), 206.

33 On the announcement of the end of major combat operations, see CNN World, "Rumsfeld: Major Combat Over in Afghanistan," CNN, May 1, 2003, http://www.cnn.com/2003/WORLD/asiapcf/central/05/01/afghan.combat/.

34 On Bush's decision to increase troop levels, see Bush, *Decision Points*,

207; Amy Belasco, "The Cost of Iraq, Afghanistan, and Other Global War on Terror Operations Since 9/11," Congressional Research Service, December 8, 2014, https://fas.org/sgp/crs/natsec/RL33110.pdf.

35 Coll, *Directorate S*, 458–59. For example, after a Taliban attack on a peace Loya Jirga, Karzai summoned National Director of Security Amrullah Saleh and Minister of the Interior Hanif Atmar to his office. The president declared that the attack had been planned by the United States to undermine his peace initiative with the Taliban. Saleh and Atmar, two of the most talented men in Karzai's cabinet, disagreed. Both men resigned at the end of the meeting.

36 For more on President Karzai and Ambassador Holbrooke's relationship, see George Packer, *Our Man: Richard Holbrooke and the End of the American Century* (New York: Alfred A. Knopf, 2019), 4–6.

37 Alissa J. Rubin, "Karzai's Antagonism Corners the West; Afghan President Is Seen as Only Viable Option, Even as He Alienates Allies," *New York Times International Edition*, April 6, 2010, Nexis Uni.

38 Frud Bezhan, "Karzai to Move Up After Stepping Down," Radio Free Europe/Radio Liberty, October 13, 2013, https://www.rferl.org/a/karzai-finances/25135480.html.

39 Coll, *Directorate S*, 409–10.

40 Peter Baker and Eric Schmitt, "Afghan War Debate Now Leans to Focus on Al Qaeda," *New York Times*, October 7, 2009, https://www.nytimes.com/2009/10/08/world/asia/08prexy.html. Administration officials talking off the record in May 2010 stated that there were fewer than one hundred Al-Qaeda fighters in Afghanistan. Joshua Partlow, "In Afghanistan, Taliban Leaving al-Qaeda Behind," *Washington Post*, November 11, 2009, http://www.washingtonpost.com/wp-dyn/content/article/2009/11/10/AR2009111019644.html.

41 On the start of the peace negotiations with the Taliban, see Rathnam Indurthy, "The Obama Administration's Strategy in Afghanistan," *International Journal on World Peace* 28, no. 3 (September 2011): 7–52, https://www.jstor.org/stable/23266718?seq=1#metadata_info_tab_contents. BBC, "How Qatar Came to Host the Taliban," BBC News, June 22, 2013, https://www.bbc.com/news/world-asia-23007401; Karen DeYoung, "U.S. to Launch Peace Talks with Taliban," *Washington Post*, June 18, 2013, https://www.washingtonpost.com/world/national-security/us-to-relaunch-peace-talks-with-taliban/2013/06/18/bd8c7f38-d81e-11e2-a016-92547bf094cc_story.html.

42 Rob Nordland, "For Swapped Taliban Prisoners from Guantánamo Bay, Few Doors to Exit Qatar," *New York Times*, May 31, 2015, https://www.nytimes.com/2015/06/01/world/middleeast/us-presses-qatar-on-travel-ban-for-swapped-taliban-prisoners.html.

43 Peter Tomsen, *The Wars of Afghanistan: Messianic Terrorism, Tribal Conflicts, and the Failures of Great Powers* (New York: PublicAffairs, 2013), 105–14.

44 John F. Burns, "Afghan President, Pressured, Reshuffles Cabinet," *New York Times*, October 11, 2008, https://www.nytimes.com/2008/10/12/world/asia/12afghan.html.

Chapter 6: Fighting for Peace

1 George Packer, "Afghanistan's Theorist-in-Chief," *The New Yorker*, July 9, 2019, www.newyorker.com/magazine/2016/07/04/ashraf-ghani-afghanistans-theorist-in-chief.

2 Taraki was a leader in the Khalqi faction of the Communist People's Democratic Party of Afghanistan. He was part of the Communist coup following the Saur Revolution, which resulted in the murder of Mohammed Daoud Khan and the rule of a Communist faction. Taraki was overthrown and assassinated shortly after he rose to power. These events set the stage for the Soviet Union.

3 Ashraf Ghani and Clare Lockhart, *Fixing Failed States: A Framework for Rebuilding a Fractured World* (New York: Oxford University Press, 2008).

4 Politico Staff, "Full Text: Trump's Speech on Afghanistan," Politico, August 22, 2017, https://www.politico.com/story/2017/08/21/trump-afghanistan-speech-text-241882; Central Intelligence Agency, "Field Listing: Terrorist Groups," CIA, February 1, 2018, https://www.cia.gov/library/publications/the-world-factbook/fields/397.html)

5 Thomas Joscelyn, forthcoming chapter, "Chapter 9, Al Qaeda Survived the War in Afghanistan." Modern-day Afghanistan had tremendous symbolic value to Al-Qaeda and other jihadist terrorists because it provided a cornerstone on which the caliphate could be built. Geographically, Afghanistan was an ideal place to organize and prepare for Al-Qaeda's campaign of terror against its "near enemy," Israel and the governments of Muslim-majority countries across the Middle East, and its "far enemy," the United States, Europe, and the West.

6 Human Rights Watch, "Pakistan Coercion, UN Complicity," February 13, 2017, https://www.hrw.org/report/2017/02/13/pakistan-coercion-un-complicity/mass-forced-return-afghan-refugees; World Population Review, "Kabul Population 2020," http://worldpopulationreview.com/world-cities/kabul-population/.

7 On the parliamentary elections, see Radio Free Afghanistan, "Voting Ends in Afghanistan's Parliamentary Elections Marred by Violence, Delays," Radio Free Europe/Radio Liberty, October 21, 2018, https://www.rferl.org/a/afghans-cast-ballots-for-second-day-in-chaotic-general-elections/29555274.html. On voter turnout in the presidential elections, see Mujib Mashal, Mohamed Fahim Abed, and Fatima Faizi, "Afghanistan Election Draws Low Turnout Amid Taliban Threats," *New York Times*, September 28, 2019, https://www.nytimes.com/2019/09/28/world/asia/afghanistan-president-election-taliban.html.

8 On enrollment, see the following: Ian S. Livingston and Michael O'Hanlon, "Afghanistan Index," Brookings Institution, September 29, 2017, https://www.brookings.edu/afghanistan-index/; Afghan Ministry of Education, "About Us: 7 Million in 2010 with Goal of 10 Million by 2015 USAID," Education—Afghanistan, USAID Afghanistan, July 22, 2019, https://www.usaid.gov/afghanistan/education.

9 According to an Asia Foundation survey of the Afghan people, access to the internet among respondents increased 400 percent from 2013 to 2018,

and 40 percent of respondents say their area has access to the internet. Dinh Thi Kieu Nhung, "Afghanistan in 2018: A Survey of the Afghan People," The Asia Foundation, https://asiafoundation.org/publication/afghanistan-in-2018-a-survey-of-the-afghan-people/, 156. For statistics on the media, see government statistics cited at TOLOnews, "Explosion Targets Media Workers in Kabul, Kills Two," August 4, 2019, https://tolonews.com/afghanistan/explosion-targets-media-workers-kabul-kills-two.

10 United Nations Assistance Mission in Afghanistan, "Protection of Civilians in Armed Conflict Annual Report 2018," United Nations Human Rights Office of the High Commissioner, February 24, 2019, 10, https://reliefweb.int/sites/reliefweb.int/files/resources/afghanistan_protection_of_civilians_annual_report_2018_final_24_feb_2019.pdf.

11 Ashraf Ghani, interview by Nikhil Kumar, *Time*, May 18, 2017, https://time.com/4781885/ashraf-ghani-afghanistan-president-interview/.

12 Civil service reform was a bright spot under a talented and principled leader, Nader Nadery.

13 For demographic statistics on Afghanistan, see "Afghanistan," CIA World Factbook, CIA.gov, https://www.cia.gov/library/publications/resources/the-world-factbook/geos/af.html. There are also many smaller minorities in Kabul, including Qizilbash (a Turkic people) and Nuristanis (also known as *kafirs* because they initially rejected Islam). Thomas J. Barfield, *Afghanistan: A Cultural and Political History* (Princeton, NJ: Princeton University Press, 2012), 53.

14 Steve Coll, *Directorate S*, 452–57.

15 Ray Rivera and Sangar Rahimi, "Afghan President Says His Country Would Back Pakistan in a Clash with the U.S." *New York Times*, October 23, 2011, https://www.nytimes.com/2011/10/24/world/asia/karzai-says-afghanistan-would-back-pakistan-in-a-conflict-with-us.html; Frud Behzan, "The Eminently Quotable Karzai," Radio Free Europe/Radio Liberty, September 29, 2014, https://www.rferl.org/a/afghanistan-karzai-quotes/26610215.html.

16 United Nations Assistance Mission in Afghanistan, "Protection of Civilians in Armed Conflict Annual Report 2010," United Nations Human Rights Office of the High Commissioner, March, 2011, 3–4, https://unama.unmissions.org/protection-of-civilians-reports.

17 Euan McKirdy and Ehsan Popalzai, "American University of Afghanistan Reopens After 2016 Attack," CNN, March 28, 2017, https://www.cnn.com/2017/03/28/asia/kabul-american-university-reopens/index.html; Mujib Mashal, Mohamed Fahim Abed, and Zahra Nader, "Attack at University in Kabul Shatters a Sense of Freedom," *New York Times*, August 25, 2016, https://www.nytimes.com/2016/08/26/world/asia/afghanistan-kabul-american-university.html.

18 United States Department of State, "Deputy Secretary Armitage's Meeting with Pakistan Intel Chief Mahmud: You're Either with Us or You're Not," unclassified, September 12, 2001, https://nsarchive2.gwu.edu/NSAEBB/NSAEBB358a/doc03-1.pdf.

19 World Bank Data, "Pakistan—Population, total," World Bank, n.d., https://

data.worldbank.org/indicator/SP.POP.TOTL?end=2018&locations=PK &name_desc=true&start=2003.

20 Peter L. Bergen, "September 11 Attacks," *Encyclopedia Britannica*, June 21, 2019, https://www.britannica.com/event/September-11-attacks.

21 Associated Press, "Pakistani Court Indicts Finance Minister on Graft Charges," Associated Press, September 26, 2017, https://apnews.com /c96efe1cc2b24a1391860c3ebd31e223. On Sharif, see interview of Nadeem Akhtar, Shamil Shams, "Why Ousted Pakistani PM Nawaz Sharif Turned Against the Powerful Military," DW, March 13, 2018, https://p.dw .com/p/2uECV; REFL, "Pakistani Finance Minister Indicted on Corruption Charges," Radio Free Europe/Radio Liberty, September 27, 2017, https://www.rferl.org/a/pakistan-finance-minister-corruption-idictment /28759837.html.

22 Two years after my visit, a car bomb killed forty Indian security personnel. The Pakistani militant group Jaish-e-Mohammed claimed responsibility, but the Indian government believed that the Pakistan Army was responsible. The countries conducted limited airstrikes. India targeted a terrorist training camp in the mountains, but Pakistan denied the camp's existence. In the process, Pakistan shot down an Indian plane and took the pilot prisoner for a period of days before he was returned to India as a "peace gesture." M. Illyas Khan, "Abhinandan: Villagers Recount Dramatic Capture of Pilot," BBC, March 1, 2019, https://www .bbc.com/news/world-asia-47397418; "Article 370: India Strips Disputed Kashmir of Special Status," BBC, August 5, 2019, https://www.bbc.com /news/world-asia-india-49231619. The move drew strong resistance from many parliamentarians and raised fears that New Delhi would encourage Hindus to move to the region in order to weaken the voice of Muslims there.

23 For more on the Pakistan Army's culture and role in society, see Christine Fair, *Fighting to the End: The Pakistan Army's Way of War* (New York: Oxford University Press, 2014); Aqil Shah, *The Army and Democracy: Military Politics in Pakistan* (Cambridge, MA: Harvard University Press, 2014).

24 Salman Masood, "More Bodies Pulled from Hotel Rubble in Pakistan," *New York Times*, September 21, 2008, https://www.nytimes.com/2008 /09/22/world/asia/22marriott.html; "Suicide Attack on Pakistani Hotel," BBC News, June 10, 2009, http://news.bbc.co.uk/2/hi/south_asia /8092147.stm.

25 Rachel Roberts, "Pakistan: Three Years after 140 Died in the Peshawar School Massacre, What Has Changed?" *The Independent*, December 16, 2017, https://www.independent.co.uk/news/world/asia/pakistan-peshawar-school-shooting-massacre-what-has-changed-happened-three-years -a8113661.html; BBC News, "Pakistan Taliban: Peshawar School Attack Leaves 141 Dead," BBC, December 16, 2014, https://www.bbc.com/news /world-asia-30491435.

26 Omar Waraich, "Pakistan Takes Fight to the Taliban," *The Independent*, December 20, 2014.

27 Naveed Mukhtar, "Afghanistan: Alternative Futures and Their Implications" (master's thesis, U.S. Army War College, 2011), 73, https://apps.dtic.mil/dtic/tr/fulltext/u2/a547182.pdf.

28 Neta C. Crawford, "Update on the Human Costs of War for Afghanistan and Pakistan, 2001 to mid-2016," Costs of War Project, Watson Institute for International and Public Affairs, Brown University, August 2016, 14, https://watson.brown.edu/costsofwar/files/cow/imce/papers/2016/War%20in%20Afghanistan%20and%20Pakistan%20UPDATE_FINAL_corrected%20date.pdf.

29 Richard P. Cronin, K. Alan Kronstadt, and Sharon Squassoni, "Pakistan's Nuclear Proliferation Activities and the Recommendations of the 9/11 Commission: U.S. Policy Constraints and Options," Congressional Research Service Report for Congress, May 24, 2005, 8, https://fas.org/sgp/crs/nuke/RL32745.pdf.

30 The U.S. special operations raid that killed Osama bin Laden in 2011 should have finally exposed the unreliability of America's nominal allies in Pakistan. The terrorist leader's compound was located near the Pakistani equivalent of the U.S. Military Academy at West Point, New York. The Pakistan Army's reaction, to feign ignorance and complain about U.S. violations of Pakistani sovereignty, was offensive.

31 Vahid Brown and Don Rassler, *Fountainhead of Jihad: The Haqqani Nexus, 1973–2012* (Oxford: Oxford University Press, 2013).

32 Brown and Rassler, *Fountainhead of Jihad.*

33 Omar Noman, *The Political Enemy of Pakistan, 1988* (New York: Routledge, 1988).

34 Ahmed Rashid, *Descent into Chaos: The U.S. and the Disaster in Pakistan, Afghanistan, and Central Asia* (New York: Penguin Books, 2009), 22.

35 On India's projected population growth, see Hannah Ritchie, "India Will Soon Overtake China to Become the Most Populous Country in the World," Our World in Data, University of Oxford, https://ourworldindata.org/india-will-soon-overtake-china-to-become-the-most-populous-country-in-the-world. On poverty in India, see World Bank, "Supporting India's Transformation," Results Briefs, October 15, 2019, https://www.worldbank.org/en/results/2019/10/15/supporting-indias-transformation; "Global Multidimensional Poverty Index 2019: Illuminating Inequalities," United Nations Development Programme, http://hdr.undp.org/sites/default/files/mpi_2019_publication.pdf. For more on India's demographics, see CIA World Factbook, "India," CIA, https://www.cia.gov/library/publications/the-world-factbook/geos/in.html.

36 Shreeya Sinha and Mark Suppes, "Timeline of the Riots in Modi's Gujarat," *New York Times*, April 6, 2014, https://www.nytimes.com/interactive/2014/04/06/world/asia/modi-gujarat-riots-timeline.html#/. By 2016, Prime Minister Modi had visited the United States four times since he was elected in 2014. Rishi Iyengar, "As India's Prime Minister Modi Visits President Obama, Both Leaders Look to Cement a Legacy," *Time*, June 7, 2016, https://time.com/4359522/india-modi-obama-visit-us/.

37 House of Representatives: Committee on Foreign Affairs, "Bad Company: Lashkar e-Tayyiba and the Growing Ambition of Islamist Militancy in Pakistan," March 11, 2010, https://www.govinfo.gov/content/pkg/CHRG-111hhrg55399/html/CHRG-111hhrg55399.htm. Mehreen Zahra-Malik, "Militant Leader Hafiz Saeed Is Released by Pakistani Court," *New York Times*, https://www.nytimes.com/2017/11/23/world/asia/hafiz-saeed-pakistan-militant.html.

38 The White House, "Remarks by the President on the Way Forward in Afghanistan," Office of the Press Secretary, June 22, 2011, https://obama whitehouse.archives.gov/the-press-office/2011/06/22/remarks-president-way-forward-afghanistan.

39 The White House, *National Security Strategy of the United States of America*, December 2017, 46, https://www.whitehouse.gov/wp-content/uploads/2017/12/NSS-Final-12-18-2017-0905.pdf.

40 U.S. Senate: Committee on Foreign Relations, "Al Qaeda, the Taliban, and Other Extremists [*sic*] Groups in Afghanistan and Pakistan," https://www.govinfo.gov/content/pkg/CHRG-112shrg67892/html/CHRG-112shrg67892.htm.

41 Clinton Thomas, "Afghanistan: Background and U.S. Policy Brief," Congressional Research Service, January 31, 2020, https://fas.org/sgp/crs/row/R45122.pdf.

42 Thomas Joscelyn, "Disconnecting the Dots," *Washington Examiner*, July 13, 2010, https://www.washingtonexaminer.com/weekly-standard/disconnecting-the-dots; Thomas Joscelyn, "Al Qaeda Is Very Much Alive," *Washington Examiner*, September 11, 2018, https://www.washingtonexaminer.com/weekly-standard/sept-11-anniversary-17-years-later-al-qaeda-is-alive.

43 Vanda Felbab-Brown, "Why Pakistan Supports Terrorist Groups, and Why the US Finds It So Hard to Induce Change," Brookings Institution, January 5, 2018, https://www.brookings.edu/blog/order-from-chaos/2018/01/05/why-pakistan-supports-terrorist-groups-and-why-the-us-finds-it-so-hard-to-induce-change/.

44 Amy Held, "Death Toll in Kabul Blast Surpasses 150, Afghan President Says," NPR, June 6, 2017, www.npr.org/sections/thetwo-way/2017/06/06/531729176/death-toll-in-kabul-blast-surpasses-150-afghan-president-says; Laura Smith-Spark and Faith Karimi. "Afghanistan Explosion: Blast Kills 90 near Diplomatic Area," CNN, June 1, 2017, www.cnn.com/2017/05/31/asia/kabul-explosion-hits-diplomatic-area/.

45 Quotes from the White House, "Remarks by President Trump on the Strategy in Afghanistan and South Asia," August 21, 2017, https://www.whitehouse.gov/briefings-statements/remarks-president-trump-strategy-afghanistan-south-asia/.

46 Michael D. Shear and Salman Masood, "Trump Tries Cooling Tensions with Pakistan to Speed Afghan Peace Talks," *New York Times*, July 22, 2019, https://www.nytimes.com/2019/07/22/world/asia/trump-pakistan-afghanistan.html.

47 This number includes total deaths in Afghanistan during Operation Enduring Freedom and Operation Freedom's Sentinel as of February 2020. U.S. Department of Defense, "Casualty Status," Department of Defense, February 3, 2020, https://www.defense.gov/casualty.pdf.

48 On the intelligence community's Worldwide Threat Assessment, see Senate Select Committee on Intelligence, Worldwide Threat Assessment, Statement for the Record (Daniel R. Coates, Director of National Intelligence), January 29, 2019, 12, https://www.dni.gov/files/ODNI/documents/2019-ATA-SFR---SSCI.pdf. For analysis of the UN policy, see Thomas Joscelyn, "The Trump Administration's Afghanistan Policy," Congressional Testimony, Foundation for Defense of Democracies, September 19, 2019, https://www.fdd.org/analysis/2019/09/19/the-trump-administrations-afghanistan-policy/.

49 On the attack in Ghazni city, see Mujib Mashal, "Afghan Talks with Taliban Reflect a Changed Nation," *New York Times*, July 7, 2019, https://www.nytimes.com/2019/07/07/world/asia/afghanistan-peace-talks-taliban.html. Michael Crowley, Lara Jakes, and Mujib Mashal, "Trump Says He's Called Off Negotiations with Taliban After Afghanistan Bombing," *New York Times*, September 10, 2019, https://www.nytimes.com/2019/09/07/us/politics/trump-taliban-afghanistan.html. On President Trump's quotes, see "Remarks by President Trump Before Marine One Departure," The White House, September 9, 2019, https://www.whitehouse.gov/briefings-statements/remarks-president-trump-marine-one-departure-63/.

50 Bill Roggio, "U.S. Military Buries Press Release that Would Announce Killing of Al Qaeda in the Indian Subcontinent's Emir," FDD's *Long War Journal*, January 15, 2020, https://www.longwarjournal.org/archives/2020/01/u-s-military-buries-press-release-that-would-announce-killing-of-al-qaeda-in-the-indian-subcontinents-emir.php.

51 Statista Research Department, "Soldiers Killed in Action in Afghanistan 2001–2019," Statista, August 22, 2019, https://www.statista.com/statistics/262894/western-coalition-soldiers-killed-in-afghanistan/; Matthew Pennington, "Pentagon: Afghan War Costing U.S. $45 Billion per Year," *Military Times*, February 6, 2018, https://www.militarytimes.com/news/pentagon-congress/2018/02/07/pentagon-afghan-war-costing-us-45-billion-per-year/.

52 Neta C. Crawford, "United States Budgetary Costs of Post-9/11 Wars Through FY 2018," Costs of War Project, Watson Institute for International and Public Affairs, Brown University, November 2017, 9, https://watson.brown.edu/costsofwar/papers/economic.

53 On the September 2019 presidential election, Pamela Constable, "Afghanistan's Ghani Wins Slim Majority in Presidential Vote, Preliminary Results Show," *Washington Post*, December 22, 2019, https://www.washingtonpost.com/world/afghanistans-ghani-wins-slim-majority-in-presidential-vote/2019/12/22/73355178-2441-11ea-b034-de7dc2b5199b_story.html. On Afghan national mood toward the Taliban, see Nhung, "Afghanistan in 2018: A Survey of the Afghan People," 43.

Chapter 7: Who Thought It Would Be Easy? From Optimism to Resignation in the Middle East

1 U.S. Department of State, "Casualty Status as of 10 a.m. EST Jan. 20, 2020," https://www.defense.gov/casualty.pdf; Leith Aboufadel, "Over 26,000 Iraqi Soldiers Killed in 4 Year War with ISIS," AMNNews, December 13, 2017, https://www.almasdarnews.com/article/26000-iraqi-soldiers-killed-4-year-war-isis/.

2 Garrett Nada and Mattisan Rowan, "Pro-Iran Militias in Iraq," Wilson Center, April 27, 2018, https://www.wilsoncenter.org/article/part-2-pro-iran-militias-iraq.

3 George Packer, "The Lesson of Tal Afar," *The New Yorker*, July 10, 2017, https://www.newyorker.com/magazine/2006/04/10/the-lesson-of-tal-afar. Confession of Abdul Ghafur Abdul Rahman Mustafa from August 28, 2008, in possession of author; Joseph L. Galloway, McClatchy Newspapers, "Regiment's Rotation out of Tal Afar Raises Questions about U.S. Strategy," McClatchy Washington Bureau, January 18, 2006, https://www.mcclatchydc.com/opinion/article24452989.html.

4 Joel Rayburn, *Iraq After America: Strongmen, Sectarians, Resistance* (Stanford, CA: Hoover Institution Press, 2014), 74–75.

5 On how Jaafari changed the Ministry of Interior, see John F. Burns, "Torture Alleged at Ministry Site Outside Baghdad," *New York Times*, November 16, 2005, https://www.nytimes.com/2005/11/16/world/middleeast/torture-alleged-at-ministry-site-outside-baghdad.html; see also Joel Rayburn, *Iraq After America: Strongmen, Sectarians, Resistance* (Stanford, CA: Hoover Institution Press, 2014), 79.

6 For a report on the abuse of Sunni prisoners, see Ned Parker, "Torture by Iraqi Militias: The Report Washington Did Not Want You to See," Reuters, December 14, 2015, https://www.reuters.com/investigates/special-report/mideast-crisis-iraq-militias/.

7 On Iranian influence in Iraqi government, see International Institute for Strategic Studies, "Iranian Influence in Iraq: Assessing Tehran's Strategy," *Strategic Comments* 13, no. 10 (December 2007):1–2, https://doi.org/10.1080/13567880701870027. See also Rayburn, Iraq After America, 80–81.

8 Rayburn, *Iraq After America*, 80.

9 Richard Spencer, "Isil Carried Out Massacres and Mass Sexual Enslavement of Yazidis, UN Confirms," *Telegraph*, October 14, 2014, https://www.telegraph.co.uk/news/worldnews/islamic-state/11160906/Isil-carried-out-massacres-and-mass-sexual-enslavement-of-Yazidis-UN-confirms.html.

10 Department of Defense, "Department of Defense Press Briefing by Secretary Mattis, General Dunford and Special Envoy McGurk on the Campaign to Defeat ISIS in the Pentagon Press Briefing Room," U.S. Department of Defense Archives, May 19, 2017, https://www.defense.gov/Newsroom/Transcripts/Transcript/Article/1188225/department-of-defense-press-briefing-by-secretary-mattis-general-dunford-and-sp/.

11 Albert Hourani, *A History of the Arab Peoples* (Cambridge, MA: Harvard University Press, 2002), 397–98.

12 Reuters, "Syria's Alawites, a Secretive and Persecuted Sect," Reuters, February 2, 2012, https://www.reuters.com/article/us-syria-alawites-sect-idUSTRE8110Q720120202.

13 Williamson Murray and Kevin M. Woods, *The Iran-Iraq War: A Military and Strategic History* (Cambridge, UK: Cambridge University Press, 2014), 242, doi:10.1017/CBO9781107449794.

14 Patrick Cockburn, *Muqtada Al-Sadr and the Battle for the Future of Iraq* (New York: Scribner, 2008), 28.

15 Khomeini's efforts were not unprecedented. The Safavid dynasty in the seventh century used Shia Islam to unify Persian society against the Sunni Ottoman Empire.

16 Central Intelligence Agency, "The Demographic Consequences of the Iran-Iraq War," May 22, 1984, released April 4, 2011, CIA-RDP85T00287R001301610001, https://www.cia.gov/library/readingroom/docs/CIA-RDP85T00287R001301610001-1.pdf.

17 Luke Harding, "Haider al-Abadi: From Exile in Britain to Iraq's Next Prime Minister," *Guardian*, August 11, 2014, https://www.theguardian.com/world/2014/aug/11/haider-al-abadi-profile-iraqs-next-prime-minister.

18 Sam Dagher, *Assad or We Burn the Country: How One Family's Lust for Power Destroyed Syria* (New York: Little, Brown and Company, 2019), 55–56.

19 Michael Knights, "Helping Iraq Take Charge of Its Command-and-Control Structure," *The Washington Institute*, September 30, 2019, https://www.washingtoninstitute.org/policy-analysis/view/helping-iraq-take-charge-of-its-command-and-control-structure.

20 The Al-Qaeda documents captured in the northern Iraqi border district of Sinjar in September 2007 (the infamous "Sinjar documents") showed that the vast majority of the mujahideen who entered Iraq—more than one hundred a month at that time—did so by way of the Damascus airport and a well-established network of safe houses and friendly Syrian officials who led them across the Iraqi frontier into Anbar or Ninewa Provinces. In a police state like Bashar al-Assad's Syria, the activity recorded in the Sinjar documents could never have taken place without the full knowledge and approval of the regime. See Brian Fishman and Joseph Felter, "Al-Qa'ida's Foreign Fighters in Iraq: A First Look at the Sinjar Records," Combatting Terrorism Center at West Point, January 2, 2007, https://ctc.usma.edu/al-qaidas-foreign-fighters-in-iraq-a-first-look-at-the-sinjar-records/.

21 Fouad Ajami, "America and the Solitude of the Syrians," *Wall Street Journal*, January 6, 2012, https://www.hoover.org/research/america-and-solitude-syrians; David Remnick, "Going the Distance: On and Off the Road with Barack Obama," *The New Yorker*, January 20, 2014, https://www.newyorker.com/magazine/2014/01/27/going-the-distance-david-remnick.

22 On the Homs riots, see Michael Weiss and Hassan Hassan, *ISIS: Inside the Army of Terror* (New York: Regan Arts, 2016), 132; Warrick, *Black Flags*, 228. For examples of this repression, see Warrick, *Black Flags*, 266; Weiss and Hassan, *ISIS*, 132.
23 Murray and Woods, *The Iran-Iraq War*, 242.
24 H. R. McMaster, "Why the U.S. Was Right in Not Trying to Take Over All of Iraq," *Philadelphia Inquirer*, June 23, 1991.
25 Conrad C. Crane and W. Andrew Terrill, "Reconstructing Iraq: Insights, Challenges, and Missions for Military Forces in a Post-Conflict Scenario," Strategic Studies Institute, U.S. Army War College, February 1, 2003, 17, https://ssi.armywarcollege.edu/pubs/display.cfm?pubID=182.
26 Stephen D. Biddle and Peter Feaver, "Assessing Strategic Choices in the War on Terror," in Beth Bailey and Richard Immerman, eds., *Understanding the U.S. Wars in Iraq and Afghanistan* (New York: NYU Press, 2015).
27 Michael R. Gordon and General Bernard E. Trainor, *The Endgame: The Inside Story of the Struggle for Iraq, from George W. Bush to Barack Obama* (London: Atlantic Books, 2013).
28 Ayman al-Zawahiri, "Knights Under the Prophet's Banner," FBIS translation of the newspaper *Asharq-Al-awsat*, 2001.
29 Gordon and Trainor, *The Endgame*, 302.
30 Kimberly Kagan, *Surge: A Military History* (New York: Encounter Books, 2009).
31 For more metrics on violence in Iraq, see Anthony H. Cordesman, "Iraq: Patterns of Violence, Casualty Trends, and Emerging Threats," Center for Strategic and International Studies, February 9, 2011, https://csis-prod.s3.amazonaws.com/s3fs-public/legacy_files/files/publication/110209_Iraq-PattofViolence.pdf.
32 Peter Baker, "Relief over U.S. Exit from Iraq Fades as Reality Overtakes Hope," *New York Times*, June 22, 2014, https://www.nytimes.com/2014/06/23/world/middleeast/relief-over-us-exit-from-iraq-fades-as-reality-overtakes-hope.html.
33 Baker, "Relief over U.S. Exit from Iraq Fades as Reality Overtakes Hope."
34 On Hashemi's death, see Jack Healy, "Arrest Order for Sunni Leader in Iraq Opens New Rift," *New York Times*, December 19, 2011. On the alienation of Iraq's Sunni populations, see Rayburn, *Iraq After America*; Emma Sky, *The Unraveling: High Hopes and Missed Opportunities in Iraq* (New York: PublicAffairs, 2015), xii.
35 Beatrice Dupuy, "President Obama Did Not Free Islamic State Leader Al-Baghdadi from Prison," *Associated Press*, October 30, 2019, https://apnews.com/afs:Content:8037620747.
36 Martin Chulov, "Gaddafi's Last Moments: 'I Saw the Hand Holding the Gun and I Saw It Fire,'" *The Guardian*, October 20, 2012.
37 Barrack Obama, "Remarks by the President on Ending the War in Iraq," The White House, October 21, 2011, transcript, https://obamawhite

house.archives.gov/the-press-office/2011/10/21/remarks-president-ending-war-iraq.

Chapter 8: Breaking the Cycle

1 Kenneth Michael Pollack, *A Path Out of the Desert: A Grand Strategy for America in the Middle East* (New York: Random House, 2008), xxxix.

2 On the Kenneth Pollack quote, see Kenneth M. Pollack, "Drowning in Riches," *New York Times,* July 13, 2008, https://www.nytimes.com/2008/07/13/opinion/13pollack.html.

3 The population of Syria pre–civil war (2011) was 21 million. As of 2018, it was 16 million. The World Bank, "Syrian Arab Republic," https://data.worldbank.org/country/syrian-arab-republic.

4 The Syrian Network for Human Rights, "Statistics of 2019," SNHR, http://sn4hr.org/.

5 United Nations, "Libya Country Profile," UN, http://data.un.org/CountryProfile.aspx/_Images/CountryProfile.aspx?crName=Libya.

6 The success of Operation Provide Comfort could help stabilize Syria. Thomas E. Ricks, "Operation Provide Comfort: A Forgotten Mission with Possible Lessons for Syria," *Foreign Policy*, February 6, 2017, https://foreignpolicy.com/2017/02/06/operation-provide-comfort-a-forgotten-mission-with-possible-lessons-for-syria/.

7 Efraim Benmelech and Esteban F. Klor, "What Explains the Flow of Foreign Fighters to ISIS?" National Bureau of Economic Research, Working Paper 22190, April 2016, 16, http://www.nber.org/papers/w22190.

8 Statistics on displacement based on Eurostat findings, Phillip Connor, "Most Displaced Syrians Are in the Middle East, and About a Million Are in Europe," FactTank, Pew Research Center, January 29, 2018, https://www.pewresearch.org/fact-tank/2018/01/29/where-displaced-syrians-have-resettled/. The UNHCR count as of October 31, 2019, was 3,680,603, UNHCR, "Syria Regional Refugee Response," UNHCR, October 31, 2019, https://data2.unhcr.org/en/situations/syria/location/113.

9 For more of the story of Omran Daqneesh, see Anne Bernard, "How Omran Daqneesh, 5, Became a Symbol of Aleppo's Suffering," *New York Times*, August 18, 2016, https://www.nytimes.com/2016/08/19/world/middleeast/omran-daqneesh-syria-aleppo.html. Tragically, 30,000 deaths in Syria were estimated between 2012 and 2016. Russian bombing in Syria began in late 2015; within that time frame, there were 7,800 deaths in Aleppo. Violations Documentation Center in Syria, "Aleppo Death Statistics: 2015/09/01–2016/12/30," VDC, http://www.vdc-sy.info/index.php/en/martyrs/1/c29ydGJ5PWEua2lsbGVkX2RhdGV8c29ydGRpcj1ERVNDfGFwcHJvdmVkPXZpc2libGV8ZXh0cmFkaXNwbGF5PTB8cHJvdmluY2U9NnxzdGFydERhdGU9MjAxNS0wOS0wMXxlbmREYXRlPTIwMTYtMTItMzB8.

10 Steve Simon and Jonathan Stevenson, "Don't Intervene in Syria," *New York Times*, October 6, 2016, https://www.nytimes.com/2016/10/06/opinion/dont-intervene-in-syria.html. But there were concerns that the threat to U.S. interests would not stay contained to Syria. In January

2014, months before President Obama intervened against ISIS, former DNI James Clapper said that Syria was becoming "in some respects, a new FATA," referring to the Federally Administered Tribal Areas of Pakistan, a long-known Al-Qaeda base, saying that the country was attracting thousands of jihadist fighters who could one day create a potential base for terrorist attacks emanating from Syria to the West.

11 @realDonaldTrump: "...almost 3 years, but it is time for us to get out of these ridiculous Endless Wars, many of them tribal, and bring our soldiers home. WE WILL FIGHT WHERE IT IS TO OUR BENEFIT, AND ONLY FIGHT TO WIN. Turkey, Europe, Syria, Iran, Iraq, Russia and the Kurds will now have to..." Twitter, October 7, 2019, https://twitter.com/realDonaldTrump/status/1181172465772482563. For the president's remarks on Syria, see President Trump, "Remarks by President Trump in Cabinet Meeting," The White House, January 3, 2019, https://www.whitehouse.gov/briefings-statements/remarks-president-trump-cabinet-meeting-12/.

12 For a perspective supporting disengagement, see Steve Simon, "After the Surge: The Case for U.S. Military Disengagement from Iraq," Council Special Report No. 23, Council on Foreign Relations Press, February 2007, https://www.cfr.org/report/after-surge.

13 Candace Dunn and Tim Hess, "The United States Is Now the Largest Global Crude Oil Producer," U.S. Energy Information Administration (EIA), Independent Statistics and Analysis, EIA, September 12, 2018, https://www.eia.gov/todayinenergy/detail.php?id=37053.

14 Michael Schwirtz, "U.N. Links North Korea to Syria's Chemical Weapons Program," *New York Times*, February 27, 2018, https://www.nytimes.com/2018/02/27/world/asia/north-korea-syria-chemical-weapons-sanctions.html. On IDF attribution, see IDF, "The Secret Operation Revealed a Decade Later," IDF Press Center, March 21, 2018, https://www.idf.il/en/articles/operations/the-secret-operation-revealed-a-decade-later/.

15 On jihadist strategies and mission, see Brian Fishman, *The Master Plan: ISIS, al-Qaeda, and the Jihadi Strategy for Final Victory* (New Haven, CT: Yale University Press, 2016), 36; Jonathan Randal, *Osama: The Making of a Terrorist* (New York: Vintage Books, 2005), 86–87, 95. On world economic growth, see Office of the Historian, "Oil Embargo, 1973–1974," U.S. Department of State, https://history.state.gov/milestones/1969-1976/oil-embargo.

16 Colin Clarke, "Expanding the ISIS Brand," RAND Corporation, February 19, 2018, https://www.rand.org/blog/2018/02/expanding-the-isis-brand.html.

17 Audrey Kurth Cronin, *Power to the People: How Open Technological Innovation Is Arming Tomorrow's Terrorists* (New York: Oxford University Press, 2020), 161–239.

18 Anna Borshchevskaya, "Will Russian-Saudi Relations Continue to Improve? What Their Recent Summit Means for the Relationship," *Foreign Affairs*, October 10, 2017, https://www.foreignaffairs.com/articles/saudi-arabia/2017-10-10/will-russian-saudi-relations-continue-improve.

19 Judah Ari Gross, "IDF Says It Has Bombed over 200 Iranian Targets in Syria Since 2017," *Times of Israel*, September 4, 2018, https://www.timesofisrael.com/idf-says-it-has-carried-out-over-200-strikes-in-syria-since-2017/.

20 Secretary of State Rex W. Tillerson, "The Way Forward in Syria," speech, Hoover Institution, Stanford, CA, January 17, 2018, https://www.hoover.org/events/tillerson_11718.

21 On the "safe zone" decision, see Julian E. Barnes and Eric Schmitt, "Trump Orders Withdrawal of U.S. Troops from Northern Syria" *New York Times*, October 13, 2019, https://www.nytimes.com/2019/10/13/us/politics/mark-esper-syria-kurds-turkey.html.

22 Kareem Khadder, Jennifer Deaton, and Sharif Paget, "Kurdish Politician and 10 Others Killed by 'Turkish-Backed Militia' in Syria, SDF Claims," CNN, October 13, 2019, https://www.cnn.com/2019/10/13/middleeast/syria-turkey-kurdish-politician-intl/index.html.

23 Alissa J. Rubin, "Iraqis Rise Against a Reviled Occupier: Iran," *New York Times*, November 4, 2019, https://www.nytimes.com/2019/11/04/world/middleeast/iraq-protests-iran.html.

24 Tim Arango and Neil MacFarquhar, "Grief and Fear in Sacramento over a Death That Set the World on Edge," *New York Times*, January 15, 2020, https://www.nytimes.com/2020/01/15/us/contractor-killed-in-iraq-sacramento.html.

25 DeirezZor 24, "Al-Sha'itat Massacre in Deir Ezzor...the 5th Anniversary," DeirezZor 24 News, September 8, 2019, https://en.deirezzor24.net/al-shaitat-massacre-in-deir-ezzor-the-5th-anniversary/.

26 Michael Shear, "Obama Administration Ends Effort to Train Syrians to Combat ISIS," *New York Times*, October 9, 2015, https://www.nytimes.com/2015/10/10/world/middleeast/pentagon-program-islamic-state-syria.html; World Bulletin News Desk, "Syrian Opp Withdraw from U.S. 'Train and Equip' Program," World Bulletin, June 23, 2015, https://www.worldbulletin.net/middle-east/syrian-opp-withdraw-from-us-train-and-equip-program-h161073.html; Ibrahim Hamidi, "Syrian Opposition Fighters Withdraw from U.S. 'Train and Equip' Program," *Syrian Observer*, June 22, 2015, https://syrianobserver.com/EN/news/29743/syrian_opposition_fighters_withdraw_from_us_train_equip_program.html.

27 Marc Lynch, "Welcome to the Syrian Jihad," *Foreign Policy*, June 6, 2013, https://foreignpolicy.com/2013/06/06/welcome-to-the-syrian-jihad/.

28 "Readout of the President's Call with Prime Minister Haider Al-Abadi of Iraq," The White House (The United States Government), March 29, 2017, https://www.whitehouse.gov/briefings-statements/readout-presidents-call-prime-minister-haider-al-abadi-iraq-2/.

29 Gonul Tol and Omer Taspinar, "Erdogan's Turn to the Kemalists," *Foreign Affairs*, November 10, 2016, https://www.foreignaffairs.com/articles/turkey/2016-10-27/erdogans-turn-kemalists.

30 On Erdogan and the AKP, see Adam Withnall @adamwithnall, "Erdogan Just Made His Most Worrying Claim Yet over the Attempted Coup

in Turkey," *Independent*, August 2, 2016, https://www.independent.co.uk/news/world/europe/erdogan-turkey-coup-latest-news-blames-us-west-terrorism-gulen-a7168271.html; Ihsan Yilmaz and Galib Bashirov, "The AKP After 15 Years: Emergence of Erdoganism in Turkey," *Third World Quarterly* 39, no. 9 (2018), https://www.tandfonline.com/doi/full/10.1080/01436597.2018.1447371#_i9. On the hostage taking, see Karen DeYoung and Kareem Fahim, "U.S.-Turkey Tensions Boil over After Arrest of Consulate Employee," *Washington Post*, October 9, 2017, https://www.washingtonpost.com/world/turkey-summons-another-us-consulate-employee-as-crisis-deepens/2017/10/09/5fbaecf6-ac7b-11e7-9b93-b97043e57a22_story.html.

31 On Turkey and Russia, see Kirişci Kemal, *Turkey and the West: Fault Lines in a Troubled Alliance* (Washington, DC: Brookings Institution Press, 2018), 175–78. On the 2020 refugee influx, see France 24, "Fighting Continues in Idlib as Turkey Talks End Inconclusively," February 11, 2020, https://www.france24.com/en/20200211-fighting-continues-in-idlib-as-turkey-russia-talks-end-inconclusively. Kareem Fahim and Robyn Dixon, "Turkey Vows to Escalate Military Action After 33 Soldiers Die in Syria," *Washington Post*, February 28, 2020.

32 Eric Schmitt, "Killing of Terrorist Leader in Yemen Is Latest Blow to Qaeda Affiliate," *New York Times*, February 10, 2020, https://www.nytimes.com/2020/02/10/us/politics/al-qaeda-yemen-qassim-al-rimi.html.

33 On the deaths of Baghdadi and al-Muhajir, see Ben Hubbard and Karam Shoumali, "Likely Successor to Dead ISIS Leader Also Reported Killed," *New York Times*, October 27, 2019, https://www.nytimes.com/2019/10/27/world/middleeast/al-baghdadi-successor-reported-killed.html. On Assad's release of jihadist terrorist prisoners, see Daniel Byman, "The Resurgence of Al Qaeda in Iraq," Brookings, December 12, 2013, https://www.brookings.edu/testimonies/the-resurgence-of-al-qaeda-in-iraq/.

34 Wahhabism was named for an eighteenth-century Najdi cleric, Muhammad Ibn Abdul Wahhab, who developed an ideology foundational to twenty-first-century radicalization and terrorist recruitment. Wahhabism and Salafism, a related nineteenth-century radical theology, reject Islam's tolerance for "people of the book"—Jews and Christians—who worship the same God based on the teachings of prophets who predated Mohammed. Wahhabism traditionally has opposed *taqlid*, or blind following of schools of law, instead asking followers to go directly to a puritanical and literalist interpretation of the Qur'an and Sunnah itself. The Great Mosque in Mecca used to have four *maqamat*, or prayer stations, each one representing one of the four schools. In 1926, King Abdulaziz and the Wahhabis got rid of the *maqamat* representing the four different Sunni schools and this practice and imposed one prayer. MBS's reference to the four schools was therefore of major significance for a Saudi leader. For a discussion of *maqamat* in Arabic, see Al Masjid al Haram, Makkawi, https://www.makkawi.com/Article/872/%D8%A7%D9%84%D9%85%D9%82%D8%A7%D9%85%D8%A7%D8%AA

-%D8%A7%D9%84%D8%A3%D8%B1%D8%A8%D8%B9-%D8%A8%D8%A7%D9%84%D9%85%D8%B3%D8%AC%D8%AF-%D8%A7%D9%84%D8%AD%D8%B1%D8%A7%D9%85.

35 On the proliferation of Saudi religious textbooks, see David Andrew Weinberg, "Textbook Diplomacy," Foundation for Defense of Democracies, October 24, 2018, https://www.fdd.org/analysis/2014/03/21/textbook-diplomacy/; David D. Kirkpatrick, "ISIS' Harsh Brand of Islam Is Rooted in Austere Saudi Creed," *New York Times*, September 24, 2014, https://www.nytimes.com/2014/09/25/world/middleeast/isis-abu-bakr-baghdadi-caliph-wahhabi.html.

36 For more on the growth of Shia extremism, see chapters 9 and 10.

37 Donald J. Trump, "President Trump's Speech to the Arab Islamic American Summit," Riyadh, Saudi Arabia, Statements and Releases, The White House, May 21, 2017, https://www.whitehouse.gov/briefings-statements/president-trumps-speech-arab-islamic-american-summit/.

38 Julian E. Barnes, "C.I.A. Concludes that Saudi Crown Prince Ordered Khashoggi Killed," *New York Times*, November 16, 2018, https://www.nytimes.com/2018/11/16/us/politics/cia-saudi-crown-prince-khashoggi.html.

Chapter 9: A Bad Deal: Iran's Forty-Year Proxy Wars and the Failure of Conciliation

1 "Nuclear," Joint Comprehensive Plan of Action, U.S. Department of State, July 14, 2015, 6–9, https://2009-2017.state.gov/documents/organization/245317.pdf.

2 Barack Obama, "Statement by the President on Iran," Speech, The White House, Office of the Press Secretary, July 14, 2015, https://obamawhitehouse.archives.gov/the-press-office/2015/07/14/statement-president-iran.

3 On inspections, see William Tobey and Judith Miller, "Are Iranian Military Bases Off-Limits to Inspection?" RealClearPolitics, September 8, 2015, https://www.realclearpolitics.com/articles/2015/09/08/are_iranian_military_bases_off-limits_to_inspection_128007.html; Olli Heinonen, "The IAEA's Right and Obligation to Inspect Military Facilities in Iran," Foundation for Defense of Democracies, April 4, 2018, https://www.fdd.org/analysis/2018/04/04/the-iaeas-right-and-obligation-to-inspect-military-facilities-in-iran/.

4 Glenn Kesler, "President Trump's Claim That Democrats Gave Iran $150 Billion," *Washington Post*, December 13, 2018, https://www.washingtonpost.com/politics/2018/12/13/president-trumps-claim-that-democrats-gave-iran-billion/.

5 Matthew Levitt, "Iran's Support for Terrorism Under the JCPOA," Washington Institute for Near East Policy, July 8, 2016, https://www.washingtoninstitute.org/policy-analysis/view/irans-support-for-terrorism-under-the-jcpoa.

6 Spencer Ackerman, "U.S. Central Command Nominee Has 'Concerns' About Progress Against Isis," *Guardian*, March 9, 2016, https://www

.theguardian.com/world/2016/mar/09/us-isis-syria-strategy-central-command-nomination-joseph-votel.

7 Lou Barletta, "H.R.1191—Iran Nuclear Agreement Review Act of 2015: 114th Congress (2015–2016)," https://www.congress.gov/bill/114th-congress/house-bill/1191/text.

8 Rex W. Tillerson, "Secretary of State Rex Tillerson. Press Availability," U.S. Embassy & Consulates in Russia, April 19, 2017, https://ru.usembassy.gov/secretary-state-rex-tillerson-press-availability/.

9 President Donald J. Trump, "President Donald J. Trump Is Ending United States Participation in an Unacceptable Iran Deal," Fact Sheets, The White House, May 8, 2018, https://www.whitehouse.gov/briefings-statements/president-donald-j-trump-ending-united-states-participation-unacceptable-iran-deal/.

10 See "Treasury Targets Persons Supporting Iranian Military and Iran's Islamic Revolutionary Guard Corps," Press Center, U.S. Department of the Treasury, July 18, 2017, https://www.treasury.gov/press-center/press-releases/Pages/sm0125.aspx; Jesse Chase-Lubitz, "Trump Slaps Sanctions on Iran While Keeping Nuclear Deal in Place—for Now," July 18, 2017, *Foreign Policy* (blog), https://foreignpolicy.com/2017/07/18/trump-slaps-sanctions-on-iran-while-keeping-nuclear-deal-in-place-for-now/.

11 "Excerpts: Donald Trump's Interview with the Wall Street Journal," *Wall Street Journal*, July 25, 2017, https://blogs.wsj.com/washwire/2017/07/25/donald-trumps-interview-with-the-wall-street-journal-edited-transcript/.

12 Donald J. Trump, "Statement by the President on the Iran Nuclear Deal," speech, Washington, DC, Statements and Releases, The White House, January 12, 2018, https://www.whitehouse.gov/briefings-statements/statement-president-iran-nuclear-deal/.

13 Trump, "Statement by the President on the Iran Nuclear Deal."

14 Donald J. Trump, "Statement from the President on the Designation of the Islamic Revolutionary Guard Corps as a Foreign Terrorist Organization," Statements and Releases, The White House, April 8, 2019, https://www.whitehouse.gov/briefings-statements/statement-president-designation-islamic-revolutionary-guard-corps-foreign-terrorist-organization/.

15 Mullahs are scholars of Islamic teaching who wield considerable political and societal power in the Islamic Republic's theocracy.

16 On the effects of the sanctions, see "How Renewed U.S. Sanctions Have Hit Iran Hard," BBC News, May 2, 2019, https://www.bbc.com/news/world-middle-east-48119109. On Europe and the sanctions, see Kenneth Katzman, "Summary," *Iran Sanctions*, Congressional Research Service, April 22, 2019, 1, https://crsreports.congress.gov/product/pdf/RS/RS20871/291.

17 On the Mr. Rouhani quote, see Alastair Gale, "Iran Presses Japan to Break with U.S. Sanctions on Tehran," *Wall Street Journal*, June 12, 2019, https://www.wsj.com/articles/japans-abe-looks-to-mediate-between-u-s

-iran-11560340410. On the letter episode, see "I Don't Consider Trump Worth Sending a Message to, We Won't Negotiate with U.S.," Khamenei.ir, June 13, 2019, http://english.khamenei.ir/news/6844/I-don-t-consider-Trump-worth-sending-a-message-to-we-won-t-negotiate.

18 On the *Kokuka Courageous* incident, see "Gulf of Oman Tanker Attacks: What We Know," BBC News, June 18, 2019, https://www.bbc.co.uk/news/world-middle-east-48627014; Amanda Macias, "U.S. Military Releases New Images of Japanese Oil Tanker Attack," CNBC, June 17, 2019, https://www.cnbc.com/2019/06/17/us-military-releases-new-images-of-japanese-oil-tanker-attack.html.

19 On President Trump's decision to halt strikes, see Patrick Wintour and Julian Borger, "Trump Says He Stopped Airstrike on Iran Because 150 Would Have Died," *Guardian*, June 21, 2019, https://www.theguardian.com/world/2019/jun/21/donald-trump-retaliatory-iran-airstrike-cancelled-10-minutes-before; Luis Martinez, Elizabeth McLaughlin, and Meredith McGraw, "Trump Says Iranian Shootdown of U.S. Military Drone May Have Been a 'Mistake,'" ABC News, June 20, 2019, https://abcnews.go.com/Politics/iran-shoots-american-drone-international-airspace-us-official/story?id=63825990.

20 On the Obama quote, see Thomas L. Friedman, "Iran and the Obama Doctrine," *New York Times*, April 5, 2015, https://www.nytimes.com/2015/04/06/opinion/thomas-friedman-the-obama-doctrine-and-iran-interview.html. On the Rhodes quote, see Gardiner Harris, "Deeper Mideast Aspirations Seen in Nuclear Deal with Iran," *New York Times*, July 31, 2015, https://www.nytimes.com/2015/08/01/world/middleeast/deeper-mideast-aspirations-seen-in-nuclear-deal-with-iran.html.

21 The Learning Network, "Jan. 20 1981: Iran Releases American Hostages as Reagan Takes Office," *New York Times*, January 20, 2012, https://learning.blogs.nytimes.com/2012/01/20/jan-20-1981-iran-releases-american-hostages-as-reagan-takes-office/.

22 On the marine barracks attack, see Lynn Maalouf, Luc Cote, and Theo Boudruche, "Lebanon's Legacy of Political Violence," International Center for Transitional Justice, September 2013, 53. The scandal known as the Iran-Contra affair began with illegal arms sales to Iran to gain the freedom of hostages and expanded when the profits from those sales were used for another purpose: buying arms for the anticommunist rebels known as the Contras, who were fighting to topple a Marxist government in Nicaragua. On hostage releases, arms deal exposure, use of arms by Contras, see David Crist, *Twilight War: The Secret History of America's Thirty-Year Conflict With Iran* (New York: Penguin House, 2012), 197–98.

23 Richard N. Haass, "The George H. W. Bush Administration," The Iran Primer, United States Institute of Peace, https://iranprimer.usip.org/resource/george-hw-bush-administration.

24 Crist, *Twilight War*, 382–85.

25 On the murder of prominent Kurdish Iranians, see Claude Moniquet, "The Recent Iranian Terrorist Plots in Europe," European Strategic In-

telligence and Security Center, February 2019, http://www.esisc.org/upload/publications/analyses/the-recent-iranian-terrorist-plots-in-europe/IRAN%20-%20RECENT%20TERRORIST%20PLOTS%20IN%20EUROPE.pdf. On the fatwa against Salman Rushdie, see Patricia Bauer, Carola Campbell, and Gabrielle Mander, "The Satanic Verses: Novel by Rushdie," *Encyclopedia Britannica*, https://www.britannica.com/topic/The-Satanic-Verses.

26 "Iran," Heritage Foundation, October 30, 2019, https://www.heritage.org/military-strength/assessing-threats-us-vital-interests/iran.

27 "Transcript of interview with Iranian president Mohammad Khatami." CNN Archive, January 7, 1998, http://www.cnn.com/WORLD/9801/07/iran/interview.html. U.S. decision making in the immediate aftermath of the bombing was complicated due to Saudi Arabia's position that Al-Qaeda was responsible for the bombing. The Saudis knew this to be false but did not want the United States to believe that a significant Saudi Hezbollah organization existed in Saudi Arabia.

28 Mir Sadat and James Hughes, "U.S.-Iran Engagement Through Afghanistan," *Middle East Policy* 17, no. 1 (2010): 35, https://mepc.org/us-iran-engagement-through-afghanistan.

29 On Khatami's speech, see Michael Rubin, "Khatami and the Myth of Reform in Iran," *The Politic* (Spring 2002), Washington Institute for Near East Policy, https://www.washingtoninstitute.org/policy-analysis/view/khatami-and-the-myth-of-reform-in-iran. On the Natanz facility, see Kelsey Davenport, "Timeline of Nuclear Diplomacy with Iran," Fact Sheets and Briefs, Arms Control Association, updated November 2018, https://www.armscontrol.org/factsheet/Timeline-of-Nuclear-Diplomacy-With-Iran.

30 Adrian Levy and Cathy Scott-Cook, "Al-Qaeda Has Rebuilt Itself—with Iran's Help," *The Atlantic*, November 11, 2017, https://www.theatlantic.com/international/archive/2017/11/al-qaeda-iran-cia/545576/.

31 Crist, *The Twilight War*, 521, 529.

32 On the Bush administration's reasoning, see Michael Rubin, "Iran's Revolutionary Guard: A Rogue Outfit," *Middle East Quarterly* 15, no. 4 (Fall 2008), https://www.meforum.org/1990/irans-revolutionary-guards-a-rogue-outfit; George W. Bush, "Press Conference by the President," The White House, February 14, 2007, https://georgewbush-whitehouse.archives.gov/news/releases/2007/02/20070214-2.html.

33 Kyle Rempfer, "Iran Killed More U.S. Troops in Iraq Than Previously Known, Pentagon Says," *MilitaryTimes*, April 4, 2019, https://www.militarytimes.com/news/your-military/2019/04/04/iran-killed-more-us-troops-in-iraq-than-previously-known-pentagon-says/.

34 On the Karbala attack, see Crist, *Twilight War*, 529. On the plot to assassinate the Saudi ambassador, see Charlie Savage and Scott Shane, "Iranians Accused of a Plot to Kill Saudis' U.S. Envoy," *New York Times*, October 11, 2011 https://www.nytimes.com/2011/10/12/us/us-accuses-iranians-of-plotting-to-kill-saudi-envoy.html?_r=1. On U.S. law enforcement disrupting plot, see "Iranian Plot to Kill Saudi Ambassador Thwarted,

U.S. Officials Say," CNN, October 12, 2011, https://www.cnn.com/2011/10/11/justice/iran-saudi-plot/index.html. On two men charged in plot, see "Two Men Charged in Alleged Plot to Assassinate Saudi Arabian Ambassador to the United States," Department of Justice Office of Public Affairs, October 11, 2011, https://www.justice.gov/opa/pr/two-men-charged-alleged-plot-assassinate-saudi-arabian-ambassador-united-states. On the British embassy attack, see Robert F. Worth and Rick Gladstone, "Iranian Protesters Attack British Embassy," *New York Times*, November 29, 2011, https://www.nytimes.com/2011/11/30/world/middleeast/tehran-protesters-storm-british-embassy.html.

35 Crist, *Twilight War*, 530.

36 Seymour M. Hersh, "The Iran Plans," *The New Yorker*, April 10, 2006, https://www.newyorker.com/magazine/2006/04/17/the-iran-plans.

37 From 2012 to 2014, the Iranian economy contracted by 9 percent per year, crude oil exports dropped precipitously, and more than $120 billion in reserves abroad were inaccessible to the regime. The regime subsequently shifted assets to South Asia and the Noor Islamic Bank in the United Arab Emirates, but Noor suspended business with Iran, which precipitated a 30 percent fall in the rial. The regime could no longer conduct dollar transactions. On these statistics, see Jay Solomon, *The Iran Wars* (New York: Random House, 2016), 167.

38 Navid Hassibi, "Why Can't Iran and Israel Be Friends?" *Guardian*, February 20, 2014, https://www.theguardian.com/world/iran-blog/2014/feb/20/why-cant-iran-and-israel-be-friends.

39 Jonathan Saul and Parisa Hafezi, "Iran Boosts Military Support in Syria to Bolster Assad," Reuters, February 21, 2014, https://www.reuters.com/article/us-syria-crisis-iran-insight/iran-boosts-military-support-in-syria-to-bolster-assad-idUSBREA1K0TV20140221; Hashmatallah Moslih, "Iran 'Foreign Legion' Leans on Afghan Shia in Syria War," Aljazeera, January 22, 2016, https://www.aljazeera.com/news/2016/01/iran-foreign-legion-leans-afghan-shia-syria-war-160122130355206.html; Margherita Stancati, "Iran Backs Taliban with Cash and Arms," *Wall Street Journal*, June 11, 2015, https://www.wsj.com/articles/iran-backs-taliban-with-cash-and-arms-1434065528.

40 Michael Doran, "Obama's Secret Iran Strategy," *Mosaic*, February 2, 2015, https://mosaicmagazine.com/essay/politics-current-affairs/2015/02/obamas-secret-iran-strategy/.

41 Katherine Bauer, "Iran on Notice," Washington Institute for Near East Policy, February 16, 2017, https://www.washingtoninstitute.org/policy-analysis/view/iran-on-notice.

42 On Hezbollah funding, see Nathan Sales, "Countering Iran's Global Terrorism," U.S. State Department, November 13, 2018, https://www.state.gov/countering-irans-global-terrorism/. On Iranian troops in Syria, see Laila Bassam, "Assad Allies, Including Iranians, Prepare Ground Attack in Syria: Sources," Reuters, October 1, 2015, https://www.reuters.com/article/us-mideast-crisis-syria-iranians-exclusi/assad-allies-in

cluding-iranians-prepare-ground-attack-in-syria-sources-idUSKCN0RV4DN20151001. On the ballistic missile tests, see Behnam Ben Taleblu, "Iranian Ballistic Missile Tests Since the Nuclear Deal," Foundation for the Defense of Democracies, February 9, 2017, https://s3.us-east-2.amazonaws.com/defenddemocracy/uploads/documents/20917_Behnam_Ballistic_Missile.pdf; "Why Iran Targets ISIS Positions in Syria's Deir Ezzor?" Iran's View, June 9, 2017, http://www.iransview.com/why-iran-targets-isis-positions-in-syrias-deir-ezzur/1729/.

43 "فشک ۳ مکبش گرزب داسف یداصتقا/ مادهنا ۱۵۷ لاناک نجهتسم," *Mehr News*, February 23, 2017, https://www.mehrnews.com/news/3915666/%da%a9%d8%b4%d9%81-%db%b3-%d8%b4%d8%a8%da%a9%d9%87-%d8%a8%d8%b2%d8%b1%da%af-%d9%81%d8%b3%d8%a7%d8%af-%d8%a7%d9%82%d8%aa%d8%b5%d8%a7%d8%af%db%8c-%d8%a7%d9%86%d9%87%d8%af%d8%a7%d9%85-%db%b1%db%b5%db%b7-%da%a9%d8%a7%d9%86%d8%a7%d9%84-%d9%85%d8%b3%d8%aa%d9%87%d8%ac%d9%86

44 "Victory for a Religious Hardliner in Iran," *The Economist*, June 27, 2005, https://www.economist.com/news/2005/06/27/victory-for-a-religious-hardliner-in-iran.

45 Bill Chappell, "Iranians Vote in Parliamentary Election, After 1 Week of Campaigning," NPR, February 21, 2020, https://www.npr.org/2020/02/21/807857001/iranians-vote-in-parliamentary-election-after-1-week-of-campaigning.

46 Hamid Dabashi, "What Happened to the Green Movement in Iran?" Al Jazeera, June 12, 2013, https://www.aljazeera.com/indepth/opinion/2013/05/201351661225981675.html.

47 On the IRGC's finances, see Bradley Bowman and Andrew Gabel, "Hold IRGC Accountable for Targeting U.S. Troops," Foundation for Defense of Democracies, April 12, 2019, https://www.fdd.org/analysis/2019/04/12/hold-irgc-accountable-for-targeting-u-s-troops/. On state-controlled companies and the Iranian economy, see Yeganeh Torbati, Bozorgmehr Sharafedin, and Babak Dehghanpisheh, "After Iran's Nuclear Pact, State Firms Win Most Foreign Deals," Reuters, January 19, 2017, https://www.reuters.com/article/us-iran-contracts-insight/after-irans-nuclear-pact-state-firms-win-most-foreign-deals-idUSKBN15328S.

48 Wendy R. Sherman, *Not for the Faint of Heart: Lessons in Courage, Power, and Persistence* (New York: PublicAffairs, 2018), 13.

49 For the Mosaddeq documents, see James C. Van Hook, ed., *Foreign Relations of the United States, 1952–1954: Iran, 1951–1954*, Office of the Historian, U.S. State Department, June 15, 2017, https://history.state.gov/historicaldocuments/frus1951-54Iran; Ali M. Ansari, *Modern Iran: The Pahlavis and After*, 2nd ed. (Harlow, UK: Pearson Longman, 2008), 164–67. On U.S. university narrative on the 1953 coup, see Gregory Brew, "The Collapse Narrative: The United States, Mohammed Mossadegh, and the Coup Decision of 1953," *Texas National Security Review*, August 2019, https://2llqix3cnhb21kcxpr2u9o1k-wpengine.netdna-ssl.com/wp

-content/uploads/2019/11/Brew_TNSR-Vol-2-Issue-4.pdf. On the media narrative on the 1953 coup, see Lawrence Wu and Michelle Lanz, "How the CIA Overthrew Iran's Democracy in 4 Days," NPR, February 7, 2019, https://www.npr.org/2019/01/31/690363402/how-the-cia-overthrew-irans-democracy-in-four-days.

50 "CIA's Role in 1953 Iran Coup Detailed," Politico, August 20, 2013, https://www.politico.com/story/2013/08/cias-role-in-1953-iran-coup-detailed-095731.

51 Ray Takeyh, "What Really Happened in Iran: The CIA, the Ouster of Mosaddeq, and the Restoration of the Shah," *Foreign Affairs*, July/August 2014, https://www.foreignaffairs.com/articles/middle-east/2014-06-16/what-really-happened-iran.

52 On Ben Rhodes quotes, see Rhodes, *The World as It Is: A Memoir of the Obama White House* (New York: Random House, 2018), 329–30. On the dilemma of war or the JCPOA, see "Ben Rhodes on Iran: 'You Either Have a Diplomatic Agreement with These Guys or There's Something That Can Escalate into a War," MSNBC, June 22, 2019, https://www.msnbc.com/saturday-night-politics/watch/ben-rhodes-on-iran-you-either-have-a-diplomatic-agreement-with-these-guys-or-there-s-something-that-can-escalate-this-into-a-war-62498885877.

53 Ali Ansari, *Modern Iran Since 1797: Reform and Revolution* (New York: Routledge, 2019).

54 Rhodes, *The World as It Is*, 325.

Chapter 10: Forcing a Choice

1 On Obama quote, see Firouz Sedarat and Lin Noueihed, "Obama Says Ready to Talk to Iran," Reuters, January 27, 2009, https://www.reuters.com/article/us-obama-arabiya/obama-says-ready-to-talk-to-iran-idUSTRE50Q23220090127.

2 On tanker attacks, see "Gulf of Oman Tanker Attacks: What We Know," BBC, June 18, 2019, https://www.bbc.com/news/world-middle-east-48627014. On the drone strike on an oil facility, see Ben Hubbard, Palko Karasz, and Stanley Reed, "Two Major Saudi Oil Installations Hit by Drone Strike, and U.S. Blames Iran," *New York Times*, September 14, 2019, https://www.nytimes.com/2019/09/14/world/middleeast/saudi-arabia-refineries-drone-attack.html. On the drone shooting, see Helene Cooper, "What We Know About Iran Shooting Down a U.S. Drone," *New York Times*, June 20, 2019, https://www.nytimes.com/2019/06/20/us/politics/drone-shot-down-iran-us.html.

3 "Iran During World War II," United States Holocaust Museum, https://www.ushmm.org/m/pdfs/Iran-During-World-War-II.pdf.

4 On France stay, see Elaine Ganley, "Khomeini Launched a Revolution from a Sleepy French Village," AP News, February 1, 2019, https://apnews.com/d154664bcfed47e49b0ae0ff3648779c. On interviews, see Crist, *Twilight War*, 14. On Khomeini's characterization of the Shah, see Suzanne Maloney, "1979: Iran and America," Brookings, January 24, 2019, https://www.brookings.edu/opinions/1979-iran-and-america/.

On Khomeini's return, see "1979: Exiled Ayatollah Khomeini Returns to Iran," On This Day, BBC, http://news.bbc.co.uk/onthisday/hi/dates/stories/february/1/newsid_2521000/2521003.stm. On slogans, see Associated Press, "AP WAS THERE: Ayatollah Ruhollah Khomeini Returns to Iran," *U.S. News & World Report*, February 1, 2019. https://www.usnews.com/news/world/articles/2019-02-01/ap-was-there-ayatollah-ruhollah-khomeini-returns-to-iran.

"The Iranian Hostage Crisis," Office of the Historian, U.S. State Department, https://history.state.gov/departmenthistory/short-history/iraniancrises.

5 On protests, see "Iran Petrol Price Hike: Protesters Warned that Security Forces May Intervene," BBC, November 17, 2019, https://www.bbc com/news/world-middle-east-50444429. On price increase, see Peter Kenyon, "Higher Gasoline Prices in Iran Fuel Demonstrations," NPR, November 19, 2019, https://www.npr.org/2019/11/19/780713507/higher-gasoline-prices-in-iran-fuel-demonstrations.

6 "IRGC Head Calls Iran Protests 'World War,'" Al-Monitor, November 25, 2019, https://www.al-monitor.com/pulse/originals/2019/11/iran-protests-number-deaths-mp-irgc.html.

7 Ali Ansari, *Confronting Iran: The Failure of American Foreign Policy and the Next Great Crisis in the Middle East* (Basic Books: New York, 2006), 153–54.

8 Farnaz Calafi, Ali Dadpay, and Pouyan Mashayekh, "Iran's Yankee Hero," *New York Times*, April 18, 2009, https://www.nytimes.com/2009/04/18/opinion/18calafi.html.

9 On demographics, see Bijan Dabell, "Iran Minorities 2: Ethnic Diversity," United States Institute of Peace, September 3, 2013, https://iranprimer.usip.org/blog/2013/sep/03/iran-minorities-2-ethnic-diversity.

10 "Iran: Targeting of Dual Citizens, Foreigners," Human Rights Watch, September 26, 2018, https://www.hrw.org/news/2018/09/26/iran-targeting-dual-citizens-foreigners.

11 Ansari, *Modern Iran Since 1797*, 407–9.

12 Ansari, *Modern Iran Since 1797*, 274.

13 On Khamenei quotes, see Ayatollah Ruhollah Khamenei, "The Election of Donald Trump Is a Clear Sign of the Political and Moral Decline of the U.S.," Khamenei.ir, June 4, 2019, http://english.khamenei.ir/news/6834/The-election-of-Donald-Trump-is-a-clear-sign-of-the-political.

14 On Europe and the JCPOA, see Davenport, "Timeline of Nuclear Diplomacy with Iran."

15 On Salami quote, see "Tehran," Flashpoint, International Crisis Group, November 28, 2019, https://www.crisisgroup.org/trigger-list/iran-us-trigger-list/flashpoints/tehran.

16 Editorial Board, "Justice Arrives for Soleimani," *Wall Street Journal*, January 3, 2020, https://www.wsj.com/articles/justice-arrives-for-soleimani-11578085286?emailToken=e11bad7a48ad072ad8c3a7b409690538UNGHGeR+Gsa+R3fS5fT6VKBXwaoQHV/gUgQIC3GVRFNRnoUquqzK1B+0GtEll5XH8b2y5QxMjRIcJX3kI8UzLA%3D%3D&reflink=article_email_share.

17 On expenditure figures, see John E. Pike, "Iran—Military Spending," Globalsecurity.org, updated July 20, 2019, https://www.globalsecurity.org/military/world/iran/budget.htm; "SIPRI Military Expenditure Database," Stockholm International Peace Research Institute, 2018, https://www.sipri.org/databases/milex.

18 David Adesnik and Behnam Ben Taleblu, "Burning Bridge: The Iranian Land Corridor to the Mediterranean," Foundation for Defense of Democracies, June 18, 2019, https://www.fdd.org/analysis/2019/06/18/burning-bridge/.

19 This funding includes up to $800 million for Hezbollah and $100 million combined for Hamas and Islamic Jihad annually. See Yaya J. Fanusie and Alex Entz, "Hezbollah Financial Assessment," *Terror Finance Briefing Book*, Center on Sanctions and Illicit Finance, Foundation for Defense of Democracies, September 2017; David Adesnik, "Iran Spends $16 Billion Annually to Support Terrorists and Rogue Regimes," Foundation for Defense of Democracies, January 10, 2018, https://www.fdd.org/analysis/2018/01/10/iran-spends-16-billion-annually-to-support-terrorists-and-rogue-regimes/; Daniel Levin, "Iran, Hamas and Palestinian Islamic Jihad," The Iran Primer, United States Institute of Peace, July 9, 2018, https://iranprimer.usip.org/blog/2018/jul/09/iran-hamas-and-palestinian-islamic-jihad. On missile strikes, see Shawn Snow, "Drone and Missile Attacks Against Saudi Arabia Underscore Need for More Robust Air Defenses," *Military Times*, October 25, 2019, https://www.militarytimes.com/flashpoints/2019/10/25/drone-and-missile-attacks-against-saudi-arabia-underscore-need-for-more-robust-air-defenses/. On ship attacks, see Michael Knights and Farzin Nadimi, "Curbing Houthi Attacks on Civilian Ships in the Bab Al-Mandab," Washington Institute, July 27, 2018, https://www.washingtoninstitute.org/policy-analysis/view/curbing-houthi-attacks-on-civilian-ships-in-the-bab-al-mandab.

20 "Saudi Crown Prince Calls Iran Leader 'New Hitler': NYT," Reuters, November 23, 2017, https://www.reuters.com/article/us-saudi-security-iran/saudi-crown-prince-calls-iran-leader-new-hitler-nyt-idUSKBN1DO0G3.

21 On the connection to PLO camps, see Tony Badran, "The Secret History of Hezbollah," Foundation for Defense of Democracies, November 18, 2013, https://www.fdd.org/analysis/2013/11/18/the-secret-history-of-hezbollah/. On Hezbollah provision of social services, see Daniel Byman, "Understanding Proto-Insurgencies: RAND Counterinsurgency Study—Paper 3", RAND Corporation, 2007, https://www.rand.org/pubs/occasional_papers/OP178.html.

22 On Hezbollah fighting in Israel and against ISIL, see Nicholas Blanford, "Lebanon: The Shiite Dimension," Wilson Center, August 27, 2015, https://www.wilsoncenter.org/article/lebanon-the-shiite-dimension. On Sunni attacks on Shia, see "Lebanon: Extremism and Counter-Extremism," Counter Extremism Project, November 1, 2019, https://www.counterextremism.com/countries/lebanon. On Hezbollah pro-

tection of Shia sites, see Joseph Daher, "Hezbollah, the Lebanese Sectarian State, and Sectarianism," Middle East Institute, April 13, 2017, https://www.mei.edu/publications/hezbollah-lebanese-sectarian-state-and-sectarianism. For map of Hezbollah fighting, see "Hezbollah: A Recognized Terrorist Organization," Israel Defense Forces, https://www.idf.il/en/minisites/hezbollah/hezbollah/hezbollah-a-recognized-terrorist-organization/.

23 On recommendations, see Yaya J. Fanusie and Alex Entz, "Hezbollah Financial Assessment," *Terror Finance Briefing Book*, Center on Sanctions and Illicit Finance, Foundation for Defense of Democracies, September 2017.

24 On Hezbollah casualties, see Aryeh Savir, "Study: 1,139 Hezbollah Terrorists Killed While Fighting in Syria," *Jewish Press*, March 28, 2019, https://www.jewishpress.com/news/us-news/study-1139-hezbollah-terrorists-killed-while-fighting-in-syria/2019/03/28/.

25 On protests, see Rebecca Collard, "Untouchable No More: Hezbollah's Fading Reputation," Foreign Policy, November 27, 2019 https://foreignpolicy.com/2019/11/27/lebanon-protests-hezbollah-fading-reputation/.

26 On Hezbollah and the IDF, see William M. Arkin, "Divine Victory for Whom? Airpower in the 2006 Israel-Hezbollah War," *Strategic Studies Quarterly* 1, no. 2 (Winter 2007): 104–5. On the IRGC and Israel, see Daniel Levin, "Iran, Hamas and Palestinian Islamic Jihad," The Iran Primer, United States Institute of Peace, July 9, 2018, https://iranprimer.usip.org/blog/2018/jul/09/iran-hamas-and-palestinian-islamic-jihad; Michael Bachner and Toi Staff, "Iran Said Increasing Hamas Funding to $30m per Month, Wants Intel on Israel," *Times of Israel*, August 5, 2019, https://www.timesofisrael.com/iran-agrees-to-increase-hamas-funding-to-30-million-per-month-report/. On IRGC threat, see Ahmad Majidyar, "IRGC General: Any Future War Will Result in Israel's Annihilation." Middle East Institute, April 20, 2018, https://www.mei.edu/publications/irgc-general-any-future-war-will-result-israels-annihilation.

27 For more on the ideology of the IRGC, see Ali Ansari and Kasra Aarabi, "Ideology and Iran's Revolution: How 1979 Changed the World," Tony Blair Institute, February 11, 2019, https://institute.global/insight/co-existence/ideology-and-irans-revolution-how-1979-changed-world. On losses in the Iran-Iraq War, see "Iran's Networks of Influence in the Middle East," International Institute for Strategic Studies, November 2019, chap. 1: "Tehran's Strategic Intent," https://www.iiss.org/publications/strategic-dossiers/iran-dossier/iran-19-03-ch-1-tehrans-strategic-intent.

28 On IDF strikes, see Joseph Hincks, "Israel Is Escalating Its Shadow War with Iran. Here's What to Know," *Time*, August 29, 2019, https://time.com/5664654/israel-iran-shadow-war/. For Israeli perspectives on this point, see TOI Staff, "Gantz Positive on Gaza Disengagement in First Interview, Drawing Right's Ire," *Times of Israel*, February 6, 2019, https://www.timesofisrael.com/gantz-positive-on-gaza-disengagement-in-first-interview-drawing-rights-ire/.

29 John Kifner, "400 Die as Iranian Marchers Battle Saudi Police in Mecca; Embassies Smashed in Tehran," *New York Times*, August 2, 1987, https://www.nytimes.com/1987/08/02/world/400-die-iranian-marchers-battle-saudi-police-mecca-embassies-smashed-teheran.html?pagewanted=all.

30 On Sheikh's execution, see Florence Gaub, "War of Words: Saudi Arabia v Iran," European Union Institute for Security Studies, February 2016, https://www.iss.europa.eu/sites/default/files/EUISSFiles/Brief_2_Saudi_Arabia___Iran_01.pdf. On Sheikh's charges, "Saudi Arabia Executes 47 on Terrorism Charges," Al Jazeera, January 2, 2016, https://www.aljazeera.com/news/2016/01/saudi-announces-execution-47-terrorists-160102072458873.html.

31 On U.S. efforts to mediate, see Jackie Northam, "Saudi Arabia Sought Dialogue with Iran. Then the U.S.-Iranian Conflict Escalated," NPR, January 9, 2020, https://www.npr.org/2020/01/09/794519810/saudi-arabia-sought-dialogue-with-iran-then-the-u-s-iranian-conflict-escalated.

32 On the foiled 1987 attack, see John E. Pike, "Iran Ajr Class Landing Ship," Globalsecurity.org, updated July 12, 2019, https://www.globalsecurity.org/military/world/iran/ajr.htm; Bradley Peniston, "Capturing the Iran Ajr," Navybook.com, http://www.navybook.com/no-higher-honor/timeline/capturing-the-iran-ajr. On the Aramco cyber attack, see David E. Sanger, *The Perfect Weapon: War, Sabotage, and Fear in the Cyber Age* (New York: Crown, 2018), 51–52.

33 On Osama bin Laden, see Thomas Joscelyn and Bill Roggio, "Analysis: CIA Releases Massive Trove of Osama bin-Laden Files," FDD's *Long War Journal*, Foundation for Defense of Democracies, November 2017, https://www.longwarjournal.org/archives/2017/11/analysis-cia-releases-massive-trove-of-osama-bin-ladens-files.php; Osama bin Laden, "Letter to Karim," Files, Office of the Director of National Intelligence, October 18, 2007, https://www.dni.gov/files/documents/ubl2016/english/Letter%20to%20Karim.pdf. On the Chabahar attack, see "Deadly Bomb Attack in Iran City of Chabahar," BBC News, December 15, 2010, https://www.bbc.com/news/world-middle-east-11997679. On the 2017 attacks, see "Islamic State Claims Stunning Attacks in Heart of Iran," Associated Press, June 7, 2017, https://apnews.com/510f0af4615443c08ff7f52c2657bb76/Islamic-State-claims-attacks-on-Iran-parliament,-shrine. On the parade shooting, see Erin Cunningham and Bijan Sabbagh, "Gunmen Kill at Least 2 Dozen in Attack on Military Parade in Iran," *Washington Post*, September 22, 2018, https://www.washingtonpost.com/world/several-killed-at-least-20-injured-in-attack-on-military-parade-in-iran/2018/09/22/ec016b97-a889-4a7d-b402-479bd6858e0a_story.html.

34 For billboard quote, see Murray, *The Iran-Iraq War*, 263.

35 Kenneth D. Ward, "Statement by Ambassador Kenneth D. Ward," Organization for the Prohibition of Chemical Weapons, November 2018, https://www.opcw.org/sites/default/files/documents/2018/11/USA_0.pdf.

36 For more on the nuclear time line, see Davenport, "Timeline of Nuclear Diplomacy with Iran"; Kelsey Davenport, "Official Proposals on the Ira-

nian Nuclear Issue, 2003–2013," Fact Sheets & Briefs, Arms Control Association, August 2017, https://www.armscontrol.org/factsheets/Iran_Nuclear_Proposals.

37 For Bush quote, see Crist, *Twilight War*, 538.

38 Ellen Nakashima and Joby Warrick, "Stuxnet Was Work of U.S. and Israeli Experts, Officials Say," *Washington Post*, June 2, 2012, https://www.washingtonpost.com/world/national-security/stuxnet-was-work-of-us-and-israeli-experts-officials-say/2012/06/01/gJQAlnEy6U_story.html.

39 On the missile test explosion, see Crist, *Twilight War*, 552–53. On the nuclear scientist assassinations, see Sanger, *The Perfect Weapon*, 26.

40 On Khamenei's vow of revenge, see "Qasem Soleimani: U.S. Kills Top Iranian General in Baghdad Air Strike," BBC News, January 3, 2020, https://www.bbc.co.uk/news/world-middle-east-50979463. On the retaliatory rocket attack, see "Iraq," International Crisis Group, January 12, 2020, https://www.crisisgroup.org/trigger-list/iran-us-trigger-list/flashpoints/iraq; Associated Press, "Military Contractor Slain in Iraq Buried in California," *New York Times*, January 7, 2020. https://www.nytimes.com/aponline/2020/01/07/us/ap-us-iraq-attack-contractor.html.

41 Estimates range from 106 to 1,500. See Farnaz Fassihi and Rick Gladston, "With Brutal Crackdown, Iran Convulsed by Worst Unrest in 40 Years," *New York Times*, December 3, 2019 https://www.nytimes.com/2019/12/01/world/middleeast/iran-protests-deaths.html; "Iran: Thousands Arbitrarily Detained and at Risk of Torture in Chilling Post-Protest Crackdown," Amnesty International, December 16, 2019, https://www.amnesty.org/en/latest/news/2019/12/iran-thousands-arbitrarily-detained-and-at-risk-of-torture-in-chilling-post-protest-crackdown/. On protest slogans, see Farnaz Fassihi, "Iran Blocks Nearly All Internet Access," *New York Times*, December 5, 2019, https://www.nytimes.com/2019/11/17/world/middleeast/iran-protest-rouhani.html; Lenah Hassaballah and Leen Alfaisal, "'Death to the Dictator': Iran Protests Intensify After Petrol Price Hike," Al Arabiya English, November 16, 2019, http://english.alarabiya.net/en/News/middle-east/2019/11/16/-Severe-protests-erupt-in-Iran-after-petrol-price-hike-State-media.html.

42 Michael Safi, "Iran: Protests and Teargas as Public Anger Grows Over Aircraft Downing," *Guardian*, January 13, 2020, https://www.theguardian.com/world/2020/jan/12/iran-riot-police-anti-government-backlash-ukraine.

43 Office of Public Affairs, U.S. Department of Justice, "Seven Iranians Working for Islamic Revolutionary Guard Corps–Affiliated Entities Charged for Conducting Coordinated Campaign of Cyber Attacks Against U.S. Financial Sector," Justice News, United States Department of Justice, March 24, 2016, https://www.justice.gov/opa/pr/seven-iranians-working-islamic-revolutionary-guard-corps-affiliated-entities-charged.

44 On the *Vincennes* incident, see Crist, *Twilight War*, 369.

45 On the Iranian brain drain, see Ali Ansari, *Modern Iran Since 1797*, 407–9; Cincotta and Karim Sadjadpour, "Iran in Transition: The Implications of the Islamic Republic's Changing Demographics," Carnegie Endowment

for International Peace, December 18, 2017, https://carnegieendowment.org/2017/12/18/iran-in-transition-implications-of-islamic-republic-s-changing-demographics-pub-75042. On Iranian military spending, see John E. Pike, "Iran—Military Spending," Globalsecurity.org, updated July 20, 2019, https://www.globalsecurity.org/military/world/iran/budget.htm.

Chapter 11: The Definition of Insanity

1 On the 2002 visit, see James Kelly, "Dealing with North Korea's Nuclear Programs," U.S. Department of State Archive, July 15, 2004, https://2001-2009.state.gov/p/eap/rls/rm/2004/34395.htm. On the light-water reactor, see International Atomic Energy Agency, "Agreed Framework of 21 October 1994 Between the United States of America and the Democratic People's Republic of Korea," Information Circular, November 2, 1994, https://www.iaea.org/sites/default/files/publications/documents/infcircs/1994/infcirc457.pdf.

2 Victor Cha, *The Impossible State: North Korea, Past and Future* (New York: Ecco, 2013), 292.

3 "Hyundai Chief Admits to N. Korean Summit Payoff—2003-02-16," Voice of America, October 29, 2009, https://www.voanews.com/archive/hyundai-chief-admits-n-korean-summit-payoff-2003-02-16.

4 Adam Taylor, "Analysis: Why the Olympics Matter When It Comes to North Korea," *Washington Post*, January 3, 2018, https://www.washingtonpost.com/news/worldviews/wp/2018/01/03/why-the-olympics-matter-when-it-comes-to-north-korea/.

5 Andrei Lankov, *The Real North Korea: Life and Politics in the Failed Stalinist Utopia* (New York: Oxford University Press, 2013), 202–3. Lankov concludes that less than 35 percent of the salary went to the workers, and with an estimated annual revenue of up to 40 million, the KIZ was a major cash cow for the North Korean government.

6 International Atomic Energy Agency, "IAEA and DPRK: Chronology of Key Events," July 25, 2014, www.iaea.org/newscenter/focus/dprk/chronology-of-key-events.

7 Choe Sang-Hun, "North Korea Claims to Conduct 2nd Nuclear Test," *New York Times*, May 25, 2009, www.nytimes.com/2009/05/25/world/asia/25nuke.html.

8 On the submarine attack, see Victor Cha, "The Sinking of Cheonan," Center for Strategic and International Studies, April 22, 2010, https://www.csis.org/analysis/sinking-cheonan. On Yeonpyeong, see "North Korea Shells Southern Island, Two Fatalities Reported," *Korea JoongAng Daily*, November 23, 2010, https://www.bbc.com/news/world-asia-pacific-11818005.

9 Siegfried Hecker, "A Return Trip to North Korea's Yongbyon Nuclear Complex," NAPSNet Special Report, Nautilus Institute, November 22, 2010, https://nautilus.org/napsnet/napsnet-special-reports/a-return-trip-to-north-koreas-yongbyon-nuclear-complex/.

10 Michael Rubin, *Dancing with the Devil: The Perils of Engaging Rogue Regimes* (New York: Encounter Books, 2014), 129–30.
11 Jimmy Carter, "Listen to North Korea," Carter Center, November 23, 2010, https://www.cartercenter.org/news/editorials_speeches/jc-listen-to-north-korea.html.
12 Carter, "Listen to North Korea." For President Obama's remarks, see "Obama, Barack H., Public Papers," *Presidents of the United States: Barack Obama, 2011* (Washington, DC: Office of the Federal Register, National Archives and Records Administration, 2015), 2:1265.
13 Jong Kun Choi, "The Perils of Strategic Patience with North Korea," *Washington Quarterly* 38, no. 4 (2016): 57–72.
14 Gerald F. Seib, Jay Solomon, and Carol E. Lee, "Barack Obama Warns Donald Trump on North Korea Threat," *Wall Street Journal*, November 22, 2016, https://www.wsj.com/articles/trump-faces-north-korean-challenge-1479855286.
15 Benjamin Haas, "South Korea: Former President Park Geun-Hye Sentenced to 24 Years in Jail," *Guardian*, April 6, 2018, www.theguardian.com/world/2018/apr/06/former-south-korea-president-park-geun-hye-guilty-of-corruption.
16 Uri Friedman, "The 'God Damn' Tree that Nearly Brought America and North Korea to War," *The Atlantic*, June 12, 2018, www.theatlantic.com/international/archive/2018/06/axe-murder-north-korea-1976/562028/.
17 See Anna Fifield, *The Great Successor: The Divinely Perfect Destiny of Brilliant Comrade Kim Jong Un* (New York: PublicAffairs, 2019), 16–19.
18 Central Intelligence Agency, "Consequences of U.S. Troop Withdrawal from Korea in Spring, 1949," CIA, February 28, 1949, https://www.cia.gov/library/readingroom/docs/DOC_0000258388.pdf.
19 James Forrestal in his diary entry on April 25, 1947, as quoted in Nadia Schadlow, *War and the Art of Governance: Consolidating Combat Success into Political Victory* (Washington, DC: Georgetown University Press, 2017). 178-179. In 1947, the Joint Chiefs had already assessed Korea as unworthy of a protracted American presence. See William Stueck and Boram Yi, "'An Alliance Forged in Blood': The American Occupation of Korea, the Korean War, and the U.S.–South Korean Alliance," *Journal of Strategic Studies* 33, no. 2 (2010), 177–209.
20 In one of Kim Il-sung's appeals to Stalin for support of the North's aggression toward the South the North Korean leader gave as a reason that the war would end rapidly before the United States could intervene. Kathryn Weathersby, "Soviet Aims in Korea and the Origins of the Korean War, 1945–1950: New Evidence from Russian Archives," Cold War International History Project Working Paper Series (1993): 28–31.
21 CIA, "Consequences."
22 John Quincy Adams, "An Address Delivered at the Request of a Commission of Citizens of Washington; on the Occasion of Reading the Declaration of Independence" (Washington, DC: Davis and Force, 1821), 29. It reads: "Wherever the standard of freedom and Independence has

been or shall be unfurled, there will her heart, her benedictions and her prayers be. But she goes not abroad, in search of monsters to destroy. She is the well-wisher to the freedom and independence of all. She is the champion and vindicator only of her own."

23 United Nations Department of Economic and Social Affairs, Statistics Division, "Country Profile: Democratic People's Republic of Korea" and "Country Profile: Republic of Korea," https://unstats.un.org/UNSD/snaama/CountryProfile?ccode=408 and https://unstats.un.org/UNSD/snaama/CountryProfile?ccode=408. South Korea held a GDP of $1.5 trillion in 2017, compared to the North's $13 billion.
24 Observatory of Economic Complexity, "Country Profile: North Korea," OEC, https://oec.world/en/profile/country/prk/
25 Summarized from Lankov, *The Real North Korea*, 32–33.
26 Andrei Lankov, "Fiasco of 386 Generation," *Korea Times*, February 5, 2008, https://www.koreatimes.co.kr/www/news/special/2008/04/180_18529.html.
27 Fifield, *The Great Successor*, 88.
28 Julian Ryall, "Kim Jong Un Was Child Prodigy Who Could Drive at Age of Three, Claims North Korean School Curriculum." *Telegraph*, April 10, 2015, https://www.telegraph.co.uk/news/worldnews/asia/northkorea/11526831/Kim-Jong-un-was-child-prodigy-who-could-drive-at-age-of-three-claims-North-Korean-school-curriculum.html.
29 Fifield, *The Great Successor*, 203–5.
30 Institute for National Security Strategy, "The Misgoverning of Kim Jong Un's Five Years in Power (김정은 집권 5년 실정 백서)," December 2016, http://www.inss.re.kr/contents/publications_yc.htm
31 Choe Sang-Hun, "In Hail of Bullets and Fire, North Korea Killed Official Who Wanted Reform," *New York Times*, March 12, 2016, https://www.nytimes.com/2016/03/13/world/asia/north-korea-executions-jang-song-thaek.html.
32 Justin McCurry, "North Korea Defence Chief Reportedly Executed with Anti-aircraft Gun," *Guardian*, May 13, 2015, https://www.theguardian.com/world/2015/may/13/north-korean-defence-minister-executed-by-anti-aircaft-gun-report.
33 Lankov, *The Real North Korea*, 43–44.
34 Fifield, *The Great Successor*, 124–27.
35 Summary based on Davenport, "Chronology of U.S.–North Korean Nuclear and Missile Diplomacy," Fact Sheets & Briefs, Arms Control Association, November 2019, https://www.armscontrol.org/factsheets/dprkchron#2016.
36 Ministry of Justice, "Immigration and Foreigner Policy Monthly Statistics (출입국·외국인정책 통계월보)," December 2018, http://www.korea.kr/archive/expDocView.do?docId=38330&call_from=rsslink.
37 Later, China would extort concessions from South Korea in exchange for lifting those sanctions imposed over THAAD. See David Voldzko, "China Wins Its War Against South Korea's U.S. THAAD Missile Shield—Without Firing a Shot," *South China Morning Post*, November 18, 2017,

https://www.scmp.com/week-asia/geopolitics/article/2120452/china-wins-its-war-against-south-koreas-us-thaad-missile.

38 Ankit Panda, "U.S. Intelligence: North Korea's Sixth Test Was a 140 Kiloton 'Advanced Nuclear' Device,'" *The Diplomat*, September 6, 2017, https://thediplomat.com/2017/09/us-intelligence-north-koreas-sixth-test-was-a-140-kiloton-advanced-nuclear-device/.

39 Donald Trump, @realDonaldTrump, "I told Rex Tillerson, our wonderful Secretary of State, that he is wasting his time trying to negotiate with Little Rocket Man... Save your energy Rex, we'll do what has to be done!" Twitter, October 1, 2017, 6:31 a.m. https://twitter.com/realDonaldTrump/status/914497947517227008.

Chapter 12: Making Him Safer Without Them

1 Choe Sang-Hun, "Happy Birthday, Trump Tells Kim. Not Enough, North Korea Says," *New York Times*, January 11, 2020, https://www.nytimes.com/2020/01/11/world/asia/trump-kim-jong-un-birthday.html.

2 For more on Japan's rapid economic expansion, see Ezra Vogel, *Japan as Number 1: Lessons for America* (Cambridge, MA: Harvard University Press, 1979), 9–10.

3 "Full Text of Abe's Speech before U.S. Congress," *Japan Times*. April 30, 2015, https://www.japantimes.co.jp/news/2015/04/30/national/politics-diplomacy/full-text-abes-speech-u-s-congress/#.XhQ_P0dKiMo.

4 On the San Francisco system and the emergence of the post–World War II security architecture in Asia, see Victor Cha, *Powerplay: The Origins of the American Alliance System in Asia* (Princeton, NJ: Princeton University Press, 2016).

5 Macrotrends, "South Korea GDP 1960–2020," https://www.macrotrends.net/countries/KOR/south-korea/gdp-gross-domestic-product. See also, Macrotrends, "South Korea Life Expectancy 1950–2020," https://www.macrotrends.net/countries/KOR/south-korea/life-expectancy.

6 World Bank, "Access to Electricity (% of Population)—Korea, Dem. People's Rep," https://data.worldbank.org/indicator/EG.ELC.ACCS.ZS?locations=KP. See also, Rick Newman, "Here's How Lousy Life Is in North Korea," *U.S. News*, April 12, 2013, https://www.usnews.com/news/blogs/rick-newman/2013/04/12/heres-how-lousy-life-is-in-north-korea

7 Elizabeth Shim, "Stunted Growth, Acute Anemia Persists in North Korean Children, Says Report," United Press International, September 18, 2015, https://www.upi.com/Top_News/World-News/2015/09/18/Stunted-growth-acute-anemia-persists-in-North-Korean-children-says-report/4351442628108/.

8 As Xi Jinping confidant Liu Mingfu told Japanese reporter Kenji Minemura, "Now is the time for Japan to escape from an excessive dependence on the United States and 'return to Asia.' With China breaking through the efforts by the United States to contain it, Japan should move away from being controlled by the United States and cooperate with China to create a new order in East Asia," Kenji Minemura, "Interview:

Liu Mingfu: China Dreams of Overtaking U.S. in Thirty Years," *Asahi Shimbun*, May 28, 2019, http://www.asahi.com/ajw/articles/AJ201905280016.html.

9 David Lai and Alyssa Blair, "How to Learn to Live with a Nuclear North Korea," *Foreign Policy*, August 7, 2017, https://foreignpolicy.com/2017/08/07/how-to-learn-to-live-with-a-nuclear-north-korea/.

10 See Kim Jong-un, "Let Us March Forward Dynamically Towards Final Victory, Holding Higher the Banner of Songun," April 15, 2012, 9, http://www.korean-books.com.kp/KBMbooks/ko/work/leader3/1202.pdf.

11 Chong Bong-uk, *Uneasy, Shaky Kim Jong-il Regime* (Seoul, South Korea: Naewoe Press, 1997), 17. See also Kim Tae-woon et al., "Analysis on the Practical Characteristics of Kim Jong-Il Era's Major Ruling Narratives (김정일 시대 주요 통치담론의 실천상 특징에 관한 고찰), *Unification Policy Studies* (2006) 27-31, http://repo.kinu.or.kr/bitstream/2015.oak/1610/1/0001423170.pdf.

12 Jeffrey Lewis, "North Korea Is Practicing for Nuclear War," *Foreign Policy*, March 9, 2017. https://foreignpolicy.com/2017/03/09/north-korea-is-practicing-for-nuclear-war/.

13 Cha, *The Impossible State*, 216.

14 United States Congress, "U.S. Congress Resolution Condemning North Korea for the Abductions and Continued Captivity of Citizens of the ROK and Japan as Acts of Terrorism and Gross Violations of Human Rights," 109th Congress, 2005, https://www.congress.gov/bill/109th-congress/house-concurrent-resolution/168.

15 See Robert S. Boynton, "North Korea's Abduction Project," *The New Yorker*, December 21, 2015, https://www.newyorker.com/news/news-desk/north-koreas-abduction-project.

16 See Cha, *The Impossible State*, 238–39.

17 More on North Korea's nuclear blackmail: Tristan Volpe, "The Unraveling of North Korea's Proliferation Blackmail Strategy," Kim Sung Chull et al., eds. *North Korea and Nuclear Weapons: Entering the New Era of Deterrence* (Washington, DC: Georgetown University Press, 2017), 73–88. See also Patrick McEachern, "More than Regime Survival," *North Korean Review* 14, no. 1 (2018): 115–18.

18 Amos Harel and Aluf Benn, "No Longer a Secret: How Israel Destroyed Syria's Nuclear Reactor," *Haaretz*, March 23, 2018, https://www.haaretz.com/world-news/MAGAZINE-no-longer-a-secret-how-israel-destroyed-syria-s-nuclear-reactor-1.5914407.

19 Toi Staff, "North Korea Offered Israel a Halt to Its Missile Sales to Iran for $1b—Report." *Times of Israel*, July 9, 2018, https://www.timesofisrael.com/north-korea-offered-israel-a-halt-to-its-missile-sales-to-iran-for-1b-report/.

20 Michael Schwirtz, "U.N. Links North Korea to Syria's Chemical Weapons Program," *New York Times*, February 27, 2018, https://www.nytimes.com/2018/02/27/world/asia/north-korea-syria-chemical-weapons-sanctions.html; see also Bruce E. Bechtol Jr., "North Korea's Illegal Weapons Trade: The Proliferation Threat from Pyongyang," *Foreign Af-*

fairs, June 6, 2018, https://www.foreignaffairs.com/articles/north-korea/2018-06-06/north-koreas-illegal-weapons-trade.

21 On the smuggling of weapons, see United Nations Security Council, "Report of the Panel of Experts Established Pursuant to Resolution 1874 (2009)," United Nations Security Council, 2019, 4, https://www.undocs.org/S/2019/171. For more on North Korea's Iran and Syria connections, see Bruce Bechtol Jr., "North Korea's Illegal Weapons Trade."

22 On Trump administration strategy on North Korea in 2017, see Donald J. Trump, "Remarks by President Trump to the National Assembly of the Republic of Korea—Seoul, Republic of Korea," Remarks, The White House, November 7, 2017, whitehouse.gov/briefings-statements/remarks-president-trump-national-assembly-republic-korea-seoul-republic-korea/; James Jeffrey, "What If H.R. McMaster Is Right About North Korea?" *The Atlantic*, January 18, 2018, https://www.theatlantic.com/international/archive/2018/01/hr-mcmaster-might-be-right-about-north-korea/550799/; Duane Patterson, "National Security Advisor General H. R. McMaster on MSNBC with Hugh," HughHewitt.com, August 5, 2017, https://www.hughhewitt.com/national-security-advisor-general-h-r-mcmaster-msnbc-hugh/#.

23 "Kim Yong-chol: North Korea's Controversial Olympics Delegate." BBC News, February 23, 2018, https://www.bbc.com/news/world-asia-43169604.

24 A South Korean newspaper claimed that Kim Jong-un had stated his preference for Vietnam-style economic opening over the Chinese model during his summit with President Moon in 2018. See Park Ui-myung and Choi Mira, "North Korean Leader Envisions Vietnam-like Opening for North Korea: Source," *Pulse*, May 4, 2018, https://pulsenews.co.kr/view.php?year=2018&no=285653. See also John Reed and Bryan Harris, "North Korea Turns to Vietnam for Economic Ideas," *Financial Times*, November 28, 2018, https://www.ft.com/content/c8a4fc68-f2cd-11e8-ae55-df4bf40f9d0d.

25 On the masters of money and on Dennis Rodman, see Fifield, *The Great Successor*, 142–43 and 174–80.

26 Jeong Yong-soo et al., "Donju Are Princes of North Korean Economy," *Korea JoongAng Daily*, October 18, 2019, http://koreajoongangdaily.joins.com/news/article/article.aspx?aid=3054069.

27 See Trump's tweet: https://twitter.com/realDonaldTrump/status/1160158591518674945?s=20.

28 White House, "Remarks by President Trump after Meeting with Vice Chairman Kim Yong Chol of the Democratic People's Republic of Korea," Remarks, The White House, June 1, 2018, https://www.whitehouse.gov/briefings-statements/remarks-president-trump-meeting-vice-chairman-kim-yong-chol-democratic-peoples-republic-korea/.

29 See Trump tweet referring to North Korea's potential to become an "economic powerhouse" and Kim as a "capable leader": https://twitter.com/realDonaldTrump/status/1094035813820784640?s=20.

30 Roberta Rampton, "'We Fell in Love': Trump Swoons over Letters from North Korea's Kim," Reuters, September 29, 2018, https://www.reuters

.com/article/us-northkorea-usa-trump/we-fell-in-love-trump-swoons-over-letters-from-north-koreas-kim-idUSKCN1MA03Q.

31 Jordan Fabian, "Trump Says Kim Not Responsible for Otto Warmbier's Death: 'I Will Take Him at His Word,'" *The Hill*, February 28, 2019, https://thehill.com/homenews/administration/431962-trump-says-kim-not-responsible-for-otto-warmbiers-death-i-will-take.

32 For the full speech, see "Address by President Moon Jae-in at May Day Stadium in Pyeongyang," *Chung Wa Dae*, https://english1.president.go.kr/briefingspeeches/speeches/70.

33 Fifield, *The Great Successor*, 277.

34 Eric Beech, "N. Korea Wanted Most Sanctions Lifted in Exchange for Partial Yongbyon Closure—U.S. Official," Reuters, February 28, 2019, https://www.reuters.com/article/northkorea-usa-briefing/nkorea-wanted-most-sanctions-lifted-in-exchange-for-partial-yongbyon-closure-us-official-idUSL3N20O1I7.

35 Timothy Martin, "North Korea Fires Insults at U.S., Spares Trump," *Wall Street Journal*, June 15, 2019, https://www.wsj.com/articles/north-korea-fires-insults-at-u-s-spares-trump-11560596401?mod=searchresults&page=1&pos=1.

36 KCNA Watch, "Report on 5th Plenary Meeting of the 7th Central Committee of the Workers' Party of Korea," *KNCA Watch*, January 1, 2020, https://kcnawatch.org/newstream/1577829999-473709661/report-on-5th-plenary-meeting-of-7th-c-c-wpk/.

37 The National Committee on North Korea, "Kim Jong Un's 2019 New Year Address," NCNK, January 1, 2019, https://www.ncnk.org/resources/publications/kimjongun_2019_newyearaddress.pdf/file_view.

38 Choe, "Happy Birthday, Trump Tells Kim," *New York Times* [date] https://www.nytimes.com/2020/01/11/world/asia/trump-kim-jong-un-birthday.html.

39 See David Sanger and Choe Sang-Hun, "North Korea Tests New Weapon," *New York Times*, April 17, 2019, https://www.nytimes.com/2019/04/17/world/asia/north-korea-missile-weapons-test.html; and BBC News, "North Korea: Kim Jong-un Oversees 'Strike Drill' Missile Component Test," BBC, May 5, 2019, https://www.bbc.com/news/world-asia-48165793.

40 Choe Sang-Hun, "New North Korean Missile Comes with Angry Message to South Korea's President," *New York Times*, July 26, 2019, https://www.nytimes.com/2019/07/26/world/asia/north-korea-missile-moon-jae-in.html.

41 Neil Connor and Nicola Smith, "Beijing Forced to Defend Trade with North Korea after Chinese-made Truck Used to Showcase Missiles," *Telegraph*, April 18, 2017, https://www.telegraph.co.uk/news/2017/04/18/china-made-truck-used-showcase-missiles-north-korea-parade/. Also, North Korean Economy Watch, "Report of the Panel of Experts Established Pursuant to Resolution 1874 (2009)," NKEW, 4, http://www.nkeconwatch.com/nk-uploads/UN-Panel-of-Experts-NORK-Report-May-2011.pdf

42 On training of DPRK hackers in China, see Kong Ji Young et al., "The

All-Purpose Sword: North Korea's Cyber Operations and Strategies," Eleventh International Conference on Cyber Conflict, 2019, 14–15, https://ccdcoe.org/uploads/2019/06/Art_08_The-All-Purpose-Sword.pdf. On the effect of sanctions, see Mathew Ha, "U.S. Sanctions North Korean Companies for Profiting from Overseas Slave Labor," Foundation for Defense of Democracies, January 15, 2020, https://www.fdd.org/analysis/2020/01/15/us-sanctions-north-korean-companies-for-profiting-from-overseas-slave-labor/.

43 Bruce E. Bechtol Jr., "North Korean Illicit Activities and Sanctions: A National Security Dilemma," *Cornell International Law Journal* 57 (2018): 51, https://www.lawschool.cornell.edu/research/ILJ/upload/Bechtol-final.pdf.

44 "Report on 5th Plenary Meeting of the 7th C.C. WPK," KNCA Watch, January 1, 2020, https://kcnawatch.org/newstream/1577829999-473709661/report-on-5th-plenary-meeting-of-7th-c-c-wpk/.

45 UN Human Rights Council, "Report of the Commission of Inquiry on Human Rights in the Democratic People's Republic of Korea," United Nations Human Rights Council, 15, https://www.ohchr.org/EN/HRBodies/HRC/CoIDPRK/Pages/ReportoftheCommissionofInquiryDPRK.aspx

46 Alexander George, *Forceful Persuasion: Coercive Diplomacy as an Alternative to War* (Washington, DC: United States Institute of Peace, 1991), 76–81.

47 For more, see Ben Dooley and Choe Sang-Hun, "Japan Imposes Broad New Trade Restrictions on South Korea," *New York Times*, August 1, 2019, https://www.nytimes.com/2019/08/01/business/japan south korea-trade.html.

48 Andy Greenberg, "Silicon Valley Has a Few Ideas for Undermining Kim Jong-un," *Wired*, March 10, 2015, https://www.wired.com/2015/03/silicon-valley-ideas-undermining-kim-jong-un/.

49 Lankov, *The Real North Korea*, 252–54.

50 Lankov, *The Real North Korea*, 254–58.

Chapter 13: Entering the Arena

1 Damon Wilson and Maks Czuperski, *Digital Resilience, Hybrid Threats*, Digital Forensic Research Lab, Atlantic Council, December 20, 2017, in possession of author. This report details the sustained Twitter campaign #FireMcMaster, and concludes that the majority of participants in this campaign were bots coming from alt-right sources of disinformation.

2 William Clinton, "Speech on China Trade Bill," Speech, Washington, DC, March 8, 2000, https://www.iatp.org/sites/default/files/Full_Text_of_Clintons_Speech_on_China_Trade_Bi.htm.

3 Adrian Shahbaz, "Freedom on the Net 2018: The Rise of Digital Authoritarianism," Freedomhouse.org, November 16, 2018, https://freedomhouse.org/report/freedom-net/freedom-net-2018/rise-digital-authoritarianism.

4 Paul Mozur, "A Genocide Incited on Facebook, with Posts from

Myanmar's Military," *New York Times*, October 15, 2018, https://www.nytimes.com/2018/10/15/technology/myanmar-facebook-genocide.html.

5 On technology's effect on children, see Nicholas Kardaras, *Glow Kids: How Screen Addiction Is Hijacking Our Kids—and How to Break the Trance* (New York: St. Martin's Griffin, 2017). On the rise of addictive technology, see Adam L. Alter, *Irresistible: The Rise of Addictive Technology and the Business of Keeping Us Hooked* (New York: Penguin Books, 2018).

6 Emerson T. Brooking and Suzanne Kianpour, "Iranian Digital Influence Efforts: Guerrilla Broadcasting for the Twenty-First Century," Atlantic Council, 2020, https://www.atlanticcouncil.org/wp-content/uploads/2020/02/IRAN-DIGITAL.pdf.

7 Tae-jun Kang, "North Korea's Influence Operations, Revealed," *The Diplomat*, July 25, 2018, https://thediplomat.com/2018/07/north-koreas-influence-operations-revealed/.

8 On Hong Kong, see Louise Matsakis, "China Attacks Hong Kong Protesters with Fake Social Posts," *Wired*, October 19, 2019, https://www.wired.com/story/china-twitter-facebook-hong-kong-protests-disinformation/. On Taiwan, see Raymond Zhong, "Awash in Disinformation Before Vote, Taiwan Points Finger at China," *New York Times*, January 6, 2020, https://www.nytimes.com/2020/01/06/technology/taiwan-election-china-disinformation.html.

9 Madeleine Carlisle, "New Orleans Declared a State of Emergency and Took Down Servers After Cyber Attack," *Time*, December 14, 2019, https://time.com/5750242/new-orleans-cyber-attack/.

10 James Clapper, "The Battle for Cybersecurity," Keynote Presentation, ICF CyberSci Symposium 2017, Fairfax, VA, September 28, 2017.

11 Todd C. Lopez, "Cyber Command Expects Lessons from 2018 Midterms to Apply in 2020," U.S. Department of Defense. February 14, 2019, https://www.defense.gov/Explore/News/Article/Article/1758488/cyber-command-expects-lessons-from-2018-midterms-to-apply-in-2020/.

12 United States Congress, House of Representatives, Hearing Before the Armed Services Committee, "Cyber Warfare in the 21st Century: Threats, Challenges, and Opportunities," 115th Congress, 75 (statement of Jason Healey, Columbia University's School of International and Public Affairs, 2017), https://govinfo.gov/content/pkg/CHRG-115hhrg24680/pdf/CHRG-115hhrg24680.pdf.

13 Mia Shuang Li, "Google's Dragonfly Will Intensify Surveillance on Journalists in China," *Columbia Journalism Review*, December 11, 2018, https://www.cjr.org/tow_center/dragonfly-censorship-google-china.php.

14 John Noble Wilford, "With Fear and Wonder in Its Wake, Sputnik Lifted Us into the Future," *New York Times*, September 25, 2007, https://www.nytimes.com/2007/09/25/science/space/25sput.html; Larry Abramson, "Sputnik Left Legacy for U.S. Science Education," NPR, September 30, 2007, https://www.npr.org/templates/story/story.php?storyId=14829195.

15 Smithsonian National Air and Space Museum, "Reflections on Post–Cold War Issues for International Space Cooperation," Smithsonian,

May 23, 2010, https://airandspace.si.edu/stories/editorial/reflections-post-cold-war-issues-international-space-cooperation.

16 "Challenges to Security in Space," Defense Intelligence Agency, January 2019, https://www.dia.mil/Portals/27/Documents/News/Military%20Power%20Publications/Space_Threat_V14_020119_sm.pdf.

17 On anti-satellite weapons, see "Counterspace Capabilities," United Nations Institute for Disarmament Research, August 6, 2018, https://www.unidir.org/files/medias/pdfs/counterspace-capabilities-backgrounder-eng-0-771.pdf.

18 Comments from the author at the National Space Council's inaugural meeting, October 5, 2017.

19 Sean Kelly, "China Is Infiltrating U.S. Space Industry with Investments," The Hill, Peter Greenberger, December 26, 2018, https://thehill.com/opinion/international/422870-chinese-is-infiltrating-us-space-industry-with-investments-and.

20 Yaakov Lappin, "Chinese Company Set to Manage Haifa's Port, Testing U.S.-Israeli Alliance," *South Florida Sun Sentinel*, January 29, 2019, https://www.sun-sentinel.com/florida-jewish-journal/fl-jj-chinese-company-set-manage-haifa-port-20190206-story.html.

21 Samm Sacks and Justin Sherman, "Global Data Governance: Concepts, Obstacles, and Prospects," New America, https://www.newamerica.org/cybersecurity-initiative/reports/global-data-governance/.

22 Department of Defense, "Missile Defense Review," 2019, https://www.defense.gov/Portals/1/Interactive/2018/11-2019-Missile-Defense-Review/The%202019%20MDR_Executive%20Summary.pdf.

23 "Statement by President Trump on the Paris Climate Accord" Remarks, The White House, June 1, 2017, https://www.whitehouse.gov/briefings-statements/statement-president-trump-paris-climate-accord/.

24 Patrick Herhold and Emily Farnworth, "The Net-Zero Challenge: Global Climate Action at a Crossroads (Part 1)," World Economic Forum in collaboration with Boston Consulting Group, December 2019, https://www.weforum.org/reports/the-net-zero-challenge-global-climate-action-at-a-crossroads-part-1.

25 Richard Muller, *Energy for Future Presidents: The Science Behind the Headlines* (W. W. Norton and Company, 2012).

26 Steve Inskeep and Ashley Westerman Inskeep. "Why Is China Placing a Global Bet on Coal?" NPR, April 19, 2019, https://www.npr.org/2019/04/29/716347646/why-is-china-placing-a-global-bet-on-coal; author's calculation based on a 1000 MWe coal plant. See Jordan Hanania et al., "Energy Education—Coal Fired Power Plant," EnergyEducation.CA, February 14, 2019, https://energyeducation.ca/encyclopedia/Coal_fired_power_plant.

27 David Obura, "Kenya's Most Polluting Coal Plant Could Poison Coastline," Climate Change News, September 20, 2017, https://www.climatechangenews.com/2017/09/20/kenyas-polluting-coal-plant-poison-coastline/.

28 John Mandyck and Eric Schultz, *Food Foolish: The Hidden Connection Between Food Waste, Hunger, and Climate Change* (Carrier Corp., 2015).

29 Note: Migration is taking a psychological as well as a physical toll, encouraging a populist turn in the polities of those nations most effected.

30 Muller, *Energy for Future Presidents*, 260.

31 Note: NITI Aayog, a "prominent government think tank," claimed that "More than 600 million Indians face 'acute water shortages.'" Seventy percent of the nation's water supply has been contaminated, which results in about 200,000 deaths every year. About two dozen cities could run out of groundwater entirely by next year, and about 40 percent of India will have "no access to drinking water" by 2030. James Temple. "India's Water Crisis Is Already Here. Climate Change Will Compound It," *MIT Technology Review*, April 24, 2019, https://www.technologyreview.com/s/613344/indias-water-crisis-is-already-here-climate-change-will-compound-it/.

32 Muller, *Energy for Future Presidents*.

33 Mandyck and Schultz, *Food Foolish*.

34 Eliza Barclay and Brian Resnick, "How Big Was the Global Climate Strike? 4 Million People, Activists Estimate," Vox, September 22, 2019, https://www.vox.com/energy-and-environment/2019/9/20/20876143/climate-strike-2019-september-20-crowd-estimate.

35 "'No Planet B': Millions Take to Streets in Global Climate Strike," Al Jazeera, September 20, 2019, https://www.aljazeera.com/news/2019/09/planet-thousands-join-global-climate-strike-asia-190920040636503.html.

36 Elizabeth Weise, "On World Environment Day, Everything You Know About Energy in the U.S. Might Be Wrong," *USA Today*, June 4, 2019, https://www.usatoday.com/story/news/2019/06/04/climate-change-coal-now-more-expensive-than-wind-solar-energy/1277637001/.

37 Alison St. John, "A Better Nuclear Power Plant?," KPBS, May 21, 2012, https://www.kpbs.org/news/2012/may/21/better-nuclear-power-plant/.

38 World Nuclear Association, "Plans for New Reactors Worldwide," updated January 2020, https://www.world-nuclear.org/information-library/current-and-future-generation/plans-for-new-reactors-worldwide.aspx.

39 Bloomberg, "Made-in-China Reactor Gains Favor at Home as U.S. Nuclear Technology Falters," *Japan Times*, April 2, 2019, https://www.japantimes.co.jp/news/2019/04/02/business/corporate-business/made-china-reactor-gains-favor-home-u-s-nuclear-technology-falters/#.Xi6NWBNKiCW.

40 Elting E. Morison, *Men, Machines, and Modern Times* (Cambridge, MA: MIT Press, 2016), 85.

Conclusion

1 Paraphrased from H. R. McMaster, *Dereliction of Duty* (New York: Harper Perennial, 1997), ix.

2 McMaster, *Dereliction of Duty*, 180–96.

3 McMaster, *Dereliction of Duty*, 260.
4 McMaster, *Dereliction of Duty*, 156.
5 My belief was consistent with Professor Richard Betts's definition of strategy as "the essential ingredient for making war either politically effective or morally tenable. It is the link between military means and political ends, the scheme for how to make one produce the other. Without strategy, there is no rationale for how force will achieve purposes worth the price in blood and treasure." See Betts, "Is Strategy an Illusion?" *International Security* (Fall 2000), http://www.columbia.edu/itc/sipa/U6800/readings-sm/strategy_betts.pdf.
6 Carl von Clausewitz, *On War* (London: Kegan Paul, Trench, Trubner & C., 1918).
7 Conrad Crane, *Avoiding Vietnam: The U.S. Army's Response to Defeat in Southeast Asia* (Carlisle, PA: Strategic Studies Institute, U.S. Army War College, 2002).
8 Hew Strachan, "Strategy and the Limitation of War," *Survival* 50, no. 1 (February/March 2008): 31–54, DOI: 10.1080/00396330801899470. See also Hew Strachan, *The Direction of War* (Cambridge, UK: Cambridge University Press, 2014), 54–55.
9 On North Vietnam's response, see Mark Moyar, *Triumph Forsaken: The Vietnam War* (Cambridge, UK: Cambridge University Press, 2006), 413.
10 Speech to the American Historical Association, December 28, 1939, https://www.marshallfoundation.org/library/speech-to-the-american-historical-association/.
11 E. J. Dionne Jr., "Kicking the Vietnam Syndrome," *Washington Post*, March 4, 1991.
12 John J. Mearsheimer, *The Great Delusion: Liberal Dreams and International Realities* (New Haven, CT: Yale University Press, 2018), 121 and 41.
13 Kelsey Piper, "George Soros and Charles Koch Team Up for a Common Cause: An End to 'Endless War,'" Vox, July 1, 2019, https://www.vox.com/2019/7/1/20677441/soros-koch-end-interventionist-wars-military; Nahal Toosi, "Koch Showers Millions on Think Tanks to Push a Restrained Foreign Policy," Politico, February 13, 2020, https://www.politico.com/news/2020/02/13/charles-koch-grants-foreign-policy-think-tanks-114898; Beverly Gage, "The Koch Foundation Is Trying to Reshape Foreign Policy. With Liberal Allies," *New York Times Magazine*, September 10, 2019, https://www.nytimes.com/interactive/2019/09/10/magazine/charles-koch-foundation-education.html.
14 Paul D. Miller, "H-Diplo/ISSF State of the Field Essay: On the Unreality of Realism in International Relations," H-Diplo, October 2, 2019, https://networks.h-net.org/node/28443/discussions/4846080/h-diploissf-state-field-essay-unreality-realism-international. For another trenchant critique of this school of thought, see Hal Brands, "Retrenchment Chic: The Dangers of Offshore Balancing," SSRN August 2015, https://papers.ssrn.com/sol3/papers.cfm?abstract_id=2737594.
15 For an essay that contains all these arguments, by a director of research

at the Soros-Koch–funded Quincy Institute for Responsible Statecraft, see Stephen Wertheim, "The Price of Primacy: Why America Shouldn't Dominate the World, *Foreign Affairs*, March/April 2020, 19–29.

16 John Stuart Mill, "On Liberty" (London: John W. Parker and Son, West Strand, 1859).

17 The White House, National Security Strategy of the United States of America, December 2017, 4, https://www.whitehouse.gov/wp-content/up loads/2017/12/NSS-Final-12-18-2017-0905.pdf.

18 Both quotations are from Ronald Granieri, "What Is Geopolitics and Why Does It Matter?" Foreign Policy Research Institute (Fall 2015), 492, https://www.fpri.org/article/2015/10/what-is-geopolitics-and-why-does-it-matter/. Audrey Kurth Cronin, *Power to the People: How Open Technological Innovation Is Arming Tomorrow's Terrorists* (New York: Oxford University Press, 2020).

19 For a succinct discussion of both forms of deterrence, see A. Wess Mitchell, "The Case for Deterrence by Denial," *The American Interest*, August 12, 2015, https://www.the-american-interest.com/2015/08/12/the-case-for-deterrence-by-denial/.

20 For benefits of alliances, see Grygiel and Mitchell, *The Unquiet Frontier*, 117–54.

21 Fukuyama, *Identity: The Demand for Dignity and the Politics of Resentment* (New York: Farrar, Straus and Giroux, 2018), 165–66.

22 Fukuyama, *Identity*, 170–71.

23 For the connection between income inequality and opportunity inequality and the importance of education, see Robert D. Putnam, *Our Kids: The American Dream in Crisis* (New York: Simon and Schuster, 2015), esp. 227–61.

24 Paul Reynolds, "History's Other Great Relief Effort," BBC, January 11, 2005, http://news.bbc.co.uk/1/hi/world/europe/4164321.stm.

25 Zachary Shore, "The Spirit of Sputnik: Will America Ever Fund Education Again?" Medium, September 3, 2018, https://medium.com/@zshore/the-spirit-of-sputnik-881b8f720736.

RECOMMENDED READING

Russia

On defending against Russia new-generation warfare: *The Lands in Between: Russia vs. the West and the New Politics of Hybrid War*, by Mitchell Alexander Orenstein.

On Putin as an operator: *Mr. Putin: Operative in the Kremlin*, by Clifford Gaddy and Fiona Hill.

On the Russian Siloviki: *All the Kremlin's Men: Inside the Court of Vladimir Putin*, by Mikhail Zygar.

On the failed Russian transition of the 1990s: *Sale of the Century: Russia's Wild Ride from Communism to Capitalism*, by Chrystia Freeland.

For a personal account of Russia's transition in the 1990s and beyond: *From Cold War to Hot Peace: An American Ambassador in Putin's Russia*, by Michael McFaul.

On Putin's rule and implications for the future: *Kremlin Winter: Russia and the Second Coming of Vladimir Putin*, by Robert Service.

China

On the U.S.-China relationship: *The Beautiful Country and the Middle Kingdom: America and China, 1776 to the Present*, by John Pomfret.

On historical memory and the Chinese Communist Party's ambitions: *Everything Under the Heavens: How the Past Helps Shape China's Push for Global Power*, by Howard French.

On the Chinese Communist Party's obsession with control: *Haunted by Chaos: China's Grand Strategy from Mao Zedong to Xi Jinping*, by Sulmaan Wasif Khan.

On Xi Jinping's reversal of reform and the development of the authoritarian surveillance police state: *The Third Revolution: Xi Jinping and the New Chinese State*, by Elizabeth Economy.

On the future of the Indo-Pacific region: *The End of the Asian Century: War, Stagnation, and the Risks to the World's Most Dynamic Region*, by Michael R. Auslin.

On China's modern transformation, *Wealth and Power: China's Long March to the Twenty-First Century*, by Orville Schell and John Delury.

South Asia

On Afghanistan's wars: *The Wars of Afghanistan: Messianic Terrorism, Tribal Conflicts, and the Failures of Great Powers*, by Peter Tomsen.

On jihadist terrorist goals and strategy: *The Master Plan: ISIS, Al-Qaeda, and the Jihadi Strategy for Final Victory*, by Brian Fishman.

On the Haqqani network as the nexus of jihad in South Asia: *The Fountainhead of Jihad: The Haqqani Nexus, 1973–2012*, by Vahid Brown and Don Rassler.

On the troubled U.S.-Pakistani relationship: *Magnificent Delusions: Pakistan, the United States, and an Epic History of Misunderstanding*, by Husain Haqqani.

On the essential elements of an effective counterterrorism strategy: *How Terrorism Ends: Understanding the Decline and Demise of Terrorist Campaigns*, by Audrey Kurth Cronin.

Middle East

On the U.S. experience in Iraq from 2003 to 2011: *The Endgame: The Inside Story of the Struggle for Iraq, from George W. Bush to Barack Obama*, by Michael R. Gordon and General Bernard E. Trainor.

On Iraq's internal political dynamics after Saddam Hussein: *Iraq After America: Strongmen, Sectarians, Resistance*, by Joel Rayburn.

On the United States' long struggle with terrorism in the Middle East: *Blood Year: The Unraveling of Western Counterterrorism*, by David Kilcullen.

On ISIS: *ISIS: Inside the Army of Terror*, by Michael Weiss and Hassan Hassan; and *Shatter the Nations: ISIS and the War for the Caliphate*, by Mike Giglio.

On the effect of the Arab Spring on the people of the region: *A Rage for Order: The Middle East in Turmoil, from Tahrir Square to ISIS*, by Robert F. Worth.

On the Syrian Civil War: *Assad or We Burn the Country: How One Family's Lust for Power Destroyed Syria*, by Sam Dagher.

Iran

On the Iranian Revolution: *The Persian Sphinx: Amir Abbas Hoveyda and the Riddle of the Iranian Revolution*, by Abbas Milani.

On Iran's four-decade proxy war in the Middle East: *The Twilight War: The Secret History of America's Thirty-Year Conflict with Iran*, by David Crist; and *Confronting Iran*, by Ali M. Ansari.

On the role of the Islamic Revolutionary Guard's Corps in exporting Iran's revolution: *Vanguard of the Imam: Religion, Politics, and Iran's Revolutionary Guards*, by Afshon Ostovar.

On the Saudi-Iranian rivalry and the drives of sectarian violence in the Middle East: *Black Wave: Saudi Arabia, Iran, and the Forty-Year Rivalry that Unraveled Culture, Religion, and Collective Memory in the Middle East*, by Kim Ghattas.

North Korea

On the history of U.S.-North Korean interactions: *The Impossible State: North Korea, Past and Future*, by Victor Cha.

On the nature of the North Korean regime: *The Real North Korea: Life and Politics in the Failed Stalinist Utopia*, by Andrei Lankov.

On the miracle of South Korea: *Nation Building in South Korea*, by Gregg A. Brazinsky.

On the U.S. relationship with Japan in the postwar period: *Japan in the American Century*, by Kenneth B. Pyle.

On Kim Jong-un: *The Great Successor: The Divinely Perfect Destiny of Brilliant Comrade Kim Jong Un*, by Anna Fifield.

Arenas

On defending against cyber-enabled information warfare and other threats to democracy: *Ill Winds: Saving Democracy from Russian Rage, Chinese Ambition, and American Complacency*, by Larry Diamond.

On the interrelated problems of climate, energy, and food security: *Food Foolish: The Hidden Connection Between Food Waste, Hunger, and Climate*, by John M. Mandyck and Eric B. Schultz.

On technology and its effect on the environment and security: *Energy for Future Presidents: The Science Behind the Headlines*, by Richard A. Muller.

On social media's detrimental effect on society: *The Square and the Tower*, by Niall Ferguson; and *LikeWar: The Weaponization of Social Media*, by P. W. Singer and Emerson T. Brooking.

On offensive cyber attacks: *Bytes, Bombs, and Spies: The Strategic Dimensions of Offensive Cyber Operations*, edited by Herbert Lin and Amy Zegart.

On technology and security: *Power to the People: How Open Technological Innovation Is Arming Tomorrow's Terrorists*, by Audrey Kurth Cronin.

Conclusion/General

On great power competition and the importance of alliances: *The Unquiet Frontier: Rising Rivals, Vulnerable Allies, and the Crisis of American Power*, by Jakub J. Grygiel and A. Wess Mitchell.

For a historic perspective on how to defeat terrorist organizations: *Return of the Barbarians: Confronting Non-State Actors from Ancient Rome to the Present*, by Jakub Grygiel.

On strategy in war: *The Direction of War: Contemporary Strategy in Historical Perspective*, by Hew Strachan.

On the need to consolidate military gains into sustainable political outcomes: *War and the Art of Governance: Consolidating Combat Success into Political Victory*, by Nadia Schadlow.

On how to think about the adversary and the importance of strategic empathy: *A Sense of the Enemy: The High Stakes History of Reading Your Rival's Mind*, by Zachary Shore.

SELECTED BIBLIOGRAPHY

Primary Sources

Belasco, Amy. "The Cost of Iraq, Afghanistan, and Other Global War on Terror Operations Since 9/11." Congressional Research Service, December 8, 2014. https://fas.org/sgp/crs/natsec/RL33110.pdf.

Bush, George W. *Decision Points*. New York: Crown Publishers, 2010.

Carter, Jimmy. "Listen to North Korea." The Carter Center, November 23, 2010. https://www.cartercenter.org/news/editorials_speeches/jc-listen-to-north-korea.html.

Department of Defence, "Defending Australia in the Asia Pacific Century: Force 2030," Australian Government, Defence White Paper, 2009.

Department of State, Office of the Historian, "Document 12: Memorandum of Conversation, Beijing, February 17–18, 1973." *Foreign Relations of the United States, 1969–1976*. Volume XVIII, *China, 1973–1976*. Washington, DC: Government Printing Office, 2007, https://history.state.gov/historicaldocuments/frus1969-76v18/d12.

Dinh Thi Kieu Nhung, "Afghanistan in 2018: A Survey of the Afghan People," Asia Foundation. https://asiafoundation.org/publication/afghanistan-in-2018-a-survey-of-the-afghan-people/.

Gerasimov, Valery. "The Value of Science Is in the Foresight: New Challenges Demand Rethinking the Forms and Methods of Carrying Out Combat Operations." *Military Review*, January–February 2016. https://www.armyupress.army.mil/Portals/7/military-review/Archives/English/MilitaryReview_20160228_art008.pdf.

Hecker, Siegfried. "A Return Trip to North Korea's Yongbyon Nuclear Complex." NAPSNet Special Reports, Nautilus Institute, November 22, 2010. https://nautilus.org/napsnet/napsnet-special-reports/a-return-trip-to-north-koreas-yongbyon-nuclear-complex/.

Mueller, Robert S. "Report on the Investigation into Russian Interference in the 2016 Presidential Election." Volume 1 of 2. U.S. Department of Justice. Washington, DC, March 2019. https://www.justice.gov/storage/report.pdf.

Obama, Barack. "Remarks by the President on Ending the War in Iraq." Office of the Press Secretary. The White House, October 21, 2011. https://obamawhitehouse.archives.gov/the-press-office/2011/10/21/remarks-president-ending-war-iraq.

———. "Statement by the President on Afghanistan." Office of the Press Secretary. The White House, October 15, 2015. https://obamawhitehouse.archives.gov/the-press-office/2015/10/15/statement-president-afghanistan.

Pence, Michael. Remarks by Vice President Pence on the Administration's Policy Toward China. Hudson Institute. Washington, DC, October 4, 2018.

Putin, Vladimir. "Russia at the Turn of the Millennium," 1999. https://pages.uoregon.edu/kimball/Putin.htm.

Rhodes, Ben. *The World as It Is: A Memoir of the Obama White House.* New York: Random House, 2018.

Rice, Condoleezza. *Democracy: Stories from the Long Road to Freedom.* New York: Hachette Book Group, 2017.

Rudd, Kevin. "China's Political Economy into 2020: Pressures on Growth, Pressures on Reform." Speech delivered at the Conference on China's Economic Future: Emerging Challenges at Home and Abroad, Chatham House, London, July 11, 2019, https://asiasociety.org/sites/default/files/2020-01/4.%20China%27s%20Political%20Economy%20into%202020_0.pdf.

Select Committee on Intelligence. "Russian Active Measures Campaigns and Interferences in the 2016 U.S. Election," Volume 2: "Russia's Use of Social Media with Additional Views." United States Senate, October 2019. https://www.intelligence.senate.gov/sites/default/files/documents/Report_Volume2.pdf.

Special Inspector General for Afghanistan Reconstruction, "January 30, 2017, Quarterly Report to the United States Congress," January 30, 2017, https://www.sigar.mil/pdf/quarterlyreports/2017-01-30qr.pdf.

Tzu, Sun. *The Art of War.* Translated by Thomas Cleary. Boston, MA: Shambhala, 2005.

United Nations Assistance Mission in Afghanistan. "Protection of Civilians in Armed Conflict Annual Report 2015." United Nations Human Rights Office of the High Commissioner, February 2016. https://unama.unmissions.org/protection-of-civilians-reports.

United Nations Security Council. "Tenth Report of the Analytical Support and Sanctions Monitoring Team Submitted Pursuant to Resolution 2255 (2015) Concerning the Taliban and Other Associated Individuals and Entities Constituting a Threat to the Peace, Stability and Security of Afghanistan." June 13, 2019. https://www.undocs.org/S/2019/481.

United States Department of State. "Deputy Secretary Armitage's Meeting with Pakistan Intel Chief Mahmud: You're Either with Us or You're Not." Unclassified September 12, 2001. https://nsarchive2.gwu.edu/NSAEBB/NSAEBB358a/doc03-1.pdf.

United States Senate Committee on Homeland Security and Governmental Affairs. "Threats to the U.S. Research Enterprise: China's Talent Recruitment Plans." November 18, 2019. https://www.hsgac.senate.gov/imo/media

/doc/2019-11-18%20PSI%20Staff%20Report%20-%20China's%20Talent%20Recruitment%20Plans.pdf.

Secondary Sources: Books

Ansari, Ali. *Confronting Iran: The Failure of American Foreign Policy and the Next Great Crisis in the Middle East*. New York: Basic Books, 2006.

———. *Modern Iran Since 1797: Reform and Revolution*. New York: Routledge, 2019.

Barfield, Thomas J. *Afghanistan: A Cultural and Political History*. Princeton, NJ: Princeton University Press, 2012.

Barrett, Roby C. *The Gulf and the Struggle for Hegemony: Arabs, Iranians, and the West in Conflict*. Washington, DC: Middle East Institute, 2016.

Beardson, Timothy. *Stumbling Giant: The Threats to China's Future*. New Haven, CT: Yale University Press, 2013.

Bousquet, Antoine J. *The Scientific Way of Warfare: Order and Chaos on the Battlefields of Modernity*. New York: Oxford University Press, 2010.

Brown, Vahid, and Don Rassler, *Fountainhead of Jihad: The Haqqani Nexus, 1973–2012*. Oxford: Oxford University Press, 2013.

Cha, Victor. *The Impossible State: North Korea, Past and Future*. New York: HarperCollins, 2012.

———. *Powerplay: The Origins of the American Alliance System in Asia*. Princeton, NJ: Princeton University Press, 2016.

Challenges to Security in Space. Washington, DC: Defense Intelligence Agency, 2019.

Chau, Donovan, and Thomas Kane. *China and International Security: History, Strategy, and 21st-Century Policy*. Westport, CT: Praeger, 2014.

Coll, Steve. *Directorate S: The C.I.A. and America's Secret Wars in Afghanistan and Pakistan*. New York: Penguin Press, 2018.

———. *Ghost Wars: The Secret History of the C.I.A., Afghanistan, and Bin Laden, from the Soviet Invasion to September 10, 2001*. New York: Penguin Press, 2004.

Conley, Heather, James Mina, Ruslan Stefanov, and Martin Vladimirov. *The Kremlin Playbook: Understanding Russian Influence in Central and Eastern Europe*. Volumes I and II. Lanham, MD: Rowman and Littlefield, 2016; Washington, DC: Center for Strategic and International Studies.

Crist, David. *The Twilight War: The Secret History of America's Thirty-Year Conflict with Iran*. New York: Penguin Books, 2013.

Dagher, Sam. *Assad or We Burn the Country: How One Family's Lust for Power Destroyed Syria*. New York: Little, Brown and Company, 2019.

Diamond, Larry, and Orville Schell, eds. *China's Influence and American Interests: Promoting Constructive Vigilance*. Stanford, CA: Hoover Institution Press, 2018.

Fifield, Anna. *The Great Successor: The Divinely Perfect Destiny of Brilliant Comrade Kim Jong Un*. New York: Hachette Book Group, 2019.

Freeland, Chrystia. *Sale of the Century: The Inside Story of the Second Russian Revolution*. London: Little, Brown and Company, 2000.

Fukuyama, Francis. *The End of History and the Last Man*. New York: Free Press, 2006.

George, Alexander. *Forceful Persuasion: Coercive Diplomacy as an Alternative to War.* Washington, DC: United States Institute of Peace, 1991.

Hersh, Seymour M. *Chain of Command: The Road from 9/11 to Abu Ghraib*. New York: HarperCollins, 2005.

Hill, Fiona, and Clifford G. Gaddy. *Mr. Putin: Operative in the Kremlin*. Washington, DC: Brookings Institution Press, 2015.

Kagan, Frederick W. *Finding the Target: The Transformation of American Military Policy.* New York: Encounter Books, 2017.

Kang, David. *East Asia Before the West: Five Centuries of Trade and Tribute*. New York: Columbia University Press, 2010.

Kennan, George. *Russia and the West Under Lenin and Stalin*. Boston: Little, Brown and Company, 1961.

Khan, Sulmaan. *Haunted by Chaos: China's Grand Strategy from Mao Zedong to Xi Jinping*. Cambridge, MA: Harvard University Press, 2018.

Kilcullen, David. *The Dragon and the Snakes: How the Rest Learned to Fight the West* (New York: Oxford University Press, 2020).

Klein, Naomi. *This Changes Everything: Capitalism vs. the Climate*. Toronto: Vintage Canada, 2015.

Kolbert, Elizabeth. *Sixth Extinction: An Unnatural History*. New York: Picador USA, 2015.

Lankov, Andrei. *The Real North Korea: Life and Politics in the Failed Stalinist Utopia*. New York: Oxford University Press, 2015.

Morison, Elting Elmore. *Men, Machines, and Modern Times*. Cambridge, MA: MIT Press, 2016.

Muller, Richard A. *Energy for Future Presidents: The Science Behind the Headlines*. New York: W. W. Norton, 2013.

Murray, Williamson, and Kevin M. Woods. *The Iran-Iraq War: A Military Strategic History*. Cambridge, UK: Cambridge University Press, 2014.

Naylor, Sean. *Not a Good Day to Die: The Untold Story of Operation Anaconda*. New York: Berkley Caliber Books, 2006.

Noman, Omar. *The Political Enemy of Pakistan, 1988*. New York: Routledge, 1988.

Ostovar, Afshon. *Vanguard of the Imam: Religion, Politics, and Iran's Revolutionary Guards*. New York: Oxford University Press, 2016.

Packer, George. *Our Man: Richard Holbrooke and the End of the American Century*. New York: Alfred A. Knopf, 2019.

Porter, Patrick. *Military Orientalism: Eastern War Through Western Eyes*. New York: Oxford University Press, 2013.

Rayburn, Joel D., and Frank K. Sobchak, eds. *The U.S. Army in the Iraq War Volume 2: Surge and Withdrawal 2007–2011*. Carlisle Barracks, PA: United States Army War College Press, 2019.

Rubin, Michael. *Dancing with the Devil: The Perils of Engaging Rogue Regimes*. New York: Encounter Books, 2014.

Schadlow, Nadia. *War and the Art of Governance: Consolidating Combat Success into Political Victory*. Washington, DC: Georgetown University Press, 2017.

Schell, Orville, and John Delury. *Wealth and Power: China's Long March to the Twenty-first Century*. New York: Random House, 2013.

Singer, P. W., and Emerson T. Brooking. *LikeWar: The Weaponization of Social Media*. Boston, MA: Mariner Books/Houghton Mifflin Harcourt, 2019.

Solomon, Jay. *Iran Wars: Spy Games, Bank Battles, and the Secret Deals that Reshaped the Middle East*. New York: Random House, 2016.

Swaine, Michael D., and Ashley J. Tellis. *Interpreting China's Grand Strategy: Past, Present, and Future*. Santa Monica, CA: RAND Corporation, 2000.

Thornton, Patricia. *Disciplining the State: Virtue, Violence, and State-making in Modern China*. Cambridge, MA: Harvard University Asia Center, 2007.

Tomsen, Peter. *The Wars of Afghanistan: Messianic Terrorism, Tribal Conflicts, and the Failures of Great Powers*. New York: PublicAffairs, 2013.

Vogel, Ezra F. *Deng Xiaoping and the Transformation of China*. Cambridge, MA: Belknap Press/Harvard University Press, 2013.

Wallace-Wells, David. *The Uninhabitable Earth: Life After Warming*. New York: Random House, 2020.

Warrick, Joby. *Black Flags: The Rise of ISIS*. New York: Doubleday, 2015.

Secondary Sources: Articles, Studies, Speeches, and Testimony

Adesnik, David, and Behnam Ben Taleblu. "Burning Bridge: The Iranian Land Corridor to the Mediterranean." Foundation for Defense of Democracies, 2019. https://www.fdd.org/analysis/2019/06/18/burning-bridge/.

Ansari, Ali, and Kasra Aarabi. "Ideology and Iran's Revolution: How 1979 Changed the World," Tony Blair Institute for Global Change, February 11, 2019. https://institute.global/policy/ideology-and-irans-revolution-how-1979-changed-world.

Bechtol, Bruce E., Jr. "North Korea's Illegal Weapons Trade: The Proliferation Threat from Pyongyang." *Foreign Affairs*, June 6, 2018. https://www.foreignaffairs.com/articles/north-korea/2018-06-06/north-koreas-illegal-weapons-trade.

Bowman, Bradley, and David Maxwell, eds. *Maximum Pressure 2.0: A Plan for North Korea*, Washington, DC: Foundation for Defense of Democracies, December 2019.

Demchak, Chris C., and Yuval Shavitt. "China's Maxim—Leave No Access Point Unexploited: The Hidden Story of China Telecom's BGP Hijacking." *Military Cyber Affairs* 3, no. 1 (2018): 5–7.

DiResta, Renee, Jonathan Albright, and Ben Johnson. "The Tactics and Tropes of the Internet Research Agency." New Knowledge, 2018. https://disinformationreport.blob.core.windows.net/disinformation-report/NewKnowledge-Disinformation-Report-Whitepaper.pdf.

Gartenstein-Ross, Daveed, and Nathaniel Barr. "How Al-Qaeda Works: The Jihadist Group's Evolving Organizational Design." Hudson Institute, June 1, 2018. https://www.hudson.org/research/14365-how-al-qaeda-works-the-jihadist-group-s-evolving-organizational-design.

Geist, Edward. "Deterrence Stability in the Cyber Age." *Strategic Studies Quarterly* 9, no. 4 (2015): 44–61.

Gronvall, Gigi. "The Security Implications of Synthetic Biology." *Survival* 60, no. 4 (2018): 165–80. DOI: 10.1080/00396338.2018.1495443.

Healey, Jason, and Robert K. Knake. "Zero Botnets: Building a Global Effort to Clean up the Internet." Council on Foreign Relations, New York, 2018.

Johnson, Jeff. "Testimony Before the U.S.-China Economic and Security Review Commission Hearing on 'Chinese Investment in the United States: Impacts and Issues for Policy Makers.'" U.S.-China Economic and Security Review Commission, January 26, 2017.

Choi, Jong Kun. "The Perils of Strategic Patience with North Korea." *Washington Quarterly* 38, no. 4 (2016): 57–72.

Kong Ji Young et al., "The All-Purpose Sword: North Korea's Cyber Operations and Strategies." Eleventh International Conference on Cyber Conflict, Tallinn, Estonia, May 28–31, 2019. https://ccdcoe.org/uploads/2019/06/Art_08_The-All-Purpose-Sword.pdf.

Lai, David, and Alyssa Blair. "How to Learn to Live with a Nuclear North Korea." *Foreign Policy*, August 7, 2017. https://foreignpolicy.com/2017/08/07/how-to-learn-to-live-with-a-nuclear-north-korea/.

Livingston, Ian S., and Michael O'Hanlon. "Afghanistan Index." Brookings Institution, September 29, 2017. https://www.brookings.edu/afghanistan-index/.

Milani, Abbas. "Islamic Republic of Iran in an Age of Global Transitions: Challenges for a Theocratic Iran." Hoover Institution, Stanford, CA, April 22, 2019. https://www.hoover.org/research/islamic-republic-iran-age-global-transitions-challenges-theocratic-iran.

Nouwens, Meia, and Helena Legarda. "China's Pursuit of Advanced Dual-Use Technologies." IISS, December 18, 2018. https://www.iiss.org/blogs/research-paper/2018/12/emerging-technology-dominance.

Paul, Christopher, and Miriam Matthews. "The Russian 'Firehose of Falsehood' Propaganda Model: Why It Might Work and Options to Counter It." RAND Corporation, 2016. https://www.rand.org/pubs/perspectives/PE198.html.

Speier, Richard, George Nacouzi, Carrie A. Lee, and Richard M. Moore. "Hypersonic Missile Nonproliferation: Hindering the Spread of a New Class of Weapons." RAND Corporation, 2017. https://www.rand.org/pubs/research_reports/RR2137.html.

United States Institute of Peace. "Syria Study Group Final Report." United States Institute of Peace. Washington, DC, 2019.

Zegart, Amy. "Cheap Fights, Credible Threats: The Future of Armed Drones and Coercion." *Journal of Strategic Studies* 43, no. 1 (2020): 6–46.

Newspapers and Periodicals

Long War Journal (Foundation for Defense of Democracies)

The New Yorker

New York Times

Philadelphia Inquirer

Time

Wall Street Journal

Washington Post

INDEX

ABOUT THE AUTHOR

H. R. MCMASTER is the Fouad and Michelle Ajami Senior Fellow at the Hoover Institution and Stanford University. He is also the Susan and Bernard Liautaud Fellow at the Freeman Spogli Institute and Lecturer at the Stanford Graduate School of Business. He serves as chairman of the advisory board of the Center for Military and Political Power at the Foundation for Defense of Democracies and the Japan Chair at the Hudson Institute. A native of Philadelphia, H.R. graduated from the United States Military Academy in 1984. He served as a U.S. Army officer for thirty-four years and retired as a lieutenant general in 2018. He remained on active duty while serving as the twenty-sixth assistant to the president for national security affairs. He taught history at West Point and holds a PhD in history from the University of North Carolina at Chapel Hill.